T0309210

Geometry of
Biharmonic Mappings
Differential Geometry of
Variational Methods

Geometry of
Biharmonic Mappings
Differential Geometry of
Variational Methods

Hajime Urakawa

Tohoku University, Japan

 World Scientific

NEW JERSEY · LONDON · SINGAPORE · BEIJING · SHANGHAI · HONG KONG · TAIPEI · CHENNAI · TOKYO

Published by

World Scientific Publishing Co. Pte. Ltd.

5 Toh Tuck Link, Singapore 596224

USA office: 27 Warren Street, Suite 401-402, Hackensack, NJ 07601

UK office: 57 Shelton Street, Covent Garden, London WC2H 9HE

Library of Congress Cataloging-in-Publication Data
Names: Urakawa, Hajime, 1946–   author.
Title: Geometry of biharmonic mappings : differential geometry of variational methods /
    by Hajime Urakawa (Tohoku University, Japan).
Description: New Jersey : World Scientific, 2018. | Includes bibliographical references.
Identifiers: LCCN 2018024556 | ISBN 9789813236394 (hardcover : alk. paper)
Subjects: LCSH: Mappings (Mathematics) | Geometry, Riemannian. | Geometry, Differential.
Classification: LCC QA613.64 .U73 2018 | DDC 514/.7--dc23
LC record available at https://lccn.loc.gov/2018024556

British Library Cataloguing-in-Publication Data
A catalogue record for this book is available from the British Library.

For any available supplementary material, please visit
https://www.worldscientific.com/worldscibooks/10.1142/10886#t=suppl

Printed in Singapore

Preface

This book aims to give a general and precise geometric theory of harmonic mappings and biharmonic mappings between two Riemannian manifolds.

The theory of biharmonic maps which was already conjectured by J. Eells and L. Lemaire in their famous lecture notes on harmonic maps. Indeed, in 1972, they defined the notion of the bienergy which is half of the integral of the square norm of the tension field over the domain manifold for a smooth mapping, and raised a problem to study its variational problem. In 1986, G.Y. Jiang proceeded the calculation of the first and the second variations of the bienergy. The epoch in the theory of biharmonic maps came in minimal submanifold theory by B. Y. Chen. He raised the B.Y. Chen's conjecture which has been unsolved until now: **Every biharmonic submanifold in the Euclidean space must be minimal.**

In Part I, we prepare the fundamental materials in differential geometry in Chapter 1, and we show the fundamental paper of G. Y. Jiang on the first and second variations of the bienergy in Chapter 2. In Chapter 2, we first give rigidity results of biharmonic submanifolds in a Riemannian manifold of non-positive curvature. In Chapter 3, we state a quite general rigidity theorem of harmonic maps, namely, every biharmonic map whose finite energy and bienergy into a Riemannian manifold of non-positive curvature must be harmonic. For a biharmonic submanifold in such a Riemannian manifold, it must be minimal which is shown in Chapter 4. In Chapter 5, a theorem which every biharmonic hypersurface in a Riemannian manifold with non-positive Ricci curvature, must be minimal, is shown.

In Chapters 7 and 8, we show the abundance of biharmonic maps into compact Lie groups and also symmetric spaces of compact type. We show, in these two chapters, the systematic ways producing and characterizing all the harmonic maps into compact Lie groups and also symmetric spaces of compact type.

In Chapter 9, the *bubbling phenomena* of biharmonic maps, we will show that the totality of biharmonic maps from a compact Riemannian manifold of dimension m whose m-energy is bounded above by a positive constant, into an arbitrary target compact Riemannian manifold is *very small*.

In Chapter 10, we answer completely an interesting problem raised by P. Baird and D. Kamissoko to produce a biharmonic but not harmonic map by conformal change of Riemannian metric of a domain Riemannian manifold.

In Part III, we treat biharmonic submanifolds of a compact Riemannian manifold of non-negative sectional curvature. In Chapter 11, we classify all the biharmonic hypersurfaces in a compact symmetric space including the standard sphere or the complex projective space. In Chapters 12, and 13, we treat biharmonic Legendrian submanifolds in a Sasaki manifold, and the one in Kähler cone manifolds, and biharmonic Lagrangian submanifolds in a Kähler manifolds.

In Chapter IV, we treat with further developments of biharmonic maps. Generalizations of the notion of harmonic map and biharmonic map, are given. The one hand is pseudo harmonic map and pseudo biharmonic map in Cauchy-Riemann geometry. The other hand is transversally harmonic map and transversally biharmonic map in foliated Riemannian geometry.

Finally, we raise two unsolved problems on biharmonic maps different from the above topics:

(1) Classification of all the biharmonic submanifolds in a compact simply connected symmetric space of higher codimension.

(2) Classification of all the biharmonic Riemannian submersions over the compact symmetric spaces including the Hopf fibering, i.e., the projection of the standard sphere of odd dimension over the complex projective space.

We express our gratitude to Professor Sigmundur Gudmundsson, Professor Shun Maeta, Professor Hisashi Naito, Professor Nobumitsu Nakauchi, Professor Shinji Ohno, and Professor Takashi Sakai, all of who were the co-authors of the papers raised in the chapters of this book.

Hajime Urakawa at Sendai, Fall 2018.

Contents

2. Preliminaries   63
3. Proof of main theorem   65

Chapter 5. Biharmonic Hypersurfaces in a Riemannian Manifold
            with Non-positive Ricci Curvature   69
1. Introduction and statement of results   69
2. Preliminaries   71
3. Some lemma for the Schrödinger type equation   72
4. Biharmonic isometric immersions   74

Chapter 6. Note on Biharmonic Map Equations   77
1. Preliminaries   77
2. Biharmonic map equations of an isometric immersion   79

Chapter 7. Harmonic Maps into Compact Lie Groups and
            Integrable Systems   85
1. Introduction and statement of results   85
2. Preliminaries   86
3. Determination of the bitension field   88
4. Biharmonic curves from $\mathbb{R}$ into compact Lie groups   94
5. Biharmonic maps from an open domain in $\mathbb{R}^2$ 99
6. Complexification of the biharmonic map equation   103
7. Determination of biharmonic maps   104

Chapter 8. Biharmonic Maps into Symmetric Spaces and
            Integrable Systems   111
1. Introduction and statement of results   111
2. Preliminaries   112
3. Determination of the bitension field   117
4. Biharmonic curves into Riemannian symmetric spaces   122
5. Biharmonic maps from plane domains   131

Chapter 9. Bubbling of Harmonic Maps and Biharmonic
            Maps   139
1. Introduction   139
2. Preliminaries   140
3. The Bochner-type estimation for the tension field of a
    biharmonic map   141
4. Moser's iteration technique and proof of Theorem 3.3   143
5. Bubbling theorem of biharmonic maps   148
6. Basic inequalities   150
7. Proof of Theorem 5.1   154

Part 1

Fundamental Materials on the Theory of Harmonic Maps and Biharmonic Maps

CHAPTER 1

Fundamental Materials of Riemannian Geometry

ABSTRACT. In this chapter, we give the fundamental materials in
Riemannian geometry. During this book, we assume basic materials on manifolds. We give, for an n dimensional manifold M with
Riemannian metric, the several notion of the length of a smooth
curve, the distance between two points, Levi-Civita connection,
the parallel transport along a curve, geodesics, the curvature tensor fields, integral, the divergence of a smooth vector field, and the
Laplace operator, Green's formula, the Laplacian for differential
forms, the first and second variational formulas of the length of
curves.

1. Riemannian manifolds

1.1. Riemannian metrics. Let us recall the definition of an n-
dimensional C^∞ manifold M. A Hausdorff topological space M is an
n-dimensional C^∞ manifold if M admits an open covering $\{U_\alpha\}_{\alpha\in\Lambda}$,
that is, each U_α ($\alpha \in \Lambda$) is an open subset satisfying $\cup_{\alpha\in\Lambda}U_\alpha = M$,
and topological homeomorphisms $\varphi_\alpha : U_\alpha \to \varphi_\alpha(U_\alpha)$ of open subset
U_α in M onto an open subset $\varphi_\alpha(U_\alpha)$ in the n-dimensional Euclidean
space $\mathbb{R}^n$ satisfying that, if $U_\alpha \cap U_\beta \neq \emptyset$ ($\alpha, \beta \in \Lambda$),

$$\varphi_\alpha \circ \varphi_\beta^{-1} : \mathbb{R}^n \supset \varphi_\beta(U_\alpha \cap U_\beta) \to \varphi_\alpha(U_\alpha \cap U_\beta) \subset \mathbb{R}^n$$

is a C^∞ diffeomorphism from an open subset $\varphi_\beta(U_\alpha \cap U_\beta)$ in $\mathbb{R}^n$ onto
another open subset $\varphi_\alpha(U_\alpha \cap U_\beta)$. A pair $(U_\alpha, \varphi_\alpha)$ ($\alpha \in \Lambda$) is called a
local chart of M.

If $(x^1, \dots, x^n)$ is the standard coordinate of the n-dimensional Eu-
clidean space $\mathbb{R}^n$, for every local chart $(U_\alpha, \varphi_\alpha)$, by means of $x_\alpha^i :=$
$x^i \circ \varphi_\alpha$ ($i = 1, \dots, n$), one can define local coordinate $(x_\alpha^1, \dots, x_\alpha^n)$ on
each open subset U_α of M . A pair $(U_\alpha, (x_\alpha^1, \dots, x_\alpha^n))$ is called **local
coordinate system**.

Next, recall the notion of a C^∞ Riemannian metric g on an n-
dimensional C^∞ manifold M.

DEFINITION 1.1. *A C^∞ **Riemannian metric** g on M is, by definition, for each point $x \in M$, g_x is a symmetric positive definite bilinear
form on the tangent space T_xM of M at x, whose g_x is C^∞ in x. That*

3

is, $g_x : T_x M \times T_x M \to \mathbb{R}$ satisfies that: for every $u, v, w \in T_x M$, $a, b \in \mathbb{R}$,

$$
\begin{cases}
g_x(au + bv, w) = a\, g_x(u, w) + b\, g_x(v, w), \\
g_x(u, v) = g_x(v, u), \\
g_x(u, u) > 0 \quad (0 \neq u \in T_x M).
\end{cases}
\tag{1.1}
$$

Then, with respect to local coordinates $(U_\alpha, (x_\alpha^1, \ldots, x_\alpha^n))$ of M, g can be written as

$$
g = \sum_{i,j=1}^n g_{ij}^\alpha \, dx_\alpha^i \otimes dx_\alpha^j \quad \text{(on } U_\alpha\text{)}.
$$

Here,

$$
g_{ij}^\alpha = g\left(\frac{\partial}{\partial x_\alpha^i}, \frac{\partial}{\partial x_\alpha^j} \right) = g\left(\frac{\partial}{\partial x_\alpha^j}, \frac{\partial}{\partial x_\alpha^i} \right) = g_{ji}^\alpha.
\tag{1.2}
$$

Then, g_x is C^∞ in $x \in M$ means that, every g_{ij}^α is C^∞ function on U_α.

For another local chart (U_β, φ_β) and local coordinate neighborhood system $(U_\beta, (x_\beta^1, \ldots, x_\beta^n))$, one can write $g = \sum_{k,\ell=1}^n g_{k\ell}^\beta \, dx_\beta^k \otimes dx_\beta^\ell$, where $g_{k\ell}^\beta = g\left(\frac{\partial}{\partial x_\beta^k}, \frac{\partial}{\partial x_\beta^\ell} \right)$. If $U_\alpha \cap U_\beta \neq \emptyset$, then it holds that, for every $k, \ell = 1, \ldots, n$,

$$
g_{k\ell}^\beta = \sum_{i,j=1}^n g_{ij}^\alpha \frac{\partial x_\alpha^i}{\partial x_\beta^k} \frac{\partial x_\alpha^j}{\partial x_\beta^\ell} \quad \text{(on } U_\alpha \cap U_\beta\text{)}.
\tag{1.3}
$$

In fact, since for a C^∞ function $f : M \to \mathbb{R}$ on M, it holds that $\frac{\partial f}{\partial x_\alpha^i} = \sum_{k=1}^n \frac{\partial x_\beta^k}{\partial x_\alpha^i} \frac{\partial f}{\partial x_\beta^k}$ on $U_\alpha \cap U_\beta$, we have

$$
\frac{\partial}{\partial x_\alpha^i} = \sum_{k=1}^n \frac{\partial x_\beta^k}{\partial x_\alpha^i} \frac{\partial}{\partial x_\beta^k} \quad (i = 1, \ldots, n).
\tag{1.4}
$$

Substituting this into (1.2), we obtain (1.3). Conversely, if we have (1.3), it holds that

$$
dx_\alpha^i = \sum_{k=1}^n \frac{\partial x_\alpha^i}{\partial x_\beta^k} dx_\beta^k \quad (i = 1, \ldots, n),
\tag{1.5}
$$

which implies that

$$
g = \sum_{i,j=1}^n g_{ij}^\alpha \, dx_\alpha^i \otimes dx_\alpha^j = \sum_{k,\ell=1}^n g_{k\ell}^\beta \, dx_\beta^k \otimes dx_\beta^\ell \quad \text{(on } U_\alpha \cap U_\beta\text{)}.
\tag{1.6}
$$

Therefore, g is determined uniquely independently on a choice of local coordinate neighborhood system $(U_\alpha, (x_\alpha^1, \ldots, x_\alpha^n))$.

Notice that $(g_{ij}^{\alpha})_{i,j=1,\dots,n}$ are C^{∞} functions on U_{α} whose values are positive definite symmetric matrices of degree n. We denote their determinants by $\det(g)$. In the following, we will sometimes denote $(U, (x^1, \dots, x^n))$ by omitting subscripts α.

1.2. Lengths of curves. A continuous curve $\sigma : [a, b] \to M$ is a C^1 **curve** if, there exists a sufficiently small positive number $\epsilon > 0$ such that, if $\sigma(t) \in M$, is defined as

$$\sigma(t) = (\sigma^1(t), \dots, \sigma^n(t)) \qquad (t \in (a - \epsilon, b + \epsilon))$$

on a local coordinate neighborhood $(U, (x^1, \dots, x^n))$ around $\sigma(t)$, each $\sigma^i(t)$ is C^1 function in t on $(a - \epsilon, b + \epsilon)$. Then, one can define the **tangent vector** $\dot{\sigma}(t) \in T_{\sigma(t)}M$ of a C^1 curve $\sigma(t)$ by

$$\dot{\sigma}(t) = \sum_{i=1}^{n} \frac{d\sigma^i(t)}{dt} \left(\frac{\partial}{\partial x^i} \right)_{\sigma(t)}. \tag{1.7}$$

Next, if we denote

$$\|\dot{\sigma}(t)\|^2 := g_{\sigma(t)}(\dot{\sigma}(t), \dot{\sigma}(t)) = \sum_{i,j=1}^{n} \frac{d\sigma^i(t)}{dt} \frac{d\sigma^j(t)}{dt} g_{ij}(\sigma(t)), \tag{1.8}$$

$[a, b] \ni t \mapsto \|\dot{\sigma}(t)\|$ is a continuous function in t, the **length** $L(\sigma)$ of a C^1 curve $\sigma : [a, b] \to M$, can be defined by

$$L(\sigma) := \int_a^b \|\dot{\sigma}(t)\| \, dt. \tag{1.9}$$

Now we will discuss the **arc length parametrization** of a C^1 curve $\sigma : [a, b] \to M$. In the following, we always assume that every C^1 curve $\sigma : [a, b] \to M$ is **regular**, i.e., $\dot{\sigma}(t) \neq 0$ ($\forall\, t \in [a, b]$). Then, we can define the length $s(t)$ of the sub-arc $\sigma : [a, t] \to M$ of a C^1 curve σ by

$$s(t) := \int_a^t \|\dot{\sigma}(r)\| \, dr. \tag{1.10}$$

Since the differentiation $s'(t)$ of $s(t)$ with respect to t is given by

$$s'(t) = \frac{ds(t)}{dt} = \|\dot{\sigma}(t)\| > 0,$$

$s(t)$ is strictly monotone increasing function in t. Thus, one can define its inverse function, by denoting as $t = t(s)$. Therefore, one can define the **parametrization in terms of the arclength** s of the curve σ by

$$\overline{\sigma}(s) := \sigma(t(s)) \qquad (0 \le s \le L(\sigma)). \tag{1.11}$$

If we denote the differentiation of $\bar{\sigma}$ with respect to s, by $\bar{\sigma}'(s)$ and let $t'(s) := \frac{dt(s)}{ds}$, then it holds that $\bar{\sigma}'(s) = \frac{d\sigma}{dt}(t(s)) \frac{dt(s)}{ds}$, and for every s,

$$\|\bar{\sigma}'(s)\| = t'(s) \left\| \frac{d\sigma}{dt}(t(s)) \right\| = \frac{dt(s)}{ds} \frac{ds(t)}{dt} = 1. \tag{1.12}$$

We usually take the parameter of a C^1 curve σ, a constant multiple of the arclength s, as $c\,s$.

1.3. Distance. One can define the distance of a connected C^∞ Riemannian manifold (M, g) by using the arclength of a C^1 curve: For every two points $x, y \in M$, let us define

$$d(x, y) := \inf\{L(\sigma)|\, \sigma \text{ is a piecewise } C^1 \text{ curve}$$

$$\text{connecting two points } x \text{ and } y\}. $$
$$\tag{1.13}$$

Here **piecewise C^1 curve** is a continuous curve connecting a finite number of C^1 curves. Since M is arc-wise connected, $d(x, y)$ is finite. Then, d satisfies the three axioms of the distance and (M, d) becomes a **metric space**:

$$\begin{array}{lll} (1) & d(x, y) = d(y, x) & (x, y \in M), \\ (2) & d(x, y) + d(y, z) \geq d(x, z) & (x, y, z \in M), \\ (3) & d(x, y) > 0 \ (x \neq y). \ d(x, y) = 0 \text{ holds if and only if } x = y. \end{array}$$

Furthermore, the topology of a metric space (M, d) coincides with the original one which defines a manifold structure of M. If the metric space (M, d) is complete, i.e., every Cauchy sequence $\{x_k\}_{k=1}^\infty$ of points in M, i.e., $d(x_k, x_\ell) \to 0 \ (k, \ell \to \infty)$ is convergent. Namely, there exists a point $x \in M$ such that $d(x_k, x) \to 0 \ (k \to \infty)$. We say a Riemannian manifold (M, g) is **complete** if (M, d) is so. Every compact Riemannian manifold is complete. We also define the **diameter** of a compact Riemannian manifold (M, g) as

$$0 < \operatorname{diam}(M, g) := \max\{d(x, y)|\, x, y \in M\} < \infty. $$
$$\tag{1.14}$$

2. Connection

2.1. Levi-Civita connection. On an n-dimensional C^∞ manifold (M, g), a **vector field** X is an assignment $X_x \in T_x M$ $(x \in M)$. By definition, a C^∞ **vector field** X is, taking a local coordinate neighborhood system of $x \in M$, $(U_\alpha, (x_\alpha^1, \ldots, x_\alpha^n))$ $(\alpha \in \Lambda)$, on U_α, it can be written as $X = \sum_{i=1}^n X_\alpha^i \frac{\partial}{\partial x_\alpha^i}$, where $X_\alpha^i \in C^\infty(U_\alpha)$ $(i = 1, \ldots, n, \alpha \in$

Λ). Taking another local coordinate system $(U_\beta, (x_\beta^1, \ldots, x_\beta^n))$, one write $X = \sum_{k=1}^n X_\beta^k \frac{\partial}{\partial x_\beta^k}$ on U_β, it holds that

$$X_\beta^k = \sum_{i=1}^n X_\alpha^i \frac{\partial x_\beta^k}{\partial x_\alpha^i} \qquad (\text{on } U_\alpha \cap U_\beta \; ; \; k = 1, \ldots, n),$$

which is called **the changing formula of local coordinates** of a vector field X.

Now, let us denote by $\mathfrak{X}(M)$, the totality of C^∞ vector fields, and by $C^\infty(M)$, the one of C^∞ functions on M. For $X \in \mathfrak{X}(M)$ and $f \in C^\infty(M)$, $Xf \in C^\infty(M)$ can be written as $(Xf)(x) = \sum_{i=1}^n X^i(x) \frac{\partial f}{\partial x^i}(x)$ $(x \in U)$ in terms of local coordinate system $(U, (x^1, \ldots, x^n))$. For every two C^∞ vector fields $X = \sum_{i=1}^n X^i \frac{\partial}{\partial x^i}$ and $Y = \sum_{i=1}^n Y^i \frac{\partial}{\partial x^i} \in \mathfrak{X}(M)$ on M, one can define the third vector field $[X, Y] \in \mathfrak{X}(M)$ on M by

$$[X, Y] = \sum_{i=1}^n \left\{ X(Y^i) - Y(X^i) \right\} \frac{\partial}{\partial x^i} = \sum_{i=1}^n \left\{ \sum_{j=1}^n \left(X^j \frac{\partial Y^i}{\partial x^j} - Y^j \frac{\partial X^i}{\partial x^j} \right) \right\} \frac{\partial}{\partial x^i}.$$

Then, it holds that

$$[X, Y] f = X(Y f) - Y(X f) \qquad (f \in C^\infty(M)),$$

where $[X, Y] \in \mathfrak{X}(M)$ is called the **bracket** of X and Y.

A **connection** ∇ on C^∞ (M, g) is a C^∞ map

$$\nabla : \mathfrak{X}(M) \times \mathfrak{X}(M) \ni (X, Y) \mapsto \nabla_X Y \in \mathfrak{X}(M)$$

satisfying the following properties:

$$\begin{cases} (1) & \nabla_X(Y + Z) = \nabla_X Y + \nabla_X Z \\ (2) & \nabla_{X+Y} Z = \nabla_X Z + \nabla_Y Z \\ (3) & \nabla_{fX} Y = f \nabla_X Y \\ (4) & \nabla_X(f Y) = (X f) Y + f \nabla_X Y, \end{cases} \qquad (2.1)$$

for $f \in C^\infty(M)$, $X, Y, Z \in \mathfrak{X}(M)$. Then, the following theorem holds.

THEOREM 2.1. *Let (M, g) be an n-dimensional C^∞ Riemannian manifold. One can define a connection, called* **Levi-Civita connection** ∇ *by the following equation:*

$$2 g(\nabla_X Y, Z) = X(g(Y, Z)) + Y(g(Z, X)) - Z(g(X, Y))$$
$$+ g(Z, [X, Y]) + g(Y, [Z, X]) - g(X, [Y, Z]),$$
$$(2.2)$$

for $X, Y, Z \in \mathfrak{X}(M)$. Then, the Levi-Civita connection ∇ satisfies

$$(1) \qquad X(g(Y,Z)) = g(\nabla_X Y, Z) + g(Y, \nabla_X Z),$$

$$(2) \qquad \nabla_X Y - \nabla_Y X - [X,Y] = 0.$$

Conversely, The only connection ∇ satisfying the properties (1) and (2) is the Levi-Civita connection.

For every $X, Y \in \mathfrak{X}(M)$, in the equation (2.2), $g(X,Y)$ is a C^∞ function on M defined by $g(X,Y)(x) := g_x(X_x, Y_x)$ $(x \in M)$. For the readers, try to prove Theorem 1.2.

If we express ∇ in terms of local coordinate system $(U, (x^1, \dots, x^n))$ of M, we have

$$\nabla_{\frac{\partial}{\partial x^i}} \frac{\partial}{\partial x^j} = \sum_{k=1}^{n} \Gamma_{ij}^{k} \frac{\partial}{\partial x^k} \qquad (\text{where } \Gamma_{ij}^{k} \in C^\infty(U), \ i, j, k = 1, \dots, n),$$

as $\left[\frac{\partial}{\partial x^i}, \frac{\partial}{\partial x^j}\right] = 0$, one can obtain

$$\Gamma_{ij}^{k} = \frac{1}{2} \sum_{\ell=1}^{n} g^{k\ell} \left(\frac{\partial g_{j\ell}}{\partial x^i} + \frac{\partial g_{i\ell}}{\partial x^j} - \frac{\partial g_{ij}}{\partial x^\ell} \right) \qquad (2.3)$$

for $X = \frac{\partial}{\partial x^i}$, $Y = \frac{\partial}{\partial x^j}$, $Z = \frac{\partial}{\partial x^k}$ in (2.2). Here, we denote $g_{ij} = g\left(\frac{\partial}{\partial x^i}, \frac{\partial}{\partial x^j}\right)$, and $(g^{k\ell})$, the inverse matrix of positive definite matrix (g_{ij}). Γ_{ij}^{k} is called **Christoffel symbol** of Levi-Civita connection ∇.

2.2. Parallel transport. For a C^1 curve $\sigma : [a,b] \to M$ in M, X is a C^1 **vector field along** σ if (1) $X(t) \in T_{\sigma(t)}M$ $(\forall\, t \in [a,b])$, and (2) in terms of local coordinates $(U, (x^1, \dots, x^n))$ at each point $\sigma(t)$, it holds that

$$X(t) = \sum_{i=1}^{n} \xi^i(t) \left(\frac{\partial}{\partial x^i} \right)_{\sigma(t)} \in T_{\sigma(t)}M,$$

where each $\xi^i(t)$ is C^1 function in t. Such a vector field X is **parallel** with respect to connection ∇ if $\nabla_{\dot{\sigma}(t)}X = 0$.

Let $\sigma(t) = (\sigma^1(t), \dots, \sigma^n(t))$ be a local expression of a C^1 curve σ. Then, it turns out that the necessary and sufficient condition to hold $\nabla_{\dot{\sigma}(t)}X = 0$ is

$$\frac{d\xi^i(t)}{dt} + \sum_{j,k=1}^{n} \Gamma_{jk}^{i}(\sigma(t)) \frac{d\sigma^j(t)}{dt} \xi^k(t) = 0 \qquad (i = 1, \dots, n),$$

$$(2.4)$$

by means of (1.7) and (2.2). For every C^1 curve $\sigma : [a,b] \to M$ and an arbitrarily given initial condition of X at $x = \sigma(a)$, i.e., the coefficients $(\xi^1(a), \ldots, \xi^n(a))$ of $X(a)$, the parallel vector field X along $\sigma : [a, b] \to M$

$$
\begin{cases}
\nabla_{\dot\sigma(t)} X = 0 \qquad (a < t < b), \\
X(a) = \displaystyle\sum_{i=1}^n \xi^i(a) \left(\frac{\partial}{\partial x^i}\right)_{\sigma(a)}
\end{cases}
\tag{2.5}
$$

is uniquely determined, because of the existence and uniqueness theorems of the first order ordinary differential system (2.4).

In particular, the correspondence

$$
P_\sigma : T_{\sigma(a)} M \ni X(a) \mapsto X(b) \in T_{\sigma(b)} M
$$

is uniquely determined. This correspondence $P_\sigma : T_{\sigma(a)} M \to T_{\sigma(b)} M$ is a linear isomorphism which satisfies

$$
g_{\sigma(b)}(P_\sigma(u), P_\sigma(v)) = g_{\sigma(a)}(u, v) \qquad (u, v \in T_{\sigma(a)} M).
\tag{2.6}
$$

This is because if we let Y and Z be parallel vector fields along σ with their initial conditions are arbitrarily given $u, v \in T_{\sigma(a)} M$, and X be $X(t) = \dot\sigma(t)$ $(t \in [a, b])$. Then, it holds that

$$
\frac{d}{dt} g_{\sigma(t)}(Y(t), Z(t)) = X\big(g(Y, Z)\big) = g(\nabla_X Y, Z) + g(Y, \nabla_X Z) = 0
$$

since $\nabla_X Y = 0$ and $\nabla_X Z = 0$. Thus, $g_{\sigma(t)}(Y(t), Z(t))$ is constant in t. $\qquad\qquad\qquad\qquad\qquad\qquad\qquad\qquad\qquad\qquad\qquad\qquad\qquad\square$

The correspondence $P_\sigma : T_{\sigma(a)} M \to T_{\sigma(b)} M$ is called **parallel transport** along a C^1 curve $\sigma : [a, b] \to M$.

2.3. Geodesic. A C^1 curve $\sigma : [a, b] \to M$ in M is **geodesic** if the tangent vector field $\dot\sigma$ is parallel, i.e., $\nabla_{\dot\sigma(t)} \dot\sigma = 0$. In terms of local coordinate system $(U, (x^1, \ldots, x^n))$ of M, if we express $\sigma(t) = (\sigma^1(t), \ldots, \sigma^n(t))$, $\dot\sigma(t) = \sum_{i=1}^n \frac{d\sigma^i(t)}{dt}\left(\frac{\partial}{\partial x^i}\right)_{\sigma(t)}$ on U, the condition $\nabla_{\dot\sigma(t)} \dot\sigma = 0$ in (1.18) is $\xi^i(t) = \frac{d\sigma^i(t)}{dt}$ $(i = 1, \ldots, n)$, it holds that

$$
\frac{d^2\sigma^i(t)}{dt^2} + \sum_{j,k=1}^n \Gamma^i_{jk}(\sigma(t)) \frac{d\sigma^j(t)}{dt} \frac{d\sigma^k(t)}{dt} = 0 \qquad (i = 1, \ldots, n)
\tag{2.7}
$$

are the second order ordinary differential system, and for arbitrarily given initial conditions $(\sigma^1(a), \ldots, \sigma^n(a))$ and $\left(\frac{d\sigma^1(t)}{dt}(a), \ldots, \frac{d\sigma^n(t)}{dt}(a)\right)$, there exist uniquely solution of (2.7) if t is close enough to a. Namely, for every point $p \in M$ and every vector $u \in T_p M$, there exists a unique geodesic $\sigma(t)$, passing through p at the initial time and having u as the initial vector at p if $|t|$ is sufficiently small. Therefore, there exists a unique geodesic satisfying $\sigma(0) = p$ and $\dot\sigma(0) = u$. Let us denote it by $\sigma(t) = \mathrm{Exp}_p(t\,u) \in M$. The **exponential map**

$$\mathrm{Exp}_p : T_pM \to M$$

can be defined locally by $T_pM \ni u \mapsto \sigma(1) = \mathrm{Exp}_p u \in M$. It is defined on a neighborhood of 0 in T_pM.

On the problem when the geodesic $t \mapsto \mathrm{Exp}_p(t\,u)$ is extended to $-\infty < t < \infty$ for every tangent vector $u \in T_pM$, the following is well known.

THEOREM 2.2 (Hopf-Rinow). *Let (M,g) be a connected C^∞ Riemannian manifold. Then, the following two conditions are equivalent:*
(1) (M,g) is complete.
(2) For every point $p \in M$, the exponential map $\mathrm{Exp}_p : T_pM \to M$ can be defined on the whole space T_pM.
Therefore, in these cases, arbitrarily given two points p and q in M can be joined by a geodesic with its length $d(p,q)$.

Due to this theorem, for every compact C^∞ Riemannian manifold (M,g), the exponential map $\mathrm{Exp}_p : T_pM \to M$ is defined on the whole space T_pM.

Thus, it is natural to define for every point $p \in M$, the **injectivity radius** at p, inj_p by

$$\mathrm{inj}_p := \sup\{r > 0 |\ \mathrm{Exp}_p \text{ is a diffeomorphism on } B_r(0_p)\}, \tag{2.8}$$

where $B_r(0_p) := \{u \in T_pM |\ g_p(u,u) < r^2\}$ is a ball with radius r, centered at the zero vector 0_p in the tangent space T_pM at p. Then, we define the **injectivity radius** of (M,g) by

$$\mathrm{inj} = \mathrm{inj}(M) := \inf\{\mathrm{inj}_p |\ p \in M\}. \tag{2.9}$$

For every compact C^∞ Riemannian manifold (M,g), $\mathrm{inj} = \mathrm{inj}(M) > 0$.

Let $\{v_i\}_{i=1}^n$ be a basis of T_pM. Then, the mapping $\mathrm{Exp}_p(\sum_{i=1}^n x^i v_i)$ $\mapsto (x^1, \dots, x^n)$ gives a local coordinate system on some neighborhood around p, called **normal coordinate system** on a neighborhood of p.

3. Curvature tensor fields

A **tensor field** T **on** M **of type** (r,s) is a C^∞ section of the vector bundle

$$\overbrace{TM \otimes \cdots \otimes TM}^{r\ \text{times}} \otimes \overbrace{T^*M \otimes \cdots \otimes T^*M}^{s\ \text{times}},$$

namely, if T is expressed in terms of C^∞ functions $T_{\alpha\,j_1\cdots j_s}^{\ \ i_1\cdots i_r}$ on U_α, with respect to the local coordinates $(U_\alpha, (x_\alpha^1, \ldots, x_\alpha^n))$ of M,

$$T = \sum T_{\alpha j_1\cdots j_s}^{\ \ i_1\cdots i_r} \frac{\partial}{\partial x_\alpha^{i_1}} \otimes \cdots \otimes \frac{\partial}{\partial x_\alpha^{i_r}} \otimes dx_\alpha^{j_1} \otimes \cdots \otimes dx_\alpha^{j_s},$$

and it has the same form for other coordinate systems $(U_\beta, (x_\beta^1, \cdots, x_\beta^n))$, then it holds that, on $U_\alpha \cap U_\beta (\neq \emptyset)$,

$$T_{\alpha\,j_1\cdots j_s}^{\ \ i_1\cdots i_r} = \sum T_{\beta\,\ell_1\cdots\ell_s}^{\ \ k_1\cdots k_r} \frac{\partial x_\alpha^{i_1}}{\partial x_\beta^{k_1}} \cdots \frac{\partial x_\alpha^{i_r}}{\partial x_\beta^{k_r}} \frac{\partial x_\beta^{\ell_1}}{\partial x_\alpha^{j_1}} \cdots \frac{\partial x_\beta^{\ell_s}}{\partial x_\alpha^{j_s}},$$

where the right hand sum is taken over all $k_1, \ldots, k_r, \ell_1, \ldots, \ell_s$ through $\{1, \ldots, n\}$.

Notice that tensor fields of type $(1,0)$ are vector fields, and alternating tensor fields of type $(0, s)$ are **differential forms** of degree s.

In terms of Levi-Civita connection ∇ of a Riemannian manifold (M, g), a tensor field R of type $(1,3)$ can be defined as follows. For vector fields $X, Y, Z \in \mathfrak{X}(M)$ on M,

$$R(X, Y)Z = \nabla_X(\nabla_Y Z) - \nabla_Y(\nabla_X Z) - \nabla_{[X,Y]}Z. \qquad (3.1)$$

Then, it holds that

$$R(X, Y)Z + R(Y, Z)X + R(Z, X)Y = 0$$

which is called the **first Bianchi identity**. Furthermore, it holds that, for $\alpha, \beta, \gamma \in C^\infty(M)$,

$$R(\alpha X, \beta Y)(\gamma Z) = \alpha \beta \gamma R(X, Y)Z. \qquad (3.2)$$

The tensor field R is called **curvature tensor field**. Due to (1.25), $(R(X, Y)Z)_x \in T_x M$ is uniquely determined only on tangent vectors $u = X_x$, $v = Y_x$, $w = Z_x \in T_x M$, so that one can write as $R(u, v)w = (R(X, Y)Z)_x \in T_x M$.

If we write R in terms of local coordinates $(U, (x^1, \ldots, x^n))$ of M, as

$$R\left(\frac{\partial}{\partial x^i}, \frac{\partial}{\partial x^j}\right) \frac{\partial}{\partial x^k} = \sum_{\ell=1} R^\ell_{\ ijk} \frac{\partial}{\partial x^\ell} \qquad (1 \leq i, j, k \leq n),$$

it holds that

$$R^\ell_{\ ijk} = \frac{\partial}{\partial x^i} \Gamma^\ell_{kj} - \frac{\partial}{\partial x^j} \Gamma^\ell_{ki} + \sum_{a=1}^n \left\{ \Gamma^a_{kj} \Gamma^\ell_{ai} - \Gamma^a_{ki} \Gamma^\ell_{aj} \right\}.$$

Taking a linearly independent system $\{u, v\}$ of the tangent space $T_x M$ at $x \in M$ of M, the quantity

$$K(u, v) := \frac{g(R(u, v)v, u)}{g(u, u)\, g(v, v) - g(u, v)^2}$$

is called the **sectional curvature** determined by $\{u, v\}$. If it holds
that $K(u,v) > 0 \, (< 0)$, for every point $x \in M$ and every linearly
independent system $\{u, v\}$ of T_xM, then (M, g) is called **positively
curved (negatively curved)**, respectively.

Let $\{e_i\}_{i=1}^n$ be an orthonormal basis of (T_xM, g_x) $(x \in M)$, one can
define a linear map $\rho : T_xM \to T_xM$ by

$$\rho(u) := \sum_{i=1}^n R(u, e_i)e_i \qquad (u \in T_xM).$$

This linear map is independent of the choice of an orthonormal basis
$\{e_i\}$ of T_xM, and ρ becomes a symmetric tensor field of type $(1, 1)$,
called the **Ricci transform**. The tensor field ρ of type $(0, 2)$ defined
by

$$\rho(u, v) = g(\rho(u), v) = g(u, \rho(v)) = \sum_{i=1}^n g(R(u, e_i)e_i, v)$$

is called the **Ricci tensor**.

Furthermore, a C^∞ function S on M defined by $S = \sum_{i=1}^n \rho(e_i, e_i)$
is called the **scalar curvature**. These definitions ρ and S are independent of the choice of an orthonormal basis $\{e_i\}_{i=1}^n$.

4. Integration

Let (M, g) be an n-dimensional compact C^∞ Riemannian manifold.
Let us define the **integral** $\int_M f\, v_g$ of a continuous function f on M.

For an n-dimensional C^∞ Riemannian manifold (M, g), let us take
a coordinate neighborhood system $\{(U_\alpha, \varphi_\alpha)| \alpha \in \Lambda\}$ which comes from
the manifold structure of M. Then, one can give a **partition of unity**
$\{\eta_\alpha | \alpha \in \Lambda\}$ subordinate to an open covering $\{U_\alpha\}_{\alpha \in \Lambda}$ of M. Namely,

(i) $\eta_\alpha \in C^\infty(M)$ $(\alpha \in \Lambda)$,
(ii) $0 \le \eta_\alpha(x) \le 1$ $(x \in M, \alpha \in \Lambda)$,
(iii) for each $\alpha \in \Lambda$, the support of η_α satisfies $\operatorname{supp}(\eta_\alpha) \subset U_\alpha$,
(iv) $\sum_{\alpha \in \Lambda} \eta_\alpha(x) = 1$ $(x \in M)$.

Here, the **support** of a continuous function f on M, $\operatorname{supp}(f)$, is by
definition the closure of $\{x \in M | f(x) \ne 0\}$.

Now, let us define the integral of a continuous function f whose
support is contained in a coordinate neighborhood $(U_\alpha, (x_\alpha^1, \dots, x_\alpha^n))$.
For such a continuous function f, let us define

$$\int_{U_\alpha} f\, v_g := \int_{\varphi_\alpha(U_\alpha)} (f \circ \varphi_\alpha^{-1}) \sqrt{\det(g)}\, dx_\alpha^1 \cdots dx_\alpha^n. \qquad (4.1)$$

Here, $\det(g) := \det\left(g\left(\frac{\partial}{\partial x_\alpha^i}, \frac{\partial}{\partial x_\alpha^j}\right)\right)$. The differential form v_g of degree
$n = \dim M$ defined by $v_g = \sqrt{\det(g)}\, dx_\alpha^1 \wedge \cdots \wedge dx_\alpha^n$ is called the

volume form of (M, g). Then, for an arbitrary continuous function f on M, the **integral** $\int_M f\, v_g$ over M is defined by

$$\int_M f\, v_g = \int_M \left\{ \sum_{\alpha \in \Lambda} \eta_\alpha \right\} f\, v_g = \sum_{\alpha \in \Lambda} \int_{U_\alpha} (\eta_\alpha f)\, v_g. \tag{4.2}$$

Here, the integral $\int_{U_\alpha} (\eta_\alpha f)\, v_g$ over each U_α in (1.27) is defined by (1.26) for $\eta_\alpha f$ since $\operatorname{supp}(\eta_\alpha f) \subset U_\alpha$.

The L^2 **inner product** $(\ ,\ )$ for two continuous functions f and h on M, and the L^2 **norm** of f are defined by

$$(f, h) = \int_M f\, h\, v_g, \qquad \|f\| = \sqrt{(f, f)}.$$

The integral of $f \equiv 1$, i.e., $\operatorname{Vol}(M, g) := \int_M v_g$ is called the **volume** of (M, g). Since we assume that M is compact, it holds that $0 < \operatorname{Vol}(M, g) < \infty$.

5. Divergence of vector fields and the Laplacian

5.1. Divergences of vector fields, gradient vector fields and the Laplacian. For every C^∞ vector field on M, $X \in \mathfrak{X}(M)$, a C^∞ function $\operatorname{div}(X)$ on M, called **divergence** of a vector field X is defined as follows: Take, first, local coordinates of M, $(U, (x^1, \ldots, x^n))$, and orthonormal frame fields $\{e_i\}_{i=1}^n$ on U, i.e., $T_x M \ni e_{i\,x}$ $(x \in U)$ satisfies $g_x(e_{i\,x}, e_{j\,x}) = \delta_{ij}$. Indeed, $\{e_i\}_{i=1}^n$ can be obtained by proceeding the Gram-Schmidt orthonormalization to n vector fields $\left\{ \frac{\partial}{\partial x^i} \right\}_{i=1}^n$ on U which are linearly independent at each point of U. Then, let $X = \sum_{i=1}^n X^i \frac{\partial}{\partial x^i}$ be a local expression of X on U. One can define $\operatorname{div}(X) \in C^\infty(M)$ by

$$\operatorname{div}(X) = \sum_{i=1}^n g(e_i, \nabla_{e_i} X)$$

$$= \frac{1}{\sqrt{\det(g)}} \sum_{i=1}^n \frac{\partial}{\partial x^i} \left(\sqrt{\det(g)}\, X^i \right). \tag{5.1}$$

This definition does not depend on choices of local coordinates $(U, (x^1, \ldots, x^n))$ and local orthonormal frame fields $\{e_i\}$.

For every C^∞ function $f \in C^\infty(M)$ on M, one can define the **gradient vector field** $X = \operatorname{grad}(f) = \nabla f \in \mathfrak{X}(M)$ is defined by

$$g(Y, X) = df(Y) = Yf \qquad (Y \in \mathfrak{X}(M)).$$

Taking local coordinates $(U, (x^1, \ldots, x^n))$ on M and local orthonormal frame fields $\{e_i\}_{i=1}^n$ on U, it holds that

$$\operatorname{grad}(f) = \nabla f = \sum_{i=1}^n e_i(f)\, e_i = \sum_{i,j=1}^n g^{ij}\, \frac{\partial f}{\partial x^j}\, \frac{\partial}{\partial x^i}, \qquad (5.2)$$

where $g_{ij} = g\!\left(\frac{\partial}{\partial x^i}, \frac{\partial}{\partial x^j}\right)$, and $(g^{k\ell})$ is the inverse matrix of (g_{ij}).

On the cotangent bundle T^*M, a natural metric, denoted by the same symbol g, can be defined in such a way that

$$g(df_1, df_2) = g(\operatorname{grad}(f_1), \operatorname{grad}(f_2)) = g(\nabla f_1, \nabla f_2) \qquad (f_1,\, f_2 \in C^\infty(M)).$$

Here, the differential 1-form $df \in \Gamma(T^*M)$ for $f \in C^\infty(M)$ is defined by $df(v) = vf$ ($v \in T_xM$), and it holds that $df = \sum_{i=1}^n \frac{\partial f}{\partial x^i}\, dx^i$.

Under these conditions, the **Laplacian** $\Delta f \in C^\infty(M)$ of every $f \in C^\infty(M)$ can be defined as follows:

$$\begin{aligned}
\Delta f &= -\operatorname{div}(\operatorname{grad} f)\\[1mm]
&= -\frac{1}{\sqrt{\det(g)}} \sum_{i,j=1}^m \frac{\partial}{\partial x^i}\!\left(\sqrt{\det(g)}\, g^{ij}\, \frac{\partial f}{\partial x^j}\right)\\[1mm]
&= -\sum_{i,j=1}^n g^{ij}\!\left(\frac{\partial^2 f}{\partial x^i \partial x^j} - \sum_{k=1}^n \Gamma_{ij}^k\, \frac{\partial f}{\partial x^k}\right)\\[1mm]
&= -\sum_{i=1}^n \left\{ e_i(e_i f) - (\nabla_{e_i} e_i)f \right\}. \qquad (5.3)
\end{aligned}$$

This linear elliptic partial differential operator $\Delta : C^\infty(M) \ni f \mapsto \Delta f \in C^\infty(M)$ acting on C^∞ functions on M is called the **Laplacian** (or the **Laplace-Beltrami operator**) which depends on a choice of g, so we denote it by Δ_g if we want to emphasize it.

5.2. Green's formula. We have

PROPOSITION 5.1. *Let (M, g) be an n-dimensional compact Riemannian manifold. For f, f_1, $f_2 \in C^\infty(M)$ and $X \in \mathfrak{X}(M)$,*

(1) $\displaystyle \int_M f\, div(X)\, v_g = -\int_M g(grad(f), X)\, v_g,$

(2) $\displaystyle \int_M (\Delta f_1)\, f_2\, v_g = \int_M g(\nabla f_1, \nabla f_2)\, v_g = \int_M f_1\, (\Delta f_2)\, v_g,$

(3) $\displaystyle \int_M div(X)\, v_g = 0$ \qquad **(Green's formula).**

Proof (1) By $\nabla_{e_i}(f\,X) = (e_i\,f)\,X + f\,\nabla_{e_i}X$, we have

$$\operatorname{div}(f\,X) = \sum_{i=1}^{n} g(e_i, \nabla_{e_i}(f\,X)) = \sum_{i=1}^{n}(e_if)g(e_i,X) + f\sum_{i=1}^{n} g(e_i,\nabla_{e_i}X)$$
$$= g(\operatorname{grad}(f),X) + f\operatorname{div}(X). \tag{5.4}$$

Integrating the both sides of (5.4) over M, by (3), we have

$$0 = \int_M \operatorname{div}(f\,X)\,v_g = \int_M g(\operatorname{grad}(f),X)\,v_g + \int_M f\operatorname{div}(X)\,v_g.$$

(2) In (1), let $f = f_1$, $X = \operatorname{grad}(f_2)$. Then, we have

$$\int_M f_1\,(\Delta f_2)\,v_g = -\int_M f_1\operatorname{div}(\operatorname{grad}(f_2))\,v_g = \int_M g(\operatorname{grad}(f_1),\operatorname{grad}(f_2))\,v_g,$$

$$g(\operatorname{grad}(f_1),\operatorname{grad}(f_2)) = g(\operatorname{grad}(f_2),\operatorname{grad}(f_1)),$$

$$\int_M g(\operatorname{grad}(f_1),\operatorname{grad}(f_2))\,v_g = \int_M (\Delta f_1)\,f_2\,v_g.$$

(3) We use partition of unity $1 = \sum_{\alpha\in\Lambda}\eta_\alpha$ subordinate to an open covering $\{U_\alpha\}_{\alpha\in\Lambda}$ of M. On each U_α, we express $X = \sum_{i=1}^{n} X_\alpha^i \frac{\partial}{\partial x_\alpha^i}$. Then,

$$\int_M \operatorname{div}(X)\,v_g = \int_M \operatorname{div}(1\cdot X)\,v_g = \int_M \operatorname{div}\left(\left(\sum_{\alpha\in\Lambda}\eta_\alpha\right)X\right)v_g$$
$$= \sum_{\alpha\in\Lambda}\int_M \operatorname{div}(\eta_\alpha\,X)\,v_g \qquad \text{(finite sum)}. \tag{5.5}$$

Here, for every $\alpha\in\Lambda$, since $\operatorname{supp}(\eta_\alpha)\subset U_\alpha$, we have

$$\int_M \operatorname{div}(\eta_\alpha\,X)\,v_g = \int_{U_\alpha}\operatorname{div}(\eta_\alpha\,X)\,v_g$$
$$= \int_{U_\alpha}\frac{1}{\sqrt{\det(g)}}\sum_{i=1}^{n}\frac{\partial}{\partial x_\alpha^i}\left(\sqrt{\det(g)}\,\eta_\alpha\,X_\alpha^i\right)\sqrt{\det(g)}\,dx_\alpha^1\cdots dx_\alpha^n$$
$$= \sum_{i=1}^{n}\int_{U_\alpha}\frac{\partial}{\partial x_\alpha^i}\left(\sqrt{\det(g)}\,\eta_\alpha\,X_\alpha^i\right)dx_\alpha^1\cdots dx_\alpha^n. \tag{5.6}$$

Here we get (1.33) $= 0$. In fact, for each $\alpha\in\Lambda$ and $i = 1,\ldots,n$, the integral $\int_{U_\alpha}\frac{\partial}{\partial x_\alpha^i}\left(\sqrt{\det(g)}\,\eta_\alpha\,X_\alpha^i\right)dx_\alpha^1\cdots dx_\alpha^n$ depends only on the boundary value of $\sqrt{\det(g)}\,\eta_\alpha\,X_\alpha^i$ at the boundary of U_α, but the boundary value must be 0 since $\operatorname{supp}(\eta_\alpha)\subset U_\alpha$. We completed the proof. $\square$

6. The Laplacian for differential forms

In this section, we treat with the Laplacian acting differential forms on an n-dimensional C^∞ compact Riemannian manifold (M, g). Let us denote by $\Gamma(E)$, the space of all C^∞ sections of a vector bundle E. For every $0 \le r \le n$, let us define $A^r(M) = \Gamma(\wedge^r T^*M)$ whose elements ω are called r-**differential forms** on M, namely, all ω satisfy

$$\omega(X_{\sigma(1)}, \ldots, X_{\sigma(r)}) = \operatorname{sgn}(\sigma)\, \omega(X_1, \ldots, X_r) \qquad (\forall\, \sigma \in \mathfrak{S}_r),$$

and are multilinear maps

$$\omega : \underbrace{TM \times \ldots \times TM}_{r \text{ times}} \ni (X_1, \cdots, X_r) \mapsto \omega(X_1, \ldots, X_r) \in C^\infty(M),$$

where $\mathfrak{S}_r$ is the permutation group of r letters $\{1, \ldots, r\}$, and $\operatorname{sgn}(\sigma)$ is the signature of a permutation σ.

Next, we define the **exterior differentiation** $d : A^r(M) \to A^{r+1}(M)$ by

$$(d\omega)(X_1, \ldots, X_{r+1}) = \sum_{i=1}^{r+1}(-1)^{i+1} X_i(\omega(X_1, \ldots, \widehat{X}_i, \cdots, X_{r+1}))$$
$$+ \sum_{i<j}(-1)^{i+j} \omega([X_i, X_j], X_1, \ldots, \widehat{X}_i, \ldots, \widehat{X}_j, \ldots, X_{r+1}) \tag{6.1}$$

for every $\omega \in A^r(M)$ and $X_1, \cdots, X_{r+1} \in \mathfrak{X}(M)$. $\widehat{X}_i$ means to delete X_i. It is known that $d(d\omega) = 0$.

Thirdly, the Riemannian metric g on M induces the natural inner product on the $\binom{n}{r}$-dimensional linear space $\wedge^r T_x^* M$ $(x \in M)$, denoted by $\langle\, ,\, \rangle_x$ $(x \in M)$. Then, the L^2-**inner product** $(\,,\,)$ on $A^r(M)$ is defined by

$$(\omega, \eta) = \int_{\{x \in M\}} \langle \omega_x, \eta_x \rangle_x \, v_g \qquad (\omega, \eta \in A^r(M)). \tag{6.2}$$

Thus, the **co-differentiation** to the exterior differentiation $d : A^r(M) \to A^{r+1}(M)$, denoted by $\delta : A^{r+1}(M) \to A^r(M)$ is the differential operator which has the property

$$(d\omega, \eta) = (\omega, \delta\eta) \qquad (\omega \in A^r(M),\ \eta \in A^{r+1}(M)).$$

Indeed, the co-differentiation $\delta\eta \in A^r(M)$ $(\eta \in A^{r+1}(M))$ is given by

$$(\delta\eta)(X_1, \ldots, X_r) = -\sum_{i=1}^{n}(\nabla_{e_i}\eta)(e_i, X_1, \ldots, X_r), \tag{6.3}$$

for $X_j \in \mathfrak{X}(M)$ $(j = 1, \ldots, r)$. Here, $\{e_i\}_{i=1}^n$ is a local orthonormal frame field on (M, g). The $\nabla_X \eta$ $(X \in \mathfrak{X}(M))$ on the right-hand side

(1.36) is the **covariant differentiation** $\nabla_X \omega \in A^r(M)$ for differential form $\omega \in A^r(M)$ of degree r which is defined by

$$(\nabla_X \omega)(X_1, \ldots, X_r) = X(\omega(X_1, \ldots, X_r))$$
$$- \sum_{i=1}^{r} \omega(X_1, \ldots, \nabla_X X_i, \ldots, X_r).$$

Then, it holds that for $\omega \in A^r(M)$,

$$(d\omega)(X_1, \ldots, X_{r+1}) = \sum_{i=1}^{r+1}(-1)^{i+1}(\nabla_{X_i}\omega)(X_1, \ldots, \widehat{X_i}, \ldots, X_{r+1}).$$

In particular, if we define for C^∞ vector field $X \in \mathfrak{X}(M)$, a 1-form $\omega \in A^1(M)$ by $\omega(Y) = g(X, Y)$ ($\forall Y \in \mathfrak{X}(M)$), then we have

$$\mathrm{div}(X) = -\delta\omega. \tag{6.4}$$

Finally, we can define the **Laplacian** acting on the space $A^r(M)$ of differential forms of degree r by

$$\Delta_r := d\delta + \delta d : A^r(M) \to A^r(M).$$

Then, it holds that

$$(\Delta_r \omega, \eta) = (d\omega, d\eta) + (\delta\omega, \delta\eta) \qquad (\omega, \eta \in A^r(M)). \tag{6.5}$$

In case of $r = 0$, $A^0(M) = C^\infty(M)$, for $f \in C^\infty(M)$, we have

$$\Delta_0 f = \delta(df) = -\sum_{i=1}^{n} \nabla_{e_i}(df)(e_i) = -\sum_{i=1}^{n}\{e_i(e_i f) - \nabla_{e_i} e_i\, f\} = \Delta f.$$

In the following, we always denote $\Delta = \Delta_0$.

7. The first and second variation formulas of the lengths of curves

In this section, we derive the well known **first variational formula** and the **second variational** of the **length** $L(c) = \int_a^b \|\dot{c}(t)\|\, dt$ of a C^∞ curve $c : [a, b] \to (M, g)$ in a Riemannian manifold (M, g), where $\|\dot{c}(t)\| = \sqrt{g(\dot{c}(t), \dot{c}(t))}$ ($\dot{c}(t) \in T_{c(t)}M$). Applications of the first and second variational formulas to the eigenvalue problem of the Laplacian will be given in Chap. 3.

Given a C^∞ curve $c : [a, b] \to M$, its **variation** is a C^∞ mapping

$$\alpha : [a, b] \times (-\epsilon, \epsilon) \ni (t, s) \mapsto \alpha(t, s) \in M$$

for sufficiently small positive number $\epsilon > 0$, which satisfies $\alpha(t, 0) = c(t)$ ($t \in [a, b]$). Then, a C^∞ family of curves $c_s : [a, b] \to M$ is given by $c_s(t) := \alpha(t, s)$ ($t \in [a, b]$), the mapping α is called **deformation** (or **variation**) of c, $\{c_s| -\epsilon < s < \epsilon\}$. Then, for each $t \in [a, b]$, the tangent vector at $s = 0$ of a C^∞ curve $(-\epsilon, \epsilon) \ni s \mapsto \alpha(t, s) \in M$ is given by

$$X(t) := \frac{d}{dt}\bigg|_{s=0} \alpha(t, s) \in T_{c(t)}M \qquad (t \in [a, b]).$$

It turns out that X is a C^∞ vector field along a curve c. We say X, a **variational vector field**.

Conversely, if a C^∞ vector field X (variational vector field) along c is given as $X(t) \in T_{c(t)}M$ $(t \in [a, b])$, one can construct a variation $\alpha : [a, b] \times (-\epsilon, \epsilon) \to M$ of c by

$$X(t) = \frac{d}{ds}\bigg|_{s=0} \alpha(t, s) \in T_{c(t)}M.$$

For example, we may define α by

$$\alpha(t, s) = \mathrm{Exp}_{c(t)}(s\, X(t)) \qquad (s \in (-\epsilon, \epsilon),\ t \in [a, b]).$$

Now let $\alpha : [a, b] \times (-\epsilon, \epsilon) \to M$ be a variation of a C^∞ curve $c : [a, b] \to M$, and $\{c_s| -\epsilon < s < \epsilon\}$ be a variation of c. Then, let us calculate the **first variation** of length,

$$\frac{d}{ds}\bigg|_{s=0} L(c_s).$$

We assume that the parameter t of a curve $c = c_0$ is a constant multiple of the arclength, and

$$g_{c(t)}(\dot{c}(t), \dot{c}(t))^{\frac{1}{2}} = \|\dot{c}(t)\| = \ell \qquad (\forall\, t \in [a, b]).$$

Here, ℓ is a constant independent of t. Then, we have the following theorem.

THEOREM 7.1 (The first variational formula).

$$\frac{d}{ds}\bigg|_{s=0} L(c_s) = \ell^{-1} \int_a^b \left\{ \frac{d}{dt} g_{c(t)}\big(X(t), \dot{c}(t)\big) - g_{c(t)}\big(X(t), \nabla_{\dot{c}(t)}\dot{c}\big) \right\} dt$$

$$= \ell^{-1} \left\{ g_{c(b)}(X(b), \dot{c}(b)) - g_{c(a)}(X(a), \dot{c}(a)) \right\}$$

$$- \ell^{-1} \int_a^b g_{c(t)}(X(t), \nabla_{\dot{c}(t)}\dot{c})\, dt. \qquad (7.1)$$

Proof We regard $\frac{\partial}{\partial t}$ and $\frac{\partial}{\partial s}$ as two vector fields T and V along $[a, b] \times (-\epsilon, \epsilon)$, respectively, namely,

$$T_{(t,s)} := \left(\frac{\partial}{\partial t}\right)_{(t,s)}, \qquad V_{(t,s)} := \left(\frac{\partial}{\partial s}\right)_{(t,s)} \qquad ((t, s) \in [a, b] \times (-\epsilon, \epsilon)).$$

Then, the tangent vector $\dot{c}_s$ of curves c_s can be written as

$$\dot{c}_s(t) = c_{s*}\left(\frac{\partial}{\partial t}\right) = \alpha_*(T)$$

in terms of the differentiation α_* of α. On the other hand, we have

$$X(t) = \frac{d}{ds}\bigg|_{s=0} \alpha(t,s) = \alpha_*(V)\bigg|_{s=0}.$$

Thus, we have

$$\frac{d}{ds}L(c_s) = \int_a^b \frac{d}{ds}\left\{g_{c(t)}(\dot{c}_s(t), \dot{c}_s(t))^{\frac{1}{2}}\right\} dt$$

$$= \frac{1}{2}\int_a^b g_{c(t)}(\dot{c}_s(t), \dot{c}_s(t))^{-\frac{1}{2}} \frac{d}{ds}g_{c(t)}(\dot{c}_s(t), \dot{c}_s(t))\, dt.$$

$$(7.2)$$

Here, notice that

$$\frac{d}{ds}g_{c(t)}(\dot{c}_s(t), \dot{c}_s(t)) = V_{(t,s)}g(\alpha_*(T), \alpha_*(T)) = 2\, g(\nabla_V \alpha_*(T), \overset{\centerdot}{\alpha}_*(T)).$$

$$(7.3)$$

For $X \in \mathfrak{X}(M)$ and a C^∞ map $\varphi : M \to N$, $\varphi_*(X)$ means $\varphi_*(X)(x) := \varphi_{*x}(X_x) \in T_{\varphi(x)}N$ $(x \in M)$. Then, due to the property (2) of Levi-Civita connection ∇ in Theorem 1.2, it holds that

$$\nabla_V \alpha_*(T) - \nabla_T \alpha_*(V) - \alpha_*([V, T]) = 0,$$

$$(7.4)$$

and, since $[\frac{\partial}{\partial s}, \frac{\partial}{\partial t}] = 0$, $[V, T] = 0$. Thus, due to (2.2), we have

$$\nabla_V \alpha_*(T) = \nabla_T \alpha_*(V).$$

Insert this into (2.2), and use the property (1) of Levi-Civita connection in Theorem 2.1, we have

$$(7.3) = 2\, g(\nabla_T \alpha_*(V), \alpha_*(T))$$

$$= 2\left\{T(g(\alpha_*(V), \alpha_*(T)) - g(\alpha_*(V), \nabla_T \alpha_*(T))\right\}.$$

$$(7.5)$$

Then, inserting (7.3) into (7.2), (7.2) can be written as follows.

$$(7.2) = \int_a^b g(\alpha_*(T), \alpha_*(T))^{-\frac{1}{2}}\left\{T(g(\alpha_*(V), \alpha_*(T)))\right.$$

$$\left. - g(\alpha_*(V), \nabla_T \alpha_*(T))\right\} dt.$$

$$(7.6)$$

Here, putting $s = 0$, we have

$$\alpha_*(T)\Big|_{s=0} = \dot{c}(t), \quad \alpha_*(V)\Big|_{s=0} = X(t), \quad \nabla_T \alpha_*(T)\Big|_{s=0} = \nabla_{\dot{c}(t)}\dot{c},$$

so the equation of (7.2) at $s = 0$ turns out that

$$\frac{d}{ds}\Big|_{s=0} L(c_s) = \ell^{-1} \int_a^b \left\{ \frac{d}{dt} g_{c(t)}(X(t), \dot{c}(t)) - g_{c(t)}(X(t), \nabla_{\dot{c}(t)}\dot{c}) \right\} dt. \tag{7.7}$$

This is the desired equation. We have Theorem 7.1. $\qquad\square$

Assume that a C^∞ curve $c : [a, b] \to M$ is a geodesic whose parameter t is a constant multiple of the arc length. Then, it is known that the following **second variational formula** is known (For proof, see [**35**], Vol. II, P. 81, or [**46**], P. 124):

$$\frac{d^2}{ds^2}\Big|_{s=0} L(c_s) = \ell^{-1} \int_a^b \left[g(\nabla_{\dot{c}(t)}X^\perp, \nabla_{\dot{c}(t)}X^\perp) - g(R(X^\perp, \dot{c}(t))\dot{c}(t), X^\perp) \right] dt$$

$$= -\ell^{-1} \int_a^b g_{c(t)}\left(\nabla_{\dot{c}(t)}(\nabla_{\dot{c}(t)}X^\perp) + R(X^\perp, \dot{c}(t))\dot{c}(t), X^\perp \right) dt$$

$$+ \ell^{-1} \left[g(\nabla_{\dot{c}(t)}X^\perp, X^\perp) \right]_{t=a}^{t=b}, \tag{7.8}$$

where $X^\perp := X - \ell^{-1} g(X, \dot{c}(t)) \dot{c}(t)$, i.e., $g(X^\perp, \dot{c}(t)) = 0$.

In particular, if $g(X(t), \dot{c}(t)) = 0$ ($t \in [a, b]$), since $X(t) = X^\perp(t)$, we have

$$\frac{d^2}{ds^2}\Big|_{s=0} L(c_s) = -\ell^{-1} \int_a^b g_{c(t)}\left(\nabla_{\dot{c}(t)}(\nabla_{\dot{c}(t)}X) + R(X, \dot{c}(t))\dot{c}(t), X \right) dt$$

$$+ \ell^{-1} \left[g(\nabla_{\dot{c}(t)}X, X) \right]_{t=a}^{t=b}. \tag{7.9}$$

A vector field $X(t)$ along a geodesic $c(t)$ satisfying that

$$\nabla_{\dot{c}(t)}(\nabla_{\dot{c}(t)}X) + R(X, \dot{c}(t))\dot{c}(t) = 0 \tag{7.10}$$

is called a **Jacobi field**.

CHAPTER 2

The First and Second Variational Formulas of the Energy

1

ABSTRACT. In [40], J. Eells and L. Lemaire introduced the notion of a k-harmonic map. In this chapter, we study the case $k = 2$, and derive the first and second variational formulas of the 2-harmonic maps. We also give non-trivial examples of 2-harmonic maps and show certain nonexistence theorems of stable 2-harmonic maps.

1. Introduction

As well known, harmonic maps between Riemannian manifolds $f : M \to N$, where M is compact, can be considered as critical maps of the energy functional $E(f) = \int_M \|df\|^2 *1$. Considering the similar ideas, in 1981, J. Eells and L. Lemaire [40], proposed the problem to consider the k-harmonic maps: critical maps of the functional

$$E_k(f) = \int_M \|(d + d^*)^k f\|^2 *1.$$

In this paper, we consider the case $k = 2$ and show the preliminary results.

We use mainly vector bundle valued differential forms and Riemannian metrics. In §2, we prepare the notation and fundamental formulas needed in the sequel.

In §3, given a compact manifold M, we derive the first variation formula of $E_2(f) = \int_M \|(d + d^*)^2 f\|^2 *1$ (Theorem 3.1) and give the definition of 2-harmonic maps $f : M \to N$ whose tension field $\tau(f)$ satisfies

$$-\overline{\nabla}^* \overline{\nabla} \tau(f) + R^N(df(e_k), \tau(f))df(e_k) = 0,$$

namely $\tau(f)$ is a solution of the Jacobi type equation.

---

[1]This chapter is due to [74]: G.Y. Jiang, *2-harmonic maps and their first and second variational formula*, Chinese Ann. Math., **7A** (1986), 388–402; Note di Matematica, **28** (2009), 209–232, translated into English by H. Urakawa.

Constant maps and harmonic maps are trivial examples of 2-harmonic maps. The main results of §4 are to give nontrivial examples of 2-harmonic maps. We consider Riemannian isometric immersions. For the isometric immersions with parallel mean curvature tensor field, we give the decomposition formula (Lemma 4.2) of the Laplacian of the tension field $\tau(f)$ with its proof: for the hypersurfaces M with non-zero parallel mean curvature tensor field in the unit sphere S^{m+1}, a necessary and sufficient condition for such isometric immersions to be 2-harmonic is that the square of the length of the second fundamental form $B(f)$ satisfies $\|B(f)\|^2 = m$. Using this, special Clifford tori in the unit sphere whose Gauss maps are studied by Y.L. Xin and Q. Chen [165], give non-trivial 2-harmonic maps which are isometric immersions in the unit sphere.

In §5, using formulas in §3, we derive the second variation formula of 2-harmonic maps (Theorem 5.1) and give the definition of stability of 2-harmonic maps (the second variation is nonnegative) and give a proof of the following (Theorem 5.2): if M is compact, and N has positive constant sectional curvature, there are no nontrivial 2-harmonic maps from M into N satisfying the conservation law. Last, when $N = \mathbb{C}P^n$ we establish nonexistence results of stable 2-harmonic maps (Lemma 5.4, Theorem 5.3, etc.). Furthermore, we give a nonexistence theorem establishing sufficient conditions that stable 2-harmonic maps be harmonic.

We would like to express our gratitude to Professors Su Bu-Chin and Hu He-Shen who introduced and helped to accomplish this paper. We also would like to express our thanks to Professors Shen Chun-Li, Xin Yuan-Long and Pan Yang-Lian who helped us during the period of our study.

2. Notation, and fundamental notions

We prepare the main materials using vector bundle valued differential forms and Riemannian metrics on bundles which are in [40, 41].

Assume that (M, g) is a m-dimensional Riemannian manifold, (N, h) a n-dimensional one, and $f : M \to N$ a C^∞ map. Given points $p \in M$ and $f(p) \in N$, under (x^i), (y^α) local coordinates around them, f can be expressed as

$$y^\alpha = f^\alpha(x^i), \tag{2.1}$$

where the indices we use run as follows

$$i, j, k, \cdots = 1, \cdots, m; \ \alpha, \beta, \gamma, \cdots = 1, \cdots, n.$$

We use the following definition: the differential df of f can be regarded as the induced bundle $f^{-1}TN$-valued 1-form

$$df(X) = f_*X, \quad \forall X \in \Gamma(TM). \tag{2.2}$$

We denote by f^*h the first fundamental form of f, which is a section of the symmetric bilinear tensor bundle $\odot^2 T^*M$; the second fundamental form $B(f)$ of f is the covariant derivative $\widetilde\nabla df$ of the 1-form df, which is a section of $\odot^2 T^*M \otimes f^{-1}TN$:

$$\forall X, Y \in \Gamma(TM):$$

$$B(f)(X,Y) = (\widetilde\nabla df)(X,Y) = (\widetilde\nabla_X df)(Y) =$$

$$= \widetilde\nabla_X df(Y) - df(\nabla_X Y) =$$

$$= \nabla'_{df(X)} df(Y) - df(\nabla_X Y). \tag{2.3}$$

Here $\nabla, \nabla', \overline\nabla, \widetilde\nabla$ are the Riemannian connections on the bundles TM, TN, $f^{-1}TN$ and $T^*M \otimes f^{-1}TN$, respectively. From $\widetilde\nabla df$, by using a local orthonormal frame field $\{e_i\}$ on M, one obtains the tension field $\tau(f)$ of f

$$\tau(f) = (\widetilde\nabla df)(e_i, e_i) = (\widetilde\nabla_{e_i} df)(e_i). \tag{2.4}$$

In the following, we use the above notations without comments, and we assume the reader is familiar with the above notation.

We say f is a harmonic map if $\tau(f) = 0$. If M is compact, we consider critical maps of the energy functional

$$E(f) = \int_M \|df\|^2 * 1, \tag{2.5}$$

where $\frac{1}{2}\|df\|^2 = \frac{1}{2}\langle df(e_i), df(e_i)\rangle_N = e(f)$ which is called the enegy density of f, and the inner product $\langle \, , \, \rangle_N$ is a Riemannian metric h, and we omit the subscript N if there is no confusion. When f is an isometric immersion, $\frac{1}{m}\tau(f)$ is the mean curvature normal vector field and harmonic maps are minimal immersions.

The curvature tensor field $\widetilde{R}(\, , \,)$ of the Riemannian metric on the bundle $T^*M \otimes f^{-1}TN$ is defined as follows

$$\forall X, Y \in \Gamma(TM):$$

$$\widetilde{R}(X,Y) = -\widetilde\nabla_X \widetilde\nabla_Y + \widetilde\nabla_Y \widetilde\nabla_X + \widetilde\nabla_{[X,Y]}. \tag{2.6}$$

Furthermore, for any $Z \in \Gamma(TM)$, we define

$$(\widetilde{R}(X,Y)df)(Z) = R^{f^{-1}TN}(X,Y)df(Z) - df(R^M(X,Y)Z) =$$

$$= R^N(df(X), df(Y))df(Z) - df(R^M(X,Y)Z), \tag{2.7}$$

where R^M, R^N, and $R^{f^{-1}TN}$ are the Riemannian curvature tensor fields on TM, TN, $f^{-1}TN$, respectively.

For 1-forms df the Weitzenböck formula is given by

$$\Delta df = \widetilde{\nabla}^* \widetilde{\nabla} df + S, \tag{2.8}$$

where $\Delta = dd^* + d^*d$ is the Hodge-Laplace operator, $-\widetilde{\nabla}^*\widetilde{\nabla} = \widetilde{\nabla}_{e_k}\widetilde{\nabla}_{e_k} - \widetilde{\nabla}_{\widetilde{\nabla}_{e_k}e_k}$ is the rough Laplacian, and the operator S is defined as follows

$$\forall X \in \Gamma(TM):$$

$$S(X) = -(\widetilde{R}(e_k, X)df)(e_k), \tag{2.9}$$

where $\{e_k\}$ is a locally defined orthonormal frame field on M.

A section of $\odot^2 T^*M$ defined by $S_f = e(f)g - f^*h$ is called the stress-energy tensor field, and f is said to satisfy the conservation law if $\mathrm{div} S_f = 0$. As in [40], $\forall X \in \Gamma(TM)$, it holds that

$$(\mathrm{div} S_f)(X) = -\langle \tau(f), df(X) \rangle. \tag{2.10}$$

Maps satisfying the conservation law are said to be *relatively harmonic* ([73]).

3. The first variation formula of 2-harmonic maps

Assume that $f : M \to N$ ia a C^∞ map, M is a compact Riemannian manifold, and N is an arbitrary Riemannian manifold. As in [40], a 2-harmonic map is a critical map of the functional

$$E_2(f) = \int_M \|(d + d^*)^2 f\|^2 * 1. \tag{3.1}$$

Here, d and d^* are the exterior differentiation and the codifferentiation on vector bundle, and $*1$ is the volume form on M.

In order to derive the analytic condition of the 2-harmonic maps, we have to calculate the first variation of $E_2(f)$ defined by (3.1). To start with let

$$f_t : M \to N, \quad t \in I_\epsilon = (-\epsilon, \epsilon), \quad \epsilon > 0, \tag{3.2}$$

be a smooth 1-parameter variation of f which yields a vector field $V \in \Gamma(f^{-1}TN)$ along f in N by

$$f_0 = f, \quad \left.\frac{\partial f_t}{\partial t}\right|_{t=0} = V. \tag{3.3}$$

Variation $\{f_t\}$ yields a C^∞ map

$$F : M \times I_\epsilon \to N,$$

$$F(p, t) = f_t(p), \quad \forall p \in M, t \in I_\epsilon. \tag{3.4}$$

If we take the local coordinates around $p \in M$, $f_t(p) \in N$, respectively, we have

$$y^\alpha = F^\alpha(x^i, t) = f_t^\alpha(x^i). \tag{3.5}$$

Taking the usual Euclidean metric on I_ϵ, with respect to the product Riemannian metric on $M \times I_\epsilon$, we denote by $\nabla, \overline{\nabla}, \widetilde{\nabla}$, the induced Riemann connections on $T(M \times I_\epsilon)$, $F^{-1}TN$, $T^*(M \times I_\epsilon) \otimes F^{-1}TN$, respectively. If $\{e_i\}$ is an orthonormal frame field defined on a neighborhood U of p, $\{e_i, \frac{\partial}{\partial t}\}$ is also an orthonormal frame field on a coordinate neighborhood $U \times I_\epsilon$ in $M \times I_\epsilon$, and it holds that

$$\nabla_{\frac{\partial}{\partial t}} \frac{\partial}{\partial t} = 0, \quad \nabla_{e_i} e_j = \nabla_{e_i} e_j, \quad \nabla_{\frac{\partial}{\partial t}} e_i = \nabla_{e_i} \frac{\partial}{\partial t} = 0. \tag{3.6}$$

It also holds that

$$\frac{\partial f_t}{\partial t} = \frac{\partial F^\alpha}{\partial t} \frac{\partial}{\partial y^\alpha} = dF\left(\frac{\partial}{\partial t}\right), \quad df_t(e_i) = dF(e_i), \tag{3.7}$$

and

$$(\widetilde{\nabla}_{e_i} df_t)(e_j) = \nabla'_{df_t(e_i)} df_t(e_j) - df_t(\nabla_{e_i} e_j) = (\widetilde{\nabla}_{e_i} dF)(e_j)$$
$$(\widetilde{\nabla}_{e_k} \widetilde{\nabla}_{e_i} df_t)(e_j) = \nabla'_{df_t(e_k)}((\widetilde{\nabla}_{e_i} df_t)(e_j)) - (\widetilde{\nabla}_{e_i} df_t)(\nabla_{e_k} e_j)$$
$$= (\widetilde{\nabla}_{e_k} \widetilde{\nabla}_{e_i} dF)(e_j)$$
$$\dotso\dotso\dotso \tag{3.8}$$

etc. Here, we used the abbreviated symbol $\widetilde{\nabla}$ on $T^*M \otimes f_t^{-1}TN$ in which we omitted t.

In the following, we need two lemmas to calculate the first variation $\frac{d}{dt} E_2(f_t)|_{t=0}$ of $E_2(f)$.

LEMMA 3.1. *Under the above notation, for any C^∞ variation $\{f_t\}$ of f, it holds that*

$$\frac{d}{dt} E_2(f_t) =$$
$$2 \int_M \left\langle (\widetilde{\nabla}_{e_i} \widetilde{\nabla}_{e_i} dF)\left(\frac{\partial}{\partial t}\right) - (\widetilde{\nabla}_{\nabla_{e_i} e_i} dF)\left(\frac{\partial}{\partial t}\right), (\widetilde{\nabla}_{e_j} dF)(e_j) \right\rangle * 1$$
$$+ 2 \int_M \left\langle R^N\left(dF(e_i), dF\left(\frac{\partial}{\partial t}\right)\right) dF(e_i), (\widetilde{\nabla}_{e_j} dF)(e_j) \right\rangle * 1. \tag{3.9}$$

PROOF. By using d, d^*, and definition of $\tau(f)$, (3.1) can be written as

$$E_2(f) = \int_M \|d^* df\|^2 * 1 = \int_M \|\tau(f)\|^2 * 1$$
$$= \int_M \langle (\widetilde{\nabla}_{e_i} df)(e_i), \widetilde{\nabla}_{e_i} df)(e_i) \rangle * 1. \qquad (3.10)$$

By noting (3.8), for variation f_t of f, it holds that

$$\frac{d}{dt} E_2(f_t) = \frac{d}{dt} \int_M \langle (\widetilde{\nabla}_{e_i} dF)(e_i), (\widetilde{\nabla}_{e_j} dF)(e_j) \rangle * 1$$
$$= 2 \int_M \langle \widetilde{\nabla}_{\frac{\partial}{\partial t}} ((\widetilde{\nabla}_{e_i} dF)(e_i)), (\widetilde{\nabla}_{e_j} dF)(e_j) \rangle * 1. \qquad (3.11)$$

By (3.6), and using the curvature tensor on $T^*(M \times I_\epsilon) \otimes F^{-1}TN$,

$$\left(\tilde{R}\left(X, \frac{\partial}{\partial t} \right) dF \right)(Y) = R^N \left(dF(X), dF\left(\frac{\partial}{\partial t} \right) \right) dF(Y)$$
$$- dF\left(R^{M \times I_\epsilon} \left(X, \frac{\partial}{\partial t} \right) Y \right)$$
$$= R^N \left(dF(X), dF\left(\frac{\partial}{\partial t} \right) \right) dF(Y), \qquad (3.12)$$

for all X, $Y \in \Gamma(TM)$. In (3.11), interchanging the order of differentiations in $\widetilde{\nabla}_{\frac{\partial}{\partial t}}((\widetilde{\nabla}_{e_i} dF)(e_i))$, we have

$$\widetilde{\nabla}_{\frac{\partial}{\partial t}}((\widetilde{\nabla}_{e_i} dF)(e_i)) = (\widetilde{\nabla}_{\frac{\partial}{\partial t}} \widetilde{\nabla}_{e_i} dF)(e_i)$$
$$= \left(\widetilde{\nabla}_{e_i} \widetilde{\nabla}_{\frac{\partial}{\partial t}} dF - \widetilde{\nabla}_{[e_i, \frac{\partial}{\partial t}]} dF + \tilde{R}\left(e_i, \frac{\partial}{\partial t} \right) dF \right)(e_i)$$
$$= \widetilde{\nabla}_{e_i}((\widetilde{\nabla}_{\frac{\partial}{\partial t}} dF)(e_i)) - (\widetilde{\nabla}_{\frac{\partial}{\partial t}} dF)(\nabla_{e_i} e_i)$$
$$+ R^N \left(dF(e_i), dF\left(\frac{\partial}{\partial t} \right) \right) dF(e_i)$$
$$= (\widetilde{\nabla}_{e_i} \widetilde{\nabla}_{e_i} dF) \left(\frac{\partial}{\partial t} \right) - (\widetilde{\nabla}_{\nabla_{e_i} e_i} dF) \left(\frac{\partial}{\partial t} \right)$$
$$+ R^N \left(dF(e_i), dF\left(\frac{\partial}{\partial t} \right) \right) dF(e_i). \qquad (3.13)$$

In the last of the above, we used the symmetry of the second fundamental form. By substituting $(3.13)^2$ into (3.11) we obtain (3.9).   □

LEMMA 3.2.

$$\int_M \left\langle (\widetilde{\nabla}_{e_i}\widetilde{\nabla}_{e_i}dF)\left(\frac{\partial}{\partial t}\right) - (\widetilde{\nabla}_{\nabla_{e_i}e_i}dF)\left(\frac{\partial}{\partial t}\right), (\widetilde{\nabla}_{e_j}dF)(e_j) \right\rangle * 1$$

$$= \int_M \left\langle dF\left(\frac{\partial}{\partial t}\right), \widetilde{\nabla}_{e_k}\widetilde{\nabla}_{e_k}((\widetilde{\nabla}_{e_j}dF)(e_j)) - \widetilde{\nabla}_{\nabla_{e_k}e_k}((\widetilde{\nabla}_{e_j}dF)(e_j)) \right\rangle * 1.$$

$$(3.14)$$

PROOF. For each $t \in I_\epsilon$, let us define a C^∞ vector field on M by

$$X = \left\langle (\widetilde{\nabla}_{e_i}dF)\left(\frac{\partial}{\partial t}\right), (\widetilde{\nabla}_{e_j}dF)(e_j) \right\rangle e_i, \qquad (3.15)$$

which is well defined because of the independence on a choice of $\{e_i\}$. The divergence of X is given by

$$\mathrm{div}X = \langle \nabla_{e_k}X, e_k \rangle_M = \nabla_{e_i}\left\langle (\widetilde{\nabla}_{e_i}dF)\left(\frac{\partial}{\partial t}\right), (\widetilde{\nabla}_{e_j}dF)(e_j) \right\rangle$$

$$+ \left\langle (\widetilde{\nabla}_{e_i}dF)\left(\frac{\partial}{\partial t}\right), (\widetilde{\nabla}_{e_j}dF)(e_j) \right\rangle \langle \nabla_{e_k}e_i, e_k \rangle_M. \qquad (3.16)$$

---

^{2}Translator's comments: to get the last equation of (3.13), we have to see that

$$(\widetilde{\nabla}_{e_i}dF)\left(\frac{\partial}{\partial t}\right) = \widetilde{\nabla}_{e_i}\left(dF\left(\frac{\partial}{\partial t}\right)\right) - dF\left(\nabla_{e_i}\frac{\partial}{\partial t}\right)$$

$$= \widetilde{\nabla}_{dF(e_i)}dF\left(\frac{\partial}{\partial t}\right)$$

$$= \widetilde{\nabla}_{dF(\frac{\partial}{\partial t})}dF(e_i) - \widetilde{\nabla}_{[dF(e_i),dF(\frac{\partial}{\partial t})]}$$

$$= (\widetilde{\nabla}_{\frac{\partial}{\partial t}}dF)(e_i),$$

and by a similar way,

$$(\widetilde{\nabla}_{\frac{\partial}{\partial t}}dF)(\nabla_{e_i}e_i) = (\widetilde{\nabla}_{\nabla_{e_i}e_i}dF)\left(\frac{\partial}{\partial t}\right).$$

Thus,

$$\widetilde{\nabla}_{e_i}((\widetilde{\nabla}_{\frac{\partial}{\partial t}}dF)(e_i))) - (\widetilde{\nabla}_{\frac{\partial}{\partial t}}dF)(\nabla_{e_i}e_i)$$

coincides with

$$\widetilde{\nabla}_{e_i}\left((\widetilde{\nabla}_{e_i}dF)\left(\frac{\partial}{\partial t}\right)\right) - (\widetilde{\nabla}_{\nabla_{e_i}e_i}dF)\left(\frac{\partial}{\partial t}\right)$$

$$= (\widetilde{\nabla}_{e_i}\widetilde{\nabla}_{e_i}dF)\left(\frac{\partial}{\partial t}\right) + (\widetilde{\nabla}_{e_i}dF)\left(\nabla_{e_i}\frac{\partial}{\partial t}\right) - (\widetilde{\nabla}_{\nabla_{e_i}e_i}dF)\left(\frac{\partial}{\partial t}\right)$$

$$= (\widetilde{\nabla}_{e_i}\widetilde{\nabla}_{e_i}dF)\left(\frac{\partial}{\partial t}\right) - (\widetilde{\nabla}_{\nabla_{e_i}e_i}dF)\left(\frac{\partial}{\partial t}\right),$$

which implies (3.13).

Noticing (3.6) and

$$\langle \nabla_{e_k} e_i, e_k \rangle_M + \langle e_i, \nabla_{e_k} e_k \rangle_M = 0, \tag{3.17}$$

we have $(3.18)^3$:

$$\operatorname{div}(X) = \langle (\widetilde{\overline{\nabla}}_{e_i} \widetilde{\overline{\nabla}}_{e_i} dF) \left(\frac{\partial}{\partial t} \right), (\widetilde{\overline{\nabla}}_{e_j} dF)(e_j) \rangle$$

$$+ \langle (\widetilde{\overline{\nabla}}_{e_i} dF) \left(\frac{\partial}{\partial t} \right), \overline{\nabla}_{e_i} ((\widetilde{\overline{\nabla}}_{e_j} dF)(e_j)) \rangle$$

$$- \langle (\widetilde{\overline{\nabla}}_{\nabla_{e_k} e_k} dF) \left(\frac{\partial}{\partial t} \right), (\widetilde{\overline{\nabla}}_{e_j} dF)(e_j) \rangle. \tag{3.18}$$

Furthermore, let us define a C^∞ vector field Y on M by

$$Y = \left\langle dF \left(\frac{\partial}{\partial t} \right), \overline{\nabla}_{e_i} ((\widetilde{\overline{\nabla}}_{e_j} dF)(e_j)) \right\rangle e_i, \tag{3.19}$$

---

^{3}Translator's comments: the first term of (3.16) coincides with

$$\langle \overline{\nabla}_{e_i} (\widetilde{\overline{\nabla}}_{e_i} dF) \left(\frac{\partial}{\partial t} \right), (\widetilde{\overline{\nabla}}_{e_j} dF)(e_j) \rangle$$

$$+ \langle (\widetilde{\overline{\nabla}}_{e_i} dF) \left(\frac{\partial}{\partial t} \right), \overline{\nabla}_{e_i} ((\widetilde{\overline{\nabla}}_{e_j} dF)(e_j)) \rangle$$

$$= \langle \widetilde{\overline{\nabla}}_{e_i} \widetilde{\overline{\nabla}}_{e_i} dF \left(\frac{\partial}{\partial t} \right) + (\widetilde{\overline{\nabla}}_{e_i} dF) \left(\nabla_{e_i} \frac{\partial}{\partial t} \right), (\widetilde{\overline{\nabla}}_{e_j} dF)(e_j) \rangle$$

$$+ \langle (\widetilde{\overline{\nabla}}_{e_i} dF) \left(\frac{\partial}{\partial t} \right), \overline{\nabla}_{e_i} ((\widetilde{\overline{\nabla}} dF)(e_j)) \rangle$$

$$= \langle \widetilde{\overline{\nabla}}_{e_i} \widetilde{\overline{\nabla}}_{e_i} dF \left(\frac{\partial}{\partial t} \right), (\widetilde{\overline{\nabla}}_{e_j} dF)(e_j) \rangle$$

$$+ \langle (\widetilde{\overline{\nabla}}_{e_i} dF) \left(\frac{\partial}{\partial t} \right), \overline{\nabla}_{e_i} ((\widetilde{\overline{\nabla}} dF)(e_j)) \rangle,$$

and the second term of (3.16) coincides with

$$\langle (\widetilde{\overline{\nabla}}_{e_i} dF) \left(\frac{\partial}{\partial t} \right), (\widetilde{\overline{\nabla}}_{e_j} dF)(e_j) \rangle \langle \nabla_{e_k} e_i, e_k \rangle_M$$

$$= - \langle (\widetilde{\overline{\nabla}}_{e_i} dF) \left(\frac{\partial}{\partial t} \right), (\widetilde{\overline{\nabla}}_{e_j} dF)(e_j) \rangle \langle e_i, \nabla_{e_k} e_k \rangle_M$$

$$= - \langle (\widetilde{\overline{\nabla}}_{\nabla_{e_k} e_k} dF) \left(\frac{\partial}{\partial t} \right), (\widetilde{\overline{\nabla}}_{e_j} dF)(e_j) \rangle.$$

Thus, we have (3.18).

which is also well defined. Then, by a similar way, we have

$$\text{div}Y = \langle \nabla_{e_k} Y, e_k \rangle_M$$

$$= \left\langle (\widetilde{\nabla}_{e_k} dF)\left(\frac{\partial}{\partial t}\right), \overline{\nabla}_{e_k}((\widetilde{\nabla}_{e_j} dF)(e_j)) \right\rangle$$

$$+ \left\langle dF\left(\frac{\partial}{\partial t}\right), \overline{\nabla}_{e_k} \overline{\nabla}_{e_k}((\widetilde{\nabla}_{e_j} dF)(e_j)) \right\rangle$$

$$- \left\langle dF\left(\frac{\partial}{\partial t}\right), \overline{\nabla}_{\nabla_{e_k} e_k}((\widetilde{\nabla}_{e_j} dF)(e_j)) \right\rangle. \tag{3.20}$$

By the Green's theorem, we have

$$\int_M \text{div}(X - Y) * 1 = 0, \tag{3.21}$$

and together with (3.18) and (3.20), we have (3.14). □

THEOREM 3.1. *Assume that* $f : M \to N$ *is a* C^∞ *map from a compact Riemannian manifold* M *into an arbitrary Riemannian manifold* N, $\{f_t\}$ *is an arbitrary* C^∞ *variation generating* V. *Then,*

$$\frac{d}{dt} E_2(f_t)\bigg|_{t=0} =$$

$$= 2 \int_M \langle V, -\overline{\nabla}^* \overline{\nabla} \tau(f) + R^N(df(e_i), \tau(f))df(e_i) \rangle * 1. \tag{3.22}$$

PROOF. Substituting (3.14) into (3.9), we have

$$\frac{d}{dt} E_2(f_t) = 2 \int_M \left\langle dF\left(\frac{\partial}{\partial t}\right), \overline{\nabla}_{e_k} \overline{\nabla}_{e_k}((\widetilde{\nabla}_{e_j} dF)(e_j)) \right.$$

$$\left. -\overline{\nabla}_{\nabla_{e_k} e_k}((\widetilde{\nabla}_{e_j} dF)(e_j)) \right\rangle * 1$$

$$+ 2 \int_M \langle R^N\left(dF(e_i), dF\left(\frac{\partial}{\partial t}\right)\right) dF(e_i), (\widetilde{\nabla}_{e_j} dF)(e_j) \rangle * 1, \tag{3.23}$$

where putting $t = 0$, noticing (3.3), (3.7), (3.8), and the symmetry of the curvature tensor, we have (3.22). Here, we used the explicit formula of the rough Laplacian on $f^{-1}TN$, that is $-\overline{\nabla}^* \overline{\nabla} = \overline{\nabla}_{e_k} \overline{\nabla}_{e_k} - \overline{\nabla}_{\nabla_{e_k} e_k}$. □

REMARK 3.1. In the above arguments, we assumed M is a compact Riemannian manifold without boundary. For a general Riemannian manifold M, let $\mathcal{D} \subset M$ be an arbitrarily bounded domain with smooth boundary, and take a variation $\{f_t\}$ of f satisfying that

$$\left.\frac{\partial f_t}{\partial t}\right|_{\partial \mathcal{D}} = 0, \quad \left.\left(\nabla_{e_i} \frac{\partial f_t}{\partial t}\right)\right|_{\partial \mathcal{D}} = 0,$$

then, in Lemma 3.2, we obtain (3.14) by applying the Green's divergence theorem to X and Y. Then, we have the first variational formula on $\mathcal{D}$ as

$$\left.\frac{d}{dt} E_2(f_t, \mathcal{D})\right|_{t=0} = 2 \int_{\mathcal{D}} \langle V, -\overline{\nabla}^*\overline{\nabla}\tau(f) + R^N(df(e_i), \tau(f))df(e_i)\rangle * 1,$$

where $E_2(f_t, \mathcal{D})$ is the corresponding functional relative to $\mathcal{D}$.

DEFINITION 3.1. For a C^∞ map $f : M \to N$ between two Riemannian manifolds, let us define the *2-tension field* $\tau_2(f)$ of f by

$$\tau_2(f) = -\overline{\nabla}^*\overline{\nabla}\tau(f) + R^N(df(e_i), \tau(f))df(e_i). \qquad (3.24)$$

f is said to be a *2-harmonic map* is if $\tau_2(f) = 0$.

The C^∞ function

$$e_2(f) = \frac{1}{2}\|(d + d^*)^2 f\|^2 = \frac{1}{2}\|\tau(f)\|^2, \qquad (3.25)$$

is called the *2-energy density*, and

$$\frac{1}{2}E_2(f) = \int_M e_2(f) * 1 < +\infty,$$

is the *2-energy* of f. If M is compact, by the first variational formula, a 2-harmonic map f is a critical point of the 2-energy.

4. Examples of 2-harmonic maps

By the definition 3.1, we have immediately

PROPOSITION 4.1. (1) *Any harmonic map is 2-harmonic.*
(2) *Any doubly harmonic function $f : M \to \mathbb{R}$ on a Riemannian manifold M is also 2-harmonic.*

PROPOSITION 4.2. *Assume that M is compact and N has non positive curvature, i.e.* $\mathrm{Riem}^N \leq 0$. *Then every 2-harmonic map $f : M \to N$ is harmonic.*

PROOF. Computing the Laplacian of the 2-energy density $e_2(f)$, we have

$$\Delta e_2(f) = \frac{1}{2}\Delta\|\tau(f)\|^2$$
$$= \langle \overline{\nabla}_{e_k}\tau(f), \overline{\nabla}_{e_k}\tau(f)\rangle + \langle -\overline{\nabla}^*\overline{\nabla}\tau(f), \tau(f)\rangle. \qquad (4.1)$$

Taking

$$\tau_2(f) = -\overline{\nabla}^*\overline{\nabla}\tau(f) + R^N(df(e_i), \tau(f))df(e_i) = 0,$$

and noticing $\mathrm{Riem}^N \leq 0$, we have

$$\Delta e_2(f) = \langle \overline{\nabla}_{e_k}\tau(f), \overline{\nabla}_{e_k}\tau(f)\rangle - R^N(df(e_i), \tau(f))df(e_i), \tau(f)\rangle$$
$$\geq 0. \qquad (4.2)$$

By the Green's theorem $\int_M \Delta e_2(f)v_g = 0$, and (4.2), we have $\Delta e_2(f) = 0$, so that $e_2(f) = \frac{1}{2}\|\tau(f)\|^2$ is constant. Again, by (4.2), we have

$$\overline{\nabla}_{e_k}\tau(f) = 0, \quad \forall k = 1, \cdots, m.$$

Therefore, by [**40**], we have[4] $\tau(f) = 0$. □

REMARK 4.1. As we know nonexistence of compact minimal submanifolds in the Euclidean space, Proposition 4.2 shows nonexistence of 2-harmonic isometric immersions from compact Riemannian manifolds.

By Proposition 4.1 harmonic maps are trivial examples of 2-harmonic ones, and in Proposition 4.2 in the case that M is compact and the sectional curvature of N does not have nonpositive curvature, one may ask examples of nontrivial 2-harmonic maps. To do it, the following lemmas complete this.

LEMMA 4.1. *Assume that* $f : M \to N$ *is a Riemannian isometric immersion whose mean curvature vector field is parallel. Then, for a*

---

[4]Translator's comments: since $\Delta e_2(f) = 0$, both terms of (4.2) are non negative, we have $\langle \overline{\nabla}_{e_k}\tau(f), \overline{\nabla}_{e_k}\tau(f)\rangle = 0$, i.e., $\overline{\nabla}_{e_k}\tau(f) = 0$ for all $k = 1, \cdots, m$. We can define a global vector field $X_f = \langle df(e_i), \tau(f)\rangle e_i \in \mathfrak{X}(M)$, whose divergence is given as

$$\mathrm{div}(X_f) = \langle \tau(f), \tau(f)\rangle + \langle df(e_i), \overline{\nabla}_{e_i}\tau(f)\rangle = \langle \tau(f), \tau(f)\rangle.$$

Integrating this over M, we have

$$0 = \int_M \mathrm{div}(X_f)v_g = \int_M \langle \tau(f), \tau(f)\rangle v_g,$$

which implies $\tau(f) = 0$.

locally defined orthonormal frame field $\{e_i\}$, we have

$$-\overline{\nabla}^*\overline{\nabla}\tau(f) = \langle -\overline{\nabla}^*\overline{\nabla}\tau(f), df(e_i)\rangle df(e_i)$$
$$+ \langle \overline{\nabla}_{e_i}\tau(f), df(e_j)\rangle \, (\widetilde{\nabla}_{e_i}df)(e_j). \qquad (4.3)$$

PROOF. Since f is an isometric immersion, $df(e_i)$ span the tangent space of $f(M) \subset N$. Since[5] the mean curvature tensor is parallel, for all $i = 1, \cdots, m$, $\overline{\nabla}_{e_i}\tau(f) \in \Gamma(f_*TM)$. Thus

$$\overline{\nabla}_{e_i}\tau(f) = \langle \overline{\nabla}_{e_i}\tau(f), df(e_j)\rangle df(e_j). \qquad (4.4)$$

Calculating this, we have

$$-\overline{\nabla}^*\overline{\nabla}\tau(f) = \langle -\overline{\nabla}^*\overline{\nabla}\tau(f), df(e_j)\rangle df(e_j)$$
$$+ \langle \overline{\nabla}_{e_i}\tau(f), \overline{\nabla}_{e_i}df(e_j)\rangle df(e_j)$$
$$+ \langle \overline{\nabla}_{e_i}\tau(f), df(e_j)\rangle \overline{\nabla}_{e_i}df(e_j). \qquad (4.5)$$

Here, if we denote $\nabla_{e_i}e_j = \Gamma_{ij}^k e_k$, we have $\Gamma_{ki}^j + \Gamma_{kj}^i = e_k\langle e_i, e_j\rangle = 0$. Since

$$(\widetilde{\nabla}_{e_i}df)(e_j) = \widetilde{\nabla}_{e_i}(df(e_j)) - df(\nabla_{e_i}e_j) \in T^{\perp}M \subset TN,$$

and $\overline{\nabla}_{e_i}\tau(f) \in f_*(TM)$, the second term of (4.5) is

$$\langle \overline{\nabla}_{e_i}\tau(f), \overline{\nabla}_{e_i}df(e_j)\rangle df(e_j) = \langle \overline{\nabla}_{e_i}\tau(f), (\widetilde{\nabla}_{e_i}df)(e_j) + df(\nabla_{e_i}e_j)\rangle df(e_j)$$
$$= \langle \overline{\nabla}_{e_i}\tau(f), df(\nabla_{e_i}e_j)\rangle df(e_j)$$
$$= \langle \overline{\nabla}_{e_i}\tau(f), df(e_k)\rangle \, df(\Gamma_{ij}^k e_j)$$
$$= \langle \overline{\nabla}_{e_i}\tau(f), df(e_k)\rangle \, df(-\Gamma_{ik}^j e_j)$$
$$= -\langle \overline{\nabla}_{e_i}\tau(f), df(e_k)\rangle \, df(\nabla_{e_i}e_k). \qquad (4.6)$$

---

[5]Translator's comments: for all $\xi \in \Gamma(T^{\perp}N)$,

$$\overline{\nabla}_X\xi = \nabla'_{f_*X}\xi = \nabla^T_{f_*X}\xi + \nabla^{\perp}_{f_*X}\xi \in TM + T^{\perp}M,$$

respectively. The condition that the mean curvature tensor is parallel means that

$$\nabla^{\perp}_{f_*X}\tau(f) = 0, \quad \forall X \in \mathfrak{X}(M),$$

which is equivalent to the condition that

$$\overline{\nabla}_X\tau(f) = \nabla^T_{f_*X}\tau(f) \in \Gamma(f_*TM).$$

Substituting (4.6) into (4.5), we have $(4.3)^6$. $\square$

LEMMA 4.2. *For an isometric immersion $f : M \to N$ with parallel mean curvature vector field, the Laplacian of $\tau(f)$ is decomposed into:*

$$-\overline{\nabla}^*\overline{\nabla}\tau(f) = \langle \tau(f), R^N(df(e_k), df(e_j))df(e_k)\rangle df(e_j)$$
$$- \langle \tau(f), (\widetilde{\nabla}_{e_i}df)(e_j)\rangle(\widetilde{\nabla}_{e_i}df)(e_j). \tag{4.7}$$

PROOF. Calculate the right hand side of (4.3). By differentiating by e_i $\langle\tau(f), df(e_j)\rangle = 0$, we have

$$\langle\overline{\nabla}_{e_i}\tau(f), df(e_j)\rangle + \langle\tau(f), \overline{\nabla}_{e_i}df(e_j)\rangle = e_i\langle\tau(f), df(e_j)\rangle = 0. \tag{4.8}$$

Then, we have

$$\langle\overline{\nabla}_{e_i}\tau(f), df(e_j)\rangle = -\langle\tau(f), \overline{\nabla}_{e_i}df(e_j)\rangle$$
$$= -\langle\tau(f), \overline{\nabla}_{e_i}df(e_j) - df(\nabla_{e_i}e_j)\rangle$$
$$= -\langle\tau(f), (\widetilde{\nabla}_{e_i}df)(e_j)\rangle. \tag{4.9}$$

For the first term of the RHS of (4.3), by differentiating by e_i (4.8), we have

$$\langle\overline{\nabla}_{e_i}\overline{\nabla}_{e_i}\tau(f), df(e_j)\rangle + 2\langle\overline{\nabla}_{e_i}\tau(f), \overline{\nabla}_{e_i}df(e_j)\rangle$$
$$+ \langle\tau(f), \overline{\nabla}_{e_i}\overline{\nabla}_{e_i}df(e_j)\rangle = 0. \tag{4.10}$$

We also have

$$\langle\overline{\nabla}_{\nabla_{e_i}e_i}\tau(f), df(e_j)\rangle + \langle\tau(f), \overline{\nabla}_{\nabla_{e_i}e_i}df(e_j)\rangle = \nabla_{e_i}e_i\langle\tau(f), df(e_j)\rangle = 0. \tag{4.11}$$

Together with (4.10) qand (4.11), we have

$$\langle-\overline{\nabla}^*\overline{\nabla}\tau(f), df(e_j)\rangle + 2\langle\overline{\nabla}_{e_i}\tau(f), \overline{\nabla}_{e_i}df(e_j)\rangle$$
$$+ \langle\tau(f), -\overline{\nabla}^*\overline{\nabla}df(e_j)\rangle = 0. \tag{4.12}$$

---

[6]Translator's comments: for (4.3), we only have to see

$$-\overline{\nabla}^*\overline{\nabla}\tau(f) = \langle-\overline{\nabla}^*\overline{\nabla}\tau(f), df(e_i)\rangle df(e_i)$$
$$+ \langle\overline{\nabla}_{e_i}\tau(f), df(e_j)\rangle\{\overline{\nabla}_{e_i}df(e_j) - df(\nabla_{e_i}e_j)\}$$
$$= \langle-\overline{\nabla}^*\overline{\nabla}\tau(f), df(e_i)\rangle df(e_i)$$
$$+ \langle\overline{\nabla}_{e_i}\tau(f), df(e_j)\rangle(\widetilde{\nabla}_{e_i}df)(e_j)$$

by (2.3), which is (4.3).

For the second term of (4.12), by making use of the fact that $\overline{\nabla}_{e_i}\tau(f) \in \Gamma(f_*TM)$ from the assumption that the mean curvature tensor is parallel, and (4.9), we have

$$
\begin{aligned}
\langle \overline{\nabla}_{e_i}\tau(f), \overline{\nabla}_{e_i}df(e_j)\rangle &= \langle \overline{\nabla}_{e_i}\tau(f), (\widetilde{\nabla}_{e_i}df)(e_j) + df(\nabla_{e_i}e_j)\rangle \\
&= \langle \overline{\nabla}_{e_i}\tau(f), df(\nabla_{e_i}e_j)\rangle \\
&= -\langle \tau(f), (\widetilde{\nabla}_{e_i}df)(\nabla_{e_i}e_j)\rangle.
\end{aligned}
\tag{4.13}
$$

For the third term of (4.12), we have

$$
\begin{aligned}
\langle \tau(f), -\overline{\nabla}^*\overline{\nabla}df(e_j)\rangle &= \langle \tau(f), \overline{\nabla}_{e_k}\overline{\nabla}_{e_k}df(e_j) - \overline{\nabla}_{e_k}df(e_j)\rangle \\
&= \langle \tau(f), \overline{\nabla}_{e_k}((\widetilde{\nabla}_{e_k}df)(e_j) + df(\nabla_{e_k}e_j)) \\
&\quad - (\widetilde{\nabla}_{\nabla_{e_k}e_k}df)(e_j) - df(\nabla_{\nabla_{e_k}e_k}e_j)\rangle \\
&= \langle \tau(f), (\widetilde{\nabla}_{e_k}\widetilde{\nabla}_{e_k}df)(e_j) \\
&\quad + 2(\widetilde{\nabla}_{e_k}df)(\nabla_{e_k}e_j) - (\widetilde{\nabla}_{\nabla_{e_k}e_k}e_j)\rangle \\
&= \langle \tau(f), (-\widetilde{\nabla}^*\widetilde{\nabla}df)(e_j)\rangle \\
&\quad + 2\langle \tau(f), (\widetilde{\nabla}_{e_k}df)(\nabla_{e_k}e_j)\rangle \\
&= \langle \tau(f), -\Delta df(e_j) + S(e_j)\rangle \\
&\quad + 2\langle \tau(f), (\widetilde{\nabla}_{e_k}df)(\nabla_{e_k}e_j)\rangle
\end{aligned}
$$

since $\langle \tau(f), df(X)\rangle = 0$ for all $X \in \mathfrak{X}(M)$ and Weitzenböck formula (2.8). Here, we have

$$
-\Delta df(e_j) = -dd^* df(e_j) = d\tau(f)(e_j) = \overline{\nabla}_{e_j}df,
$$

and by (2.9) and (2.7),

$$
\begin{aligned}
S(e_j) &= -\sum_{k=1}^{m}(\widetilde{R}(e_k, e_j)df)(e_k) \\
&= -\sum_{k=1}^{m}\{R^N(df(e_k), df(e_j))df(e_k) - df(R^M(e_k, e_j)e_k))\}.
\end{aligned}
$$

Thus, we have

$$
\begin{aligned}
\langle \tau(f), \overline{\nabla}^*\overline{\nabla}df(e_j)\rangle &= 2\langle \tau(f), (\widetilde{\nabla}_{e_k}df)(\nabla_{e_k}e_j)\rangle \\
&\quad + \langle \tau(f), \overline{\nabla}_{e_j}\tau(f) - R^N(df(e_k), df(e_j))df(e_k) + df(R^M(e_k, e_j)e_k)\rangle \\
&= 2\langle \tau(f), (\widetilde{\nabla}_{e_k}df)(\nabla_{e_k}e_j)\rangle \\
&\quad - \langle \tau(f), R^N(df(e_k), df(e_j))df(e_k)\rangle
\end{aligned}
\tag{4.14}
$$

since $\overline{\nabla}_{e_j}\tau(f) \in \Gamma(f_*TM)$. Substituting (4.13) and (4.14) into (4.12), we have

$$\langle -\overline{\nabla}^*\overline{\nabla}\tau(f), df(e_j)\rangle = \langle \tau(f), R^N(df(e_k), df(e_j))df(e_k)\rangle.$$
(4.15)

Finally, substituting (4.9) and (4.15) into (4.3), we have (4.7).   □

Taking for M to be a hypersurface in the unit sphere $N = S^{m+1}$ with $n = m+1$, we have

THEOREM 4.1.   Let $f : M \to S^{m+1}$ be an isometric immersion having parallel mean curvature vector field with non-zero mean curvature. Then, the necessary and sufficient condition for f to be 2-harmonic is $\|B(f)\|^2 = m = \dim M$.

PROOF. Since S^m has constant sectional curvature, the normal component of $R^N(df(e_k), df(e_j))df(e_k)$ is zero, (4.7) in Lemma 4.2 becomes

$$\overline{\nabla}^*\overline{\nabla}\tau(f) = -\langle \tau(f), (\widetilde{\nabla}_{e_i}df)(e_j)\rangle (\widetilde{\nabla}_{e_i}df)(e_j).$$

Noticing $R^N(df(e_k), \tau(f))df(e_k) = m\tau(f)$, the condition for f to be 2-harmonic becomes

$$-\langle \tau(f), (\widetilde{\nabla}_{e_i}df)(e_j)\rangle (\widetilde{\nabla}_{e_i}df)(e_j) + m\tau(f) = 0.$$
(4.16)

Denoting by ξ, the unit normal vector field on $f(M)$, and

$$(\widetilde{\nabla}_{e_i}df)(e_j) = B(f)(e_i, e_j) = H_{ij}\xi$$

in (2.3), we have $\tau(f) = H_{ii}\xi$ which implies

$$\|\tau(f)\|^2 = H_{ii}H_{jj}, \quad \|B(f)\|^2 = B(f)(e_i, e_j) = H_{ij}H_{ij}.$$

Substituting these into (4.16), we have

$$(mH_{kk} - H_{kk}H_{ij}H_{ij})\xi = 0,$$

which is equivalent to

$$(m - \|B(f)\|^2)\|\tau(f)\| = 0.$$
(4.17)

Since $\|\tau(f)\| \neq 0$, the condition $\|B(f)\|^2 = m$ is equivalent to 2-harmonicity.   □

Example 4.1   Due to Theorem 4.1, we can obtain non-trivial examples of 2-harmonic maps. Consider the Clifford torus in the unit sphere S^{m+1}:

$$M_k^m(1) = S^k\left(\sqrt{\frac{1}{2}}\right) \times S^{m-k}\left(\sqrt{\frac{1}{2}}\right),$$

where the integer k satisfies $0 \le k \le m$ ([32]). The isometric embeddings $f : M_k^m(1) \to S^{m+1}$ with $k \ne \frac{m}{2}$ are non-trivial 2-harmonic maps. Indeed, f has the parallel second fundamental form, and parallel mean curvature vector field, and by direct computation, we have $\|B(f)\|^2 = k+m-k = m$, $\|\tau(f)\| = |k-(m-k)| = |2k-m| \ne 0$, so by Theorem 4.1, f is a nontrivial 2-harmonic map.

5. The second variation of 2-harmonic maps

Assume that M is compact, $f : M \to N$ is a 2-harmonic map. We will compute the second variation formula. By using the variation formula in §3 and notation, we continue to calculate (3.23):

$$\frac{1}{2}\frac{d}{dt}E_2(f_t) = \int_M \left\langle dF\left(\frac{\partial}{\partial t}\right), \widetilde{\nabla}_{e_k}\widetilde{\nabla}_{e_k}((\widetilde{\nabla}_{e_j}dF)(e_j))\right.$$

$$- \widetilde{\nabla}_{\nabla_{e_k}e_k}((\widetilde{\nabla}_{e_j}dF)(e_j))$$

$$+ \left.R^N(dF(e_i), (\widetilde{\nabla}_{e_j}dF)(e_j)dF(e_i)\right\rangle * 1. \qquad (5.1)$$

Differentiating (5.1) by t, we have

$$\frac{1}{2}\frac{d^2}{dt^2}E_2(f_t) = \int_M \left\langle \widetilde{\nabla}_{\frac{\partial}{\partial t}}dF\left(\frac{\partial}{\partial t}\right), \widetilde{\nabla}_{e_k}\widetilde{\nabla}_{e_k}((\widetilde{\nabla}_{e_j}dF)(e_j))\right.$$

$$- \widetilde{\nabla}_{\nabla_{e_k}e_k}((\widetilde{\nabla}_{e_j}dF)(e_j))$$

$$+ \left.R^N(dF(e_i), (\widetilde{\nabla}_{e_j}dF)(e_j)dF(e_i)\right\rangle * 1$$

$$+ \int_M \left\langle dF\left(\frac{\partial}{\partial t}\right), \nabla_{\frac{\partial}{\partial t}}\left[\widetilde{\nabla}_{e_k}\widetilde{\nabla}_{e_k}((\widetilde{\nabla}_{e_j}dF)(e_j))\right.\right.$$

$$- \widetilde{\nabla}_{\nabla_{e_k}e_k}((\widetilde{\nabla}_{e_j}dF)(e_j))$$

$$+ \left.\left.R^N(dF(e_i), (\widetilde{\nabla}_{e_j}dF)(e_j)dF(e_i)\right]\right\rangle * 1. \qquad (5.2)$$

We need two Lemmas to calculate the covariant differentiation with respect to $\frac{\partial}{\partial t}$ the second term of RHS of (5.2).

LEMMA 5.1.

$$\overline{\nabla}_{\frac{\partial}{\partial t}}\overline{\nabla}_{e_k}\overline{\nabla}_{e_k}((\widetilde{\nabla}_{e_j}dF)(e_j)) = \overline{\nabla}_{e_k}\overline{\nabla}_{e_k}\left[(\widetilde{\nabla}_{e_i}\widetilde{\nabla}_{e_i}dF)\left(\frac{\partial}{\partial t}\right)\right.$$

$$- (\widetilde{\nabla}_{\nabla_{e_i}e_i}dF)\left(\frac{\partial}{\partial t}\right)$$

$$\left.+ R^N\left(dF(e_j), dF\left(\frac{\partial}{\partial t}\right)\right)dF(e_j)\right]$$

$$+ \overline{\nabla}_{e_k}\left[R^N\left(dF(e_k), dF\left(\frac{\partial}{\partial t}\right)\right)((\widetilde{\nabla}_{e_j}dF)(e_j))\right]$$

$$+ R^N\left(dF(e_k), dF\left(\frac{\partial}{\partial t}\right)\right)\overline{\nabla}((\widetilde{\nabla}dF)(e_j)). \qquad (5.3)$$

PROOF. Let us make use of the curvature formula in $F^{-1}TN$ changing variables:

$$\overline{\nabla}_{\frac{\partial}{\partial t}}\overline{\nabla}_{e_k} = \overline{\nabla}_{e_k}\overline{\nabla}_{\frac{\partial}{\partial t}} + R^N\left(dF(e_k), dF\left(\frac{\partial}{\partial t}\right)\right). \qquad (5.4)$$

Using twice this formula, we have

$$\overline{\nabla}_{\frac{\partial}{\partial t}}\overline{\nabla}_{e_k}\overline{\nabla}_{e_k}((\widetilde{\nabla}_{e_j}dF)(e_j)) = \overline{\nabla}_{e_k}\overline{\nabla}_{\frac{\partial}{\partial t}}\overline{\nabla}_{e_k}((\widetilde{\nabla}_{e_j}dF)(e_j))$$

$$+ R^N\left(dF(e_k), dF\left(\frac{\partial}{\partial t}\right)\right)\overline{\nabla}_{e_k}((\widetilde{\nabla}_{e_j}dF)(e_j))$$

$$= \overline{\nabla}_{e_k}\left[\overline{\nabla}_{e_k}\overline{\nabla}_{\frac{\partial}{\partial t}}((\widetilde{\nabla}_{e_j}dF)(e_j))\right.$$

$$\left.+ R^N\left(dF(e_k), dF\left(\frac{\partial}{\partial t}\right)\right)((\widetilde{\nabla}_{e_j}dF)(e_j))\right]$$

$$+ R^N\left(dF(e_k), dF\left(\frac{\partial}{\partial t}\right)\right)\overline{\nabla}_{e_k}((\widetilde{\nabla}_{e_j}dF)(e_j)).$$

Here, substituting (3.13) into $\overline{\nabla}_{\frac{\partial}{\partial t}}((\widetilde{\nabla}_{e_j}dF)(e_j))$ in the first term of the RHS, we have (5.3). $\qquad\square$

LEMMA 5.2.

$$
\overline{\nabla}_{\frac{\partial}{\partial t}} \overline{\nabla}_{\nabla_{e_k} e_k} ((\widetilde{\nabla}_{e_j} dF)(e_j)) = \overline{\nabla}_{\nabla_{e_k} e_k} \left[(\widetilde{\nabla}_{e_i} \widetilde{\nabla}_{e_i} dF) \left(\frac{\partial}{\partial t} \right) \right.
$$
$$
- (\widetilde{\nabla}_{\nabla_{e_i} e_i} dF) \left(\frac{\partial}{\partial t} \right)
$$
$$
\left. + R^N \left(dF(e_j), dF \left(\frac{\partial}{\partial t} \right) \right) dF(e_j) \right]
$$
$$
+ R^N \left(dF(\nabla_{e_k} e_k), dF \left(\frac{\partial}{\partial t} \right) \right) ((\widetilde{\nabla}_{e_j} dF)(e_j)). \tag{5.5}
$$

PROOF. In a similar way as Lemma 5.1, since $[\nabla_{e_k} e_k, \frac{\partial}{\partial t}] = 0$, we have

$$
\overline{\nabla}_{\frac{\partial}{\partial t}} \overline{\nabla}_{\nabla_{e_k} e_k} = \overline{\nabla}_{\nabla_{e_k} e_k} \overline{\nabla}_{\frac{\partial}{\partial t}} + R^N \left(dF(\nabla_{e_k} e_k), dF \left(\frac{\partial}{\partial t} \right) \right).
$$

Changing variables, and substituting again (3.13), we have (5.5).   □

LEMMA 5.3.

$$
\overline{\nabla}_{\frac{\partial}{\partial t}} \left[R^N(dF(e_i), (\widetilde{\nabla} dF)(e_j)) dF(e_i) \right]
$$
$$
= (\nabla'_{dF(e_i)} R^N) \left(dF \left(\frac{\partial}{\partial t} \right), (\widetilde{\nabla}_{e_j} dF)(e_j) \right) dF(e_i)
$$
$$
+ (\nabla'_{(\nabla_{e_j} dF)(e_j)} R^N) \left(dF(e_i), dF \left(\frac{\partial}{\partial t} \right) \right) dF(e_i)
$$
$$
+ R^N \left((\widetilde{\nabla}_{e_i} dF) \left(\frac{\partial}{\partial t} \right), (\widetilde{\nabla}_{e_j} dF)(e_j) \right) dF(e_i)
$$
$$
+ R^N(dF(e_i), (\widetilde{\nabla}_{e_j} dF)(e_j)) \left((\widetilde{\nabla}_{e_i} dF) \left(\frac{\partial}{\partial t} \right) \right)
$$
$$
+ R^N \left(dF(e_i), (\widetilde{\nabla}_{e_k} \widetilde{\nabla}_{e_k} dF) \left(\frac{\partial}{\partial t} \right) \right.
$$
$$
- (\widetilde{\nabla}_{\nabla_{e_k} e_k} dF) \left(\frac{\partial}{\partial t} \right)
$$
$$
\left. + R^N(dF(e_k), dF \left(\frac{\partial}{\partial t} \right)) dF(e_k) \right) dF(e_i). \tag{5.6}
$$

PROOF. We directly compute the LHS of (5.6). By definition of $\nabla'_{dF(\frac{\partial}{\partial t})} R$, and then by using the second Bianchi identity, (3.13) and $\overline{\nabla}_{\frac{\partial}{\partial t}} dF(e_i) = \overline{\nabla}_{e_i} dF(\frac{\partial}{\partial t})$, we have

$$\overline{\nabla}_{\frac{\partial}{\partial t}} \left[R^N (dF(e_i), (\widetilde{\nabla} dF)(e_j)) dF(e_i) \right]$$
$$= (\nabla'_{dF(\frac{\partial}{\partial t})} R^N) \left(dF(e_i), (\widetilde{\nabla}_{e_j} dF)(e_j) \right) dF(e_i)$$
$$+ R^N \left(\overline{\nabla}_{\frac{\partial}{\partial t}} dF(e_i), (\widetilde{\nabla}_{e_j} dF)(e_j) \right) dF(e_i)$$
$$+ R^N \left(dF(e_i), \overline{\nabla}_{\frac{\partial}{\partial t}} ((\widetilde{\nabla}_{e_j} dF)(e_j)) \right) dF(e_i)$$
$$+ R^N \left(dF(e_i), (\widetilde{\nabla}_{e_j} dF)(e_j) \right) \overline{\nabla}_{\frac{\partial}{\partial t}} dF(e_i)$$
$$= (\nabla'_{dF(e_i)} R^N) \left(dF \left(\frac{\partial}{\partial t} \right), (\widetilde{\nabla}_{e_j} dF)(e_j) \right) dF(e_i)$$
$$+ (\nabla'_{(\widetilde{\nabla}_{e_j} dF)(e_j)} R^N) \left(dF(e_i), dF \left(\frac{\partial}{\partial t} \right) \right) dF(e_i)$$

$$(\text{continued}) + R^N \left((\widetilde{\nabla}_{e_i} dF) \left(\frac{\partial}{\partial t} \right), (\widetilde{\nabla}_{e_j} dF)(e_j) \right) dF(e_i)$$
$$+ R^N \left(dF(e_i), (\widetilde{\nabla}_{e_j} dF)(e_j) \right) \left((\widetilde{\nabla}_{e_i} dF) \left(\frac{\partial}{\partial t} \right) \right)$$
$$+ R^N \Bigg(dF(e_i), (\widetilde{\nabla}_{e_k} \widetilde{\nabla}_{e_k} dF) \left(\frac{\partial}{\partial t} \right)$$
$$- (\widetilde{\nabla}_{\nabla_{e_k} e_k} dF) \left(\frac{\partial}{\partial t} \right)$$
$$+ R^N (dF(e_k), dF \left(\frac{\partial}{\partial t} \right)) dF(e_k) \Bigg) dF(e_i).$$

We have (5.6).   □

THEOREM 5.1. *Let $f : M \to N$ be a 2-harmonic map from a compact Riemannian manifold M into an arbitrary Riemannian manifold N, and $\{f_t\}$ an arbitrary C^∞ variation of f satisfying (3.2) and (3.3).*

Then, the second variation formula of $\frac{1}{2}E_2(f_t)$ is given as follows.

$$\frac{1}{2}\frac{\partial^2}{dt^2}E_2(f_t)\Big|_{t=0} = \int_M \langle -\overline{\nabla}^*\overline{\nabla}V + R^N(df(e_i),V)df(e_i),$$

$$-\overline{\nabla}^*\overline{\nabla}V + R^N(df(e_i),V)df(e_i)\rangle * 1$$

$$+ \int_M \langle V, (\nabla'_{df(e_i)}R^N)(df(e_i),\tau(f))V$$

$$+ (\nabla'_{\tau(f)}R^N)(df(e_i),V)df(e_i)$$

$$+ R^N(\tau(f),V)\tau(f)$$

$$+ 2R^N(df(e_k),V)\overline{\nabla}_{e_k}\tau(f)$$

$$+ 2R^N(df(e_i),\tau(f))\overline{\nabla}_{e_i}V\rangle * 1. \qquad (5.7)$$

PROOF. Putting $t = 0$ in (5.2), the first term of RHS vanishes since f is 2-harmonic. It suffices to substitute (5.3), (5.4) and (5.5) in Lemmas 5.1, 5.2, 5.3 into the second term. Then, we have

$$\frac{1}{2}\frac{d^2}{dt^2}E_2(f_t)\Big|_{t=0} = \int_M \langle V, -\overline{\nabla}^*\overline{\nabla}(-\overline{\nabla}^*\overline{\nabla}V + R^N(df(e_i),V)df(e_i))$$

$$+ \overline{\nabla}_{e_k}(R^N(df(e_k),V)\tau(f))$$

$$+ R^N(df(e_k),V)\overline{\nabla}_{e_k}\tau(f)$$

$$- R^N(df(\nabla_{e_k}e_k),V)\tau(f)$$

$$+ (\nabla'_{df(e_i)}R^N)(V,\tau(f))df(e_i)$$

$$+ (\nabla'_{\tau(f)}R^N)(df(e_i),V)df(e_i)$$

$$+ R^N(\overline{\nabla}_{e_i}V,\tau(f))df(e_i)$$

$$+ R^N(df(e_i),\tau(f))\overline{\nabla}_{e_i}V$$

$$+ R^N(df(e_i),-\overline{\nabla}^*\overline{\nabla}V$$

$$+ R^N(df(e_j),V)df(e_j))df(e_i)\rangle * 1. \qquad (5.8)$$

In the first term of (5.8), we have by Green's theorem,

$$\int_M \langle V, -\overline{\nabla}^*\overline{\nabla}(-\overline{\nabla}^*\overline{\nabla}V + R^N(df(e_i),V)df(e_i)\rangle * 1$$

$$= \int_M \langle -\overline{\nabla}^*\overline{\nabla}V, -\overline{\nabla}^*\overline{\nabla}V + R^N(df(e_i),V)df(e_i)\rangle * 1. \qquad (5.9)$$

For the last term of the RHS of (5.8), by the symmetric property of the curvature

$$\int_M \langle V, R^N(df(e_i),W)df(e_i)\rangle * 1 = \int_M \langle W, R^N(df(e_i),V)df(e_i)\rangle * 1,$$

we have

$$\int_M \langle V, R^N(df(e_i), -\overline{\nabla}^*\overline{\nabla}V + R^N(df(e_j), V)df(e_j)df(e_i)\rangle * 1$$

$$= \int_M \langle R^N(df(e_i), V)df(e_i),$$

$$- \overline{\nabla}^*\overline{\nabla}V + R^N(df(e_j), V)df(e_j)\rangle * 1. \qquad (5.10)$$

For the second term of the RHS of (5.8), we have

$$\overline{\nabla}_{e_k}(R^N(df(e_k), V)\tau(f)) = (\nabla'_{df(e_k)}R^N)(df(e_k), V)\tau(f)$$

$$+ R^N(\overline{\nabla}_{e_k}df(e_k), V)\tau(f)$$

$$+ R^N(df(e_k), \overline{\nabla}_{e_k}V)\tau(f)$$

$$+ R^N(df(e_k), V)\overline{\nabla}_{e_k}\tau(f). \qquad (5.11)$$

Substituting (5.9), (5.10) and (5.11) into (5.8), we have

$$\frac{1}{2}\frac{\partial^2}{dt^2}E_2(f_t)\Big|_{t=0} = \int_M \langle -\overline{\nabla}^*\overline{\nabla}V + R^N(df(e_i), V)df(e_i),$$

$$- \overline{\nabla}^*\overline{\nabla}V + R^N(df(e_i), V)df(e_i)\rangle * 1$$

$$+ \int_M \langle V, (\nabla'_{df(e_i)}R^N)(df(e_i), \tau(f))V$$

$$+ R^N(\tau(f), V)\tau(f) + R^N(df(e_k), \overline{\nabla}_{e_k}V)\tau(f)$$

$$+ 2R^N(df(e_k), V)\overline{\nabla}_{e_k}\tau(f)$$

$$+ (\nabla'_{df(e_i)}R^N)(V, \tau(f))df(e_i)$$

$$+ (\nabla'_{\tau(f)}R^N)(df(e_i), V)df(e_i)$$

$$+ R^N(\overline{\nabla}_{e_i}V, \tau(f))df(e_i)$$

$$+ R^N(df(e_i), \tau(f))\overline{\nabla}_{e_i}V\rangle * 1. \qquad (5.12)$$

By the first Bianchi identity, we have

$$R^N(df(e_k), \overline{\nabla}_{e_k}V)\tau(f) + R^N(\overline{\nabla}_{e_i}V, \tau(f)df(e_i)$$

$$= R^N(df(e_i), \tau(f))\overline{\nabla}_{e_i}V,$$

$$(\nabla'_{df(e_k)}R^N)(df(e_k), V)\tau(f) + (\nabla'_{df(e_k)}R^N)(V, \tau(f))df(e_k)$$

$$= (\nabla'_{df(e_k)}R^N)(df(e_k), \tau(f))V.$$

Substituting these into (5.12), we have (5.7).   □

By the second variation formula, we derive the notion of stable 2-harmonic maps.

DEFINITION 5.1. *Let $f : M \to N$ be a 2-harmonic map of a compact Riemannian manifold M into any Riemannian manifold N. If the second variation of 2-energy is non-negative for every variation $\{f_t\}$ of f, i.e., the RHS of (5.7) is non-negative for every vector field V along f, f is said to be a* stable 2-harmonic map.

By definition of 2-energy, any harmonic maps are stable 2-harmonic maps. This may also be seen as follows: since $\tau(f) = 0$, for a vector field V of any variation $\{f_t\}$ we have

$$\frac{1}{2}\frac{d^2}{dt^2}E_2(f_t)\bigg|_{t=0} = \int_M \| -\overline{\nabla}^*\overline{\nabla}V + R^N(df(e_i), V)df(e_i)\|^2 * 1 \geq 0.$$

THEOREM 5.2. *Assume that M is a compact Riemannian manifold, and N is a Riemannian manifold with a positive constant sectional curvature $K > 0$. Then, there is no non-trivial stable 2-harmonic map satisfying the conservation law.*

PROOF. Since N has constant curvature, $\nabla' R^N = 0$, so that (5.7) becomes

$$\frac{1}{2}\frac{d^2}{dt^2}\bigg|_{t=0} E_2(f_t) = \int_M \| -\overline{\nabla}^*\overline{\nabla}V + R^N(df(e_i), V)df(e_i)\|^2 * 1$$
$$+ \int_M \langle V, R^N(\tau(f), V)\tau(f) + 2R^N(df(e_i), V)\overline{\nabla}_{e_k}\tau(f)$$
$$+ 2R^N(df(e_i), \tau(f))\overline{\nabla}_{e_i}V \rangle * 1. \tag{5.13}$$

Especially, if we take $V = \tau(f)$, then, the first term of the RHS of (5.13) and the first integrand of the second term vanish, so we have

$$\frac{1}{2}\frac{d^2}{dt^2}\bigg|_{t=0} E_2(f_t) = 4\int_M \langle R^N(df(e_k), \tau(f))\overline{\nabla}_{e_k}\tau(f), \tau(f)\rangle * 1$$
$$= 4K\int_M \left[\langle df(e_k), \overline{\nabla}_{e_k}\tau(f)\rangle\|\tau(f)\|^2 \right.$$
$$\left. - \langle df(e_k), \tau(f)\rangle\langle\tau(f), \overline{\nabla}_{e_k}\tau(f)\rangle\right] * 1. \tag{5.14}$$

Since f satisfies the conservation law, i.e., $-\langle\tau(f), df(X)\rangle = (\text{div} S_f)(X) = 0$ for all $X \in \mathfrak{X}(M)$, we have

$$\langle df(e_k), \tau(f)\rangle = 0,$$

and

$$\langle df(e_k), \overline{\nabla}_{e_k}\tau(f)\rangle = -\langle \overline{\nabla}_{e_k}df(e_k), \tau(f)\rangle + e_k\langle df(e_k), \tau(f)\rangle$$
$$= -\|\tau(f)\|^2 - \langle df(\nabla_{e_k}e_k), \tau(f)\rangle$$
$$= -\|\tau(f)\|^2. \tag{5.15}$$

Substituting (5.15) into (5.14), we have

$$0 \leq \frac{1}{2}\frac{d^2}{dt^2}\bigg|_{t=0} E_2(f_t) = -4K\int_M \|\tau(f)\|^4 * 1 \leq 0,$$

which implies that $\tau(f) \equiv 0$. $\square$

In order to apply the second variation formula, we take $N = \mathbb{C}P^n$.

LEMMA 5.4. *Assume that* $f : M \to \mathbb{C}P^n$ *is a stable 2-harmonic map of a compact Riemannian manifold which satisfies the conservation law and* $\|\tau(f)\|^2 > 3\sqrt{2e(f)}\,\|\nabla\tau(f)\|$ *pointwisely on* M. *Then,* f *is harmonic. Here, we denote* $\|\nabla\tau(f)\|^2 = \langle \overline{\nabla}_{e_k}\tau(f), \overline{\nabla}_{e_k}\tau(f)\rangle$.

PROOF. Assume that f satisfies all the assumption, but not harmonic. Since $\nabla'R^N = 0$, if we take $V = \tau(f)$, both the first term and the integrand of the second term of (5.13) vanish, and we use the explicit formula of the curvature tensor of $\mathbb{C}P^n$, (5.13) becomes as follows.

$$\frac{1}{2}\frac{d^2}{dt^2}\bigg| E_2(f_t) = 4\int_M \langle R^N(df(e_k), \tau(f))\overline{\nabla}_{e_k}\tau(f), \tau(f)\rangle * 1$$

$$= C\int_M \langle\langle df(e_k), \overline{\nabla}_{e_k}\tau(f)\rangle\tau(f) - \langle\tau(f), \overline{\nabla}_{e_k}\tau(f)\rangle df(e_k)$$

$$+ \langle Jdf(e_k), \overline{\nabla}_{e_k}\tau(f)\rangle J\tau(f) - \langle J\tau(f), \overline{\nabla}_{e_k}\tau(f)\rangle Jdf(e_k)$$

$$+ 2\langle Jdf(e_k), \tau(f)\rangle J\overline{\nabla}_{e_k}\tau(f), \tau(f)\rangle * 1, \tag{5.16}$$

where C is a positive constant depending only on $\mathbb{C}P^n$. By (5.15) and $\langle J\tau(f), \tau(f)\rangle = 0$, we have

$$\frac{1}{2}\frac{d^2}{dt^2}\bigg| E_2(f_t) = C\int_M [-\|\tau(f)\|^4$$

$$+ 3\langle Jdf(e_k), \tau(f)\rangle\langle J\overline{\nabla}_{e_k}\tau(f), \tau(f)\rangle] * 1. \tag{5.17}$$

For each k, by Schwarz inequality twice, we have

$$\langle Jdf(e_k), \tau(f)\rangle \langle J\bar{\nabla}_{e_k}\tau(f), \tau(f)\rangle$$

$$\leq \sqrt{\langle Jdf(e_k), Jdf(e_k)\rangle}\|\tau(f)\|\sqrt{\langle J\bar{\nabla}_{e_k}\tau(f), J\bar{\nabla}_{e_k}\tau(f)\rangle}\|\tau(f)\|$$

$$= \|\tau(f)\|^2\sqrt{\langle df(e_k), df(e_k)\rangle\langle \bar{\nabla}_{e_k}\tau(f), \bar{\nabla}_{e_k}\tau(f)\rangle}.$$

By taking the sum over k, and by Schwarz inequality, we have

$$\langle Jdf(e_k), \tau(f)\rangle \langle J\bar{\nabla}_{e_k}\tau(f), \tau(f)\rangle$$

$$\leq \|\tau(f)\|^2\sqrt{\langle df(e_i), df(e_i)\rangle\langle \bar{\nabla}_{e_j}\tau(f), \bar{\nabla}_{e_j}\tau(f)\rangle}$$

$$= \sqrt{2e(f)}\|\tau(f)\|^2\|\bar{\nabla}\tau(f)\|. \tag{5.18}$$

Substituting this into (5.17), we have

$$0 \leq \frac{1}{2}\frac{d^2}{dt^2}\Big|E_2(f_t) \leq C\int_M \|\tau(f)\|^2\left(3\sqrt{2e(f)}\|\bar{\nabla}\| - \|\tau(f)\|^2\right)*1$$

which is impossible if $\|\tau(f)\|^2 > 3\sqrt{2e(f)}\|\bar{\nabla}\tau(f)\|$. □

LEMMA 5.5. *Assume that $f : M \to N = \mathbb{C}P^n$ a 2-harmonic map from a compact Riemannian manifold into $\mathbb{C}P^n$ with constant holomorphic sectional curvature $C > 0$ which satisfies the conservation law and $\|\tau(f)\|^2 = constant$. Then, it holds that*

$$\frac{C}{2}e(f)\|\tau(f)\|^2 \leq \|\bar{\nabla}\tau(f)\|^2 \leq 2Ce(f)\|\tau(f)\|^2. \tag{5.19}$$

PROOF. Since f is 2-harmonic, we can still use the equality in (4.2), so that

$$0 = \frac{1}{2}\Delta\|\tau(f)\|^2 = \|\bar{\nabla}\tau(f)\|^2 - \langle R^N(df(e_i), \tau(f))df(e_i), \tau(f)\rangle. \tag{5.20}$$

We denote by $\mathrm{Riem}^N(df(e_i) \wedge \tau(f))$, the sectional curvature through $df(e_i)$ and $\tau(f)$. Since this plane does not degenerate, and f satisfies the conservation law, for each i,

$$\langle R^N(df(e_i), \tau(f))df(e_i), \tau(f)\rangle$$

$$= \mathrm{Riem}^N(df(e_i) \wedge \tau(f)) \cdot \langle df(e_i), df(e_i)\rangle \|\tau(f)\|^2. \tag{5.21}$$

Recall that the sectional curvature of $\mathbb{C}P^n$ satisfies

$$\frac{C}{4} \leq \mathrm{Riem}^N \leq C, \tag{5.22}$$

so that by (5.21), (5.22), we have

$$\frac{C}{2}e(f)\|\tau(f)\|^2 \leq \langle R^N(df(e_i), \tau(f))df(e_i), \tau(f)\rangle$$

$$\leq 2Ce(f)\|\tau(f)\|^2. \tag{5.23}$$

Thus, we have (5.19).   □

THEOREM 5.3. *Let* $f : M \to \mathbb{C}P^n$ *a stable 2-harmonic map from a compact Riemannian manifolds* M *into* $\mathbb{C}P^n$ *with constant holomorphic sectional curvature* $C > 0$, *which satisfies the conservation law, and* $\|\tau(f)\|^2 = constant.$ *If the density function of* f *satisfies*

$$e(f) < \frac{\|\tau(f)\|}{6\sqrt{C}}, \tag{5.24}$$

then f *is harmonic.*

PROOF. Assume that there exists such a stable 2-harmonic map but not harmonic. By Lemma 5.4, there exists a point $p \in M$ at which

$$0 < \|\tau(f)\|^2 \leq 3\sqrt{2e(f)}\|\overline{\nabla}\tau(f)\|.$$

By Lemma 5.5, it holds that, at this point,

$$\|\tau(f)\|^2 \leq 3\sqrt{2e(f)}\|\overline{\nabla}\tau(f)\| \leq 6\sqrt{C}e(f)\|\tau(f)\|.$$

Then, at this point,

$$0 \leq \|\tau(f)\|(6\sqrt{C}e(f) - \|\tau(f)\|).$$

Since $\|\tau(f)\| > 0$ at p, we have $6\sqrt{C}e(f) - \|\tau(f)\| \geq 0$ at p which contradicts the assumption (5.24).   □

COROLLARY 5.1. *Assume that* $f : M \to \mathbb{C}P^n$ *is a 2-harmonic isometric immersion from a* m-*dimensional compact Riemannian manifold* M *into* $\mathbb{C}P^n$ *with constant holomorphic sectional curvature* $C > 0$ *whose* $\|\tau(f)\|$ *is constant and satisfies* $\|\tau(f)\| > 3\sqrt{C}m.$ *Then* f *can not be stable.*

Part 2

Rigidity and Abundance of Biharmonic Maps

CHAPTER 3

Biharmonic Maps into a Riemannian Manifold of Non-positive Curvature

1

ABSTRACT. In this chapter, we study biharmonic maps between Riemannian manifolds with finite energy and finite bi-energy. We show that if the domain manifold is complete and the target space has non-positive curvature, then such a map must be harmonic. We apply it to isometric immersions and horizontally conformal submersions.

1. Introduction

Harmonic maps play a central role in geometry; they are critical points of the energy functional $E(\varphi) = \frac{1}{2} \int_M |d\varphi|^2 \, v_g$ for smooth maps φ of (M, g) into (N, h). The Euler-Lagrange equations are given by the vanishing of the tension filed $\tau(\varphi)$. In 1983, J. Eells and L. Lemaire [40] extended the notion of harmonic map to biharmonic map, which are, by definition, critical points of the bienergy functional

$$E_2(\varphi) = \frac{1}{2} \int_M |\tau(\varphi)|^2 \, v_g. \tag{1.1}$$

After G.Y. Jiang [74] studied the first and second variation formulas of E_2, extensive studies in this area have been done (for instance, see [8], [16], [89], [90], [102], [120], [131], [63], [64], [73], etc.). Notice that harmonic maps are always biharmonic by definition.

For harmonic maps, it is well known that:

If a domain manifold (M, g) is complete and has non-negative Ricci curvature, and the sectional curvature of a target manifold (N, h) is non-positive, then every energy finite harmonic map is a constant map (cf. [139]).

---

[1]This chapter is due to [111]: N. Nakauchi, H. Urakawa and S. Gudmundsson, *Biharmonic maps into a Riemannian manifold of non-positive curvature*, Geom. Dedicata, **169** (2014), 263–272. It is also related to [108] and [109].

Therefore, it is a natural question to consider biharmonic maps into a Riemannian manifold of non-positive curvature. In this connection, Baird, Fardoun and Ouakkas (cf. [**8**]) showed that:

If a non-compact Riemannian manifold (M, g) is complete and has non-negative Ricci curvature and (N, h) has non-positive sectional curvature, then every bienergy finite biharmonic map of (M, g) into (N, h) is harmonic.

In this paper, we will show that

THEOREM 1.1. (*cf.* **Theorem 2.1**) *Under only the assumptions of completeness of (M, g) and non-positivity of curvature of (N, h),*
(1) every biharmonic map $\varphi : (M, g) \to (N, h)$ with finite energy and finite bienergy must be harmonic.
(2) In the case $\mathrm{Vol}(M, g) = \infty$, under the same assumtion, every biharmonic map $\varphi : (M, g) \to (N, h)$ with finite bienergy is harmonic.

We do not need any assumption on the Ricci curvature of (M, g) in Theorem 1.1. Since (M, g) is a non-compact complete Riemannian manifold whose Ricci curvature is non-negative, then $\mathrm{Vol}(M, g) = \infty$ (cf. Theorem 7, p. 667, [**166**]). Thus, Theorem 1.1, (2) recovers the result of Baird, Fardoun and Ouakkas. Furthermore, Theorem 1.1 is sharp because one can not weaken the assumptions because the generalized Chen's conjecture does not hold if (M, g) is not complete (cf. recall the counter examples of Ou and Tang [**123**]). The two assumptions: finiteness of the energy and bienergy, are necessary. Indeed, there exists a biharmonic map φ which is not harmonic, but energy and bienergy are infinite. For example, $f(x) = r(x)^2 = \sum_{i=1}^{m}(x_i)^2, x = (x_1, \cdots, x_m) \in \mathbb{R}^m$ is biharmonic, but not harmonic, and have infinite energy and bienergy.

As the first bi-product of our method, we obtain (cf. [**108**], [**109**])

THEOREM 1.2. (*cf.* **Theorem 3.1**) *Assume that (M, g) is a complete Riemannian manifold, and let $\varphi : (M, g) \to (N, h)$ is an isometric immersion, and the sectional curvature of (N, h) is non-positive. If $\varphi : (M, g) \to (N, h)$ is biharmonic and $\int_M |\xi|^2 v_g < \infty$, then it is minimal. Here, ξ is the mean curvature normal vector field of the isometric immersion φ.*

Theorem 1.2 (cf. Theorem 3.1) gives an affirmative answer to the generalized B.Y. Chen's conjecture (cf. [**16**]) under natural conditions.
For the second bi-product, we can apply Theorem 1.1 to a horizontally conformal submersion (cf. [**7**],[**10**]). Then, we have

THEOREM 1.3. (*cf.* **Corollary 3.4**) *Let* (M^m, g) *be a non-compact complete Riemannian manifold* $(m > 2)$, *and* (N^2, h), *a Riemannian surface with non-positive curvature. Let* λ *be a positive function on* M *belonging to* $C^\infty(M) \cap L^2(M)$, *and* $\varphi : (M, g) \to (N^2, h)$, *a horizontally conformal submersion with a dilation* λ. *If* φ *is biharmonic and* $\lambda |\hat{\mathbf{H}}|_g \in L^2(M)$, *then* φ *is a harmonic morphism. Here,* $\hat{\mathbf{H}}$ *is trace of the second fundamental form of each fiber of* φ.

Acknowledgement.   The second author would like to express his sincere gratitude to Professor Sigmundur Gudmundsson for his hospitality and very intensive and helpful discussions during for the second author staying at Lund University, at 2012 May. Addition of part of Section Three to the original version of the first and second authors was based by this joint work with him during this period.

2. Preliminaries and statement of main theorem

In this section, we prepare materials for the first and second variational formulas for the bienergy functional and biharmonic maps. Let us recall the definition of a harmonic map $\varphi : (M, g) \to (N, h)$, of a compact Riemannian manifold (M, g) into another Riemannian manifold (N, h), which is an extremal of the *energy functional* defined by

$$E(\varphi) = \int_M e(\varphi)\, v_g,$$

where $e(\varphi) := \frac{1}{2}|d\varphi|^2$ is called the energy density of φ. That is, for any variation $\{\varphi_t\}$ of φ with $\varphi_0 = \varphi$,

$$\frac{d}{dt}\Big|_{t=0} E(\varphi_t) = - \int_M h(\tau(\varphi), V)v_g = 0, \tag{2.1}$$

where $V \in \Gamma(\varphi^{-1}TN)$ is a variation vector field along φ which is given by $V(x) = \frac{d}{dt}|_{t=0}\varphi_t(x) \in T_{\varphi(x)}N$, $(x \in M)$, and the *tension field* is given by $\tau(\varphi) = \sum_{i=1}^m B(\varphi)(e_i, e_i) \in \Gamma(\varphi^{-1}TN)$, where $\{e_i\}_{i=1}^m$ is a locally defined frame field on (M, g), and $B(\varphi)$ is the second fundamental form of φ defined by

$$B(\varphi)(X, Y) = (\widetilde{\nabla}d\varphi)(X, Y)$$
$$= (\widetilde{\nabla}_X d\varphi)(Y)$$
$$= \overline{\nabla}_X(d\varphi(Y)) - d\varphi(\nabla_X Y), \tag{2.2}$$

for all vector fields $X, Y \in \mathfrak{X}(M)$. Here, ∇, and ∇^N, are connections on TM, TN of (M, g), (N, h), respectively, and $\overline{\nabla}$, and $\widetilde{\nabla}$ are the induced ones on $\varphi^{-1}TN$, and $T^*M \otimes \varphi^{-1}TN$, respectively. By (2.1), φ is harmonic if and only if $\tau(\varphi) = 0$.

The second variation formula is given as follows. Assume that φ is harmonic. Then,

$$\left.\frac{d^2}{dt^2}\right|_{t=0} E(\varphi_t) = \int_M h(J(V), V)v_g, \qquad (2.3)$$

where J is an elliptic differential operator, called the *Jacobi operator* acting on $\Gamma(\varphi^{-1}TN)$ given by

$$J(V) = \overline{\Delta}V - \mathcal{R}(V), \qquad (2.4)$$

where $\overline{\Delta}V = \overline{\nabla}^*\overline{\nabla}V = -\sum_{i=1}^m \{\overline{\nabla}_{e_i}\overline{\nabla}_{e_i}V - \overline{\nabla}_{\nabla_{e_i}e_i}V\}$ is the *rough Laplacian* and $\mathcal{R}$ is a linear operator on $\Gamma(\varphi^{-1}TN)$ given by $\mathcal{R}(V) = \sum_{i=1}^m R^N(V, d\varphi(e_i))d\varphi(e_i)$, and R^N is the curvature tensor of (N, h) given by $R^N(U, V) = \nabla^N_U\nabla^N_V - \nabla^N_V\nabla^N_U - \nabla^N_{[U,V]}$ for $U, V \in \mathfrak{X}(N)$.

J. Eells and L. Lemaire [**40**] proposed polyharmonic (k-harmonic) maps and Jiang [**74**] studied the first and second variation formulas of biharmonic maps. Let us consider the *bienergy functional* defined by

$$E_2(\varphi) = \frac{1}{2}\int_M |\tau(\varphi)|^2 v_g, \qquad (2.5)$$

where $|V|^2 = h(V, V)$, $V \in \Gamma(\varphi^{-1}TN)$.

Then, the first variation formula of the bienergy functional is given (the first variation formula) by

$$\left.\frac{d}{dt}\right|_{t=0} E_2(\varphi_t) = -\int_M h(\tau_2(\varphi), V)v_g. \qquad (2.6)$$

Here,

$$\tau_2(\varphi) := J(\tau(\varphi)) = \overline{\Delta}(\tau(\varphi)) - \mathcal{R}(\tau(\varphi)), \qquad (2.7)$$

which is called the *bitension field* of φ, and J is given in (2.4).

A smooth map φ of (M, g) into (N, h) is said to be *biharmonic* if $\tau_2(\varphi) = 0$.

Then, we can state our main theorem.

THEOREM 2.1. *Assume that (M, g) is complete and the sectional curvature of (N, h) is non-positive.*

(1) *Every biharmonic map $\varphi : (M, g) \to (N, h)$ with finite energy $E(\varphi) < \infty$ and finite bienergy $E_2(\varphi) < \infty$, is harmonic.*

(2) *In the case $\mathrm{Vol}(M, g) = \infty$, every biharmonic map $\varphi : (M, g) \to (N, h)$ with finite bienergy $E_2(\varphi) < \infty$, is harmonic.*

3. Proof of main theorem and two applications

In this section we will give a proof of Theorem 2.1 which consists of four steps.

(*The first step*)   For a fixed point $x_0 \in M$, and for every $0 < r < \infty$, we first take a cut-off C^∞ function η on M (for instance, see [150], p. 167) satisfying that

$$\begin{cases} 0 \le \eta(x) \le 1 & (x \in M), \\ \eta(x) = 1 & (x \in B_r(x_0)), \\ \eta(x) = 0 & (x \notin B_{2r}(x_0)), \\ |\nabla \eta| \le \dfrac{2}{r} & (x \in M). \end{cases} \qquad (3.1)$$

For a biharmonic map $\varphi : (M, g) \to (N, h)$, the bitension field is given as

$$\tau_2(\varphi) = \overline{\Delta}(\tau(\varphi)) - \sum_{i=1}^{m} R^N(\tau(\varphi), d\varphi(e_i)) d\varphi(e_i) = 0, \qquad (3.2)$$

so we have

$$\int_M \langle \overline{\Delta}(\tau(\varphi)), \eta^2 \tau(\varphi) \rangle \, v_g = \int_M \eta^2 \sum_{i=1}^{m} \langle R^N(\tau(\varphi), d\varphi(e_i)) d\varphi(e_i), \tau(\varphi) \rangle \, v_g$$
$$\le 0, \qquad (3.3)$$

since the sectional curvature of (N, h) is non-positive.

(*The second step*) Therefore, by (3.3) and noticing that $\overline{\Delta} = \overline{\nabla}^* \overline{\nabla}$, we obtain

$$0 \ge \int_M \langle \overline{\Delta}(\tau(\varphi)), \eta^2 \tau(\varphi) \rangle \, v_g$$
$$= \int_M \langle \overline{\nabla} \tau(\varphi), \overline{\nabla}(\eta^2 \tau(\varphi)) \rangle \, v_g$$
$$= \int_M \sum_{i=1}^{m} \langle \overline{\nabla}_{e_i} \tau(\varphi), \overline{\nabla}_{e_i}(\eta^2 \tau(\varphi)) \rangle \, v_g$$
$$= \int_M \sum_{i=1}^{m} \left\{ \eta^2 \langle \overline{\nabla}_{e_i} \tau(\varphi), \overline{\nabla}_{e_i} \tau(\varphi) \rangle + e_i(\eta^2) \langle \overline{\nabla}_{e_i} \tau(\varphi), \tau(\varphi) \rangle \right\} v_g$$
$$= \int_M \eta^2 \sum_{i=1}^{m} \left| \overline{\nabla}_{e_i} \tau(\varphi) \right|^2 v_g + 2 \int_M \sum_{i=1}^{m} \langle \eta \overline{\nabla}_{e_i} \tau(\varphi), e_i(\eta) \tau(\varphi) \rangle \, v_g,$$
$$\qquad (3.4)$$

where we used $e_i(\eta^2) = 2\eta\, e_i(\eta)$ at the last equality. By moving the second term in the last equality of (3.4) to the left hand side, we have

$$\int_M \eta^2 \sum_{i=1}^m |\overline{\nabla}_{e_i}\tau(\varphi)|^2\, v_g \leq -2\int_M \sum_{i=1}^m \langle \eta\, \overline{\nabla}_{e_i}\tau(\varphi), e_i(\eta)\,\tau(\varphi)\rangle\, v_g$$

$$= -2\int_M \sum_{i=1}^m \langle V_i, W_i\rangle\, v_g, \qquad (3.5)$$

where we put $V_i := \eta\, \overline{\nabla}_{e_i}\tau(\varphi)$, and $W_i := e_i(\eta)\,\tau(\varphi)$ $(i = 1\cdots, m)$.

Now let recall the following Cauchy-Schwartz inequality:

$$\pm 2\,\langle V_i, W_i\rangle \leq \epsilon |V_i|^2 + \frac{1}{\epsilon}|W_i|^2 \qquad (3.6)$$

for all positive $\epsilon > 0$ because of the inequality $0 \leq |\sqrt{\epsilon}\, V_i \pm \frac{1}{\sqrt{\epsilon}}\, W_i|^2$. Therefore, for (3.5), we obtain

$$-2\int_M \sum_{i=1}^m \langle V_i, W_i\rangle\, v_g \leq \epsilon\int_M \sum_{i=1}^m |V_i|^2\, v_g + \frac{1}{\epsilon}\int_M \sum_{i=1}^m |W_i|^2\, v_g. \qquad (3.7)$$

If we put $\epsilon = \frac{1}{2}$, we obtain, by (3.5) and (3.7),

$$\int_M \eta^2 \sum_{i=1}^m |\overline{\nabla}_{e_i}\tau(\varphi)|^2\, v_g \leq \frac{1}{2}\int_M \sum_{i=1}^m \eta^2\, |\overline{\nabla}_{e_i}\tau(\varphi)|^2\, v_g$$

$$+ 2\int_M \sum_{i=1}^m e_i(\eta)^2\, |\tau(\varphi)|^2\, v_g. \qquad (3.8)$$

Thus, by (3.8) and (3.1), we obtain

$$\int_M \eta^2 \sum_{i=1}^m |\overline{\nabla}_{e_i}\tau(\varphi)|^2\, v_g \leq 4\int_M |\nabla\eta|^2\, |\tau(\varphi)|^2\, v_g$$

$$\leq \frac{16}{r^2}\int_M |\tau(\varphi)|^2\, v_g. \qquad (3.9)$$

(*The third step*)   Since (M, g) is complete and non-compact, we can tend r to infinity. By the assumption $E_2(\varphi) = \frac{1}{2}\int_M |\tau(\varphi)|^2\, v_g < \infty$, the right hand side goes to zero. And also, if $r \to \infty$, the left hand side of (3.9) goes to $\int_M \sum_{i=1}^m |\overline{\nabla}_{e_i}\tau(\varphi)|^2\, v_g$ since $\eta = 1$ on $B_r(x_0)$. Thus, we obtain

$$\int_M \sum_{i=1}^m |\overline{\nabla}_{e_i}\tau(\varphi)|^2\, v_g = 0. \qquad (3.10)$$

Therefore, we obtain, for every vector field X in M,

$$\overline{\nabla}_X\tau(\varphi) = 0. \qquad (3.11)$$

Then, we have, in particular, $|\tau(\varphi)|$ is constant, say c. Because, for every vector field X on M, at each point in M,

$$X\,|\tau(\varphi)|^2 = 2\langle\overline{\nabla}_X\tau(\varphi),\tau(\varphi)\rangle = 0. \tag{3.12}$$

Therefore, if $\mathrm{Vol}(M,g) = \infty$ and $c \neq 0$, then

$$\tau_2(\varphi) = \frac{1}{2}\int_M |\tau(\varphi)|^2\,v_g = \frac{c^2}{2}\,\mathrm{Vol}(M,g) = \infty \tag{3.13}$$

which yields a contradiction. Thus, we have $|\tau(\varphi)| = c = 0$, i.e., φ is harmonic. We have (2).

(*The fourth step*) For (1), assume both $E(\varphi) < \infty$ and $E_2(\varphi) < \infty$. Then, let us consider a 1-form α on M defined by

$$\alpha(X) := \langle d\varphi(X),\tau(\varphi)\rangle, \qquad (X \in \mathfrak{X}(M)). \tag{3.14}$$

Note here that

$$\int_M |\alpha|\,v_g = \int_M \left(\sum_{i=1}^m |\alpha(e_i)|^2\right)^{1/2} v_g$$

$$\leq \int_M |d\varphi|\,|\tau(\varphi)|\,v_g$$

$$\leq \left(\int_M |d\varphi|^2\,v_g\right)^{1/2} \left(\int_M |\tau(\varphi)|^2\,v_g\right)^{1/2}$$

$$= 2\sqrt{E(\varphi)\,E_2(\varphi)} < \infty. \tag{3.15}$$

Moreover, the divergent $\delta\alpha := -\sum_{i=1}^m (\nabla_{e_i}\alpha)(e_i) \in C^\infty(M)$ turns out (cf. [**40**], p. 9) that

$$-\delta\alpha = |\tau(\varphi)|^2 + \langle d\varphi,\overline{\nabla}\tau(\varphi)\rangle = |\tau(\varphi)|^2. \tag{3.16}$$

Indeed, we have

$$-\delta\alpha = \sum_{i=1}^m e_i\langle d\varphi(e_i),\tau(\varphi)\rangle - \sum_{i=1}^m \langle d\varphi(\nabla_{e_i}e_i),\tau(\varphi)\rangle$$

$$= \langle\sum_{i=1}^m \left(\overline{\nabla}_{e_i}(d\varphi(e_i)) - d\varphi(\nabla_{e_i}e_i)\right),\tau(\varphi)\rangle$$

$$+ \sum_{i=1}^m \langle d\varphi(e_i),\overline{\nabla}_{e_i}\tau(\varphi)\rangle$$

$$= \langle\tau(\varphi),\tau(\varphi)\rangle + \langle d\varphi,\overline{\nabla}\tau(\varphi)\rangle$$

which is equal to $|\tau(\varphi)|$ since $\overline{\nabla}\tau(\varphi) = 0$.

By (3.16) and $E_2(\varphi) = \frac{1}{2}\int_M |\tau(\varphi)|^2\,v_g < \infty$, the function $-\delta\alpha$ is also integrable over M. Thus, together with (3.15), we can apply Gaffney's theorem (see 5.1 in Appendices, below) for the 1-form α.

Then, by integrating (3.16) over M, and by Gaffney's theorem, we have

$$0 = \int_M (-\delta\alpha)\, v_g = \int_M |\tau(\varphi)|^2\, v_g, \qquad (3.17)$$

which yields that $\tau(\varphi) = 0$.

We have Theorem 2.1. □

Our method can be applied to an isometric immersion $\varphi : (M, g) \to (N, h)$. In this case, the 1-form α defined by (3.14) in the proof of Theorem 2.3 vanishes automatically without using Gaffney's theorem since $\tau(\varphi) = m\,\xi$ belongs to the normal component of $T_{\varphi(x)}N$ $(x \in M)$, where ξ is the mean curvature normal vector field and $m = \dim(M)$. Thus, (3.16) turns out that

$$0 = -\delta\alpha = |\tau(\varphi)|^2 + \langle d\varphi, \overline{\nabla}\tau(\varphi)\rangle = |\tau(\varphi)|^2 \qquad (3.18)$$

which implies that $\tau(\varphi) = m\,\xi = 0$, i.e., φ is minimal. Thus, we obtain

THEOREM 3.1. *Assume that (M, g) is a complete Riemannian manifold, and let $\varphi : (M, g) \to (N, h)$ is an isometric immersion, and the sectional curvature of (N, h) is non-positive. If $\varphi : (M, g) \to (N, h)$ is biharmonic and $\int_M |\xi|^2\, v_g < \infty$, then φ is minimal. Here, ξ is the mean curvature normal vector field of the isometric immersion of φ.*

We also apply Theorem 2.1 to a horizontally conformal submersion $\varphi : (M^m, g) \to (N^n, h)$ $(m > n \geq 2)$ (cf. [10], see also [54]). In the case that a Riemannian submersion from a space form of constant sectional curvature into a Riemann surface (N^2, h), Wang and Ou (cf. [161], see also [91]) showed that it is biharmonic if and only if it is harmonic. We treat with a submersion from a higher dimensional Riemannian manifold (M, g) (cf. [7]). Namely, let $\varphi : M \to N$ be a submersion, and each tangent space $T_x M$ $(x \in M)$ is decomposed into the orthogonal direct sum of the *vertical space* $\mathcal{V}_x = \mathrm{Ker}(d\varphi_x)$ and the *horizontal space* $\mathcal{H}_x$:

$$T_x M = \mathcal{V}_x \oplus \mathcal{H}_x, \qquad (3.19)$$

and we assume that there exists a positive C^∞ function λ on M, called the *dilation*, such that, for each $x \in M$,

$$h(d\varphi_x(X), d\varphi_x(Y)) = \lambda^2(x)\, g(X, Y), \quad (X, Y \in \mathcal{H}_x).$$
$$(3.20)$$

The map φ is said to be *horizontally homothetic* if the dilation λ is constant along horizontally curves in M.

If $\varphi : (M^m, g) \to (N^n, h)$ $(m > n \geq 2)$ is a horizontally conformal submersion . Then, the tension field $\tau(\varphi)$ is given (cf. [7], [10]) by

$$\tau(\varphi) = \frac{n-2}{2} \lambda^2 \, d\varphi\Big(\mathrm{grad}_{\mathcal{H}}\Big(\frac{1}{\lambda^2}\Big)\Big) - (m-n)d\varphi\big(\hat{\mathbf{H}}\big),$$
(3.21)

where $\mathrm{grad}_{\mathcal{H}}\big(\frac{1}{\lambda^2}\big)$ is the $\mathcal{H}$-component of the decomposition according to (3.19) of $\mathrm{grad}\big(\frac{1}{\lambda^2}\big)$, and $\hat{\mathbf{H}}$ is the trace of the second fundamental form of each fiber which is given by $\hat{\mathbf{H}} = \frac{1}{m-n} \sum_{k=n+1}^{m} \mathcal{H}(\nabla_{e_k} e_k)$, where a local orthonormal frame field $\{e_i\}_{i=1}^m$ on M is taken in such a way that $\{e_{ix} | i = 1, \cdots, n\}$ belong to $\mathcal{H}_x$ and $\{e_{jx} | j = n+1, \cdots, m\}$ belong to $\mathcal{V}_x$ where x is in a neighborhood in M. Then, due to Theorems 2.1 and (3.21), we have immediately

THEOREM 3.2. *Let* (M^m, g) *be a complete non-compact Riemannian manifold, and* (N^n, h), *a Riemannian manifold with the non-positive sectional curvature* $(m > n \geq 2)$. *Let* $\varphi : (M, g) \to (N, h)$ *be a horizontally conformal submersion with the dilation* λ *satisfying that*

$$\int_M \lambda^2 \left| \frac{n-2}{2} \lambda^2 \, \mathrm{grad}_{\mathcal{H}}\Big(\frac{1}{\lambda^2}\Big) - (m-n)\,\hat{\mathbf{H}} \right|_g^2 v_g < \infty.$$
(3.22)

Assume that, either $\int_M \lambda^2 \, v_g < \infty$ *or* $\mathrm{Vol}(M, g) = \int_M v_g = \infty$. *Then, if* $\varphi : (M, g) \to (N, h)$ *is biharmonic, then it is a harmonic morphism.*

Due to Theorem 3.2, we have:

COROLLARY 3.1. *Let* (M^m, g) *be a complete non-compact Riemannian manifold, and* (N^2, h), *a Riemannian surface with the non-positive sectional curvature* $(m > n = 2)$. *Let* $\varphi : (M, g) \to (N, h)$ *be a horizontally conformal submersion with the dilation* λ *satisfying that*

$$\int_M \lambda^2 \left| \hat{\mathbf{H}} \right|_g^2 v_g < \infty.$$
(3.23)

Assume that, either $\int_M \lambda^2 \, v_g < \infty$ *or* $\mathrm{Vol}(M, g) = \int_M v_g = \infty$. *Then, if* $\varphi : (M, g) \to (N, h)$ *is biharmonic, then it is a harmonic morphism.*

Corollary 3.3 implies

COROLLARY 3.2. *Let* (M^m, g) *be a non-compact complete Riemannian manifold* $(m > 2)$, *and* (N^2, h), *a Riemannian surface with non-positive curvature. Let* λ *be a positive function in* $C^\infty(M) \cap L^2(M)$, *where* $L^2(M)$ *is the space of square integrable functions on* (M, g).

Then, every biharmonic horizontally conformal submersion $\varphi : (M^m, g)$ *$\to (N^2, h)$ with a dilation λ and a bounded $|\hat{\mathbf{H}}|_g$, exactly $\lambda |\hat{\mathbf{H}}|_g \in$ $L^2(M)$, must be a harmonic morphism.*

REMARK 3.1. *(1)   Notice that in Corollary 3.4, (1), there is no restriction to the dilation λ because of $\dim N = 2$. This implies that for every positive C^∞ function λ in $C^\infty(M) \cap L^2(M)$ satisfying (3.2), we have a harmonic morphism $\varphi : (M^m, g) \to (N^2, h)$.*
(2)   For a biharmonic map of (M, g) into (N, h), the non-positivity of (N, h) implies that

$$\langle \tau(\varphi), \overline{\Delta}\tau(\varphi) \rangle = \sum_{i=1}^{m} \langle R^N(\tau(\varphi), d\varphi(e_i))d\varphi(e_i), \tau(\varphi) \rangle \leq 0,$$

$$(3.24)$$

which is stronger than the Bochner type formula $|\tau(\varphi)| \Delta |\tau(\varphi)| \geq 0$. However, we can prove Theorem 2.1 in an alternative way by using the latter one. Here $\Delta = \sum_{i=1}^{m}(e_i{}^2 - \nabla_{e_i} e_i)$ denotes the negative Laplace operator acting on $C^\infty(M)$.

4. Appendix

In this appendix, we recall Gaffney's theorem ([**52**]):

THEOREM 4.1. *(Gaffney) Let (M, g) be a complete Riemannian manifold. If a C^1 1-form α satifies that $\int_M |\alpha| \, v_g < \infty$ and $\int_M (\delta\alpha) \, v_g < \infty$, or equivalently, a C^1 vector field X defined by $\alpha(Y) = \langle X, Y \rangle$ $(\forall Y \in \mathfrak{X}(M))$ satisfies that $\int_M |X| \, v_g < \infty$ and $\int_M \mathrm{div}(X) \, v_g < \infty$, then*

$$\int_M (-\delta\alpha) \, v_g = \int_M \mathrm{div}(X) \, v_g = 0. \qquad (4.1)$$

PROOF. For completeness, we give a proof. By integrating over M, the both hand sides of

$$\mathrm{div}(\eta^2 X) = \eta^2 \, \mathrm{div}(X) + 2\eta \, \langle \nabla\eta, X \rangle, \qquad (4.2)$$

we have

$$\int_M \mathrm{div}(\eta^2 X) \, v_g = \int_M \eta^2 \, \mathrm{div}(X) \, v_g + 2 \int_M \eta \, \langle \nabla\eta, X \rangle \, v_g. \qquad (4.3)$$

Since the support of $\eta^2 X$ is compact, the left hand side must vanish. So, we have

$$\int_M \eta^2 \, \mathrm{div}(X) \, v_g = -2 \int_M \eta \, \langle \nabla\eta, X \rangle \, v_g. \qquad (4.4)$$

Therefore, we have

$$\left| \int_{B_r(x_0)} \operatorname{div}(X) \, v_g \right| \le \left| \int_M \eta^2 \operatorname{div}(X) \, v_g \right|$$

$$= 2 \left| \int_M \eta \, \langle \nabla \eta, X \rangle \, v_g \right|$$

$$\le 2 \int_M \eta \, |\nabla \eta| \, |X| \, v_g$$

$$\le \frac{4}{r} \int_M |X| \, v_g. \tag{4.5}$$

By the assumption that $\int_M |X| \, v_g < \infty$, the right hand side goes to 0 if r tends to infinity. Since $B_r(x_0)$ goes to M as $r \to \infty$, due to completeness of (M, g), and the assumption that $\int_M \operatorname{div}(X) \, v_g < \infty$, we have $\int_M \operatorname{div}(X) \, v_g = \lim_{r \to \infty} \int_{B_r(x_0)} \operatorname{div}(X) \, v_g = 0$. $\qquad \square$

CHAPTER 4

Biharmonic Submanifolds in a Riemannian Manifold with Non-positive Curvature

1

ABSTRACT. In this chapter, we show that, for every biharmonic submanifold (M, g) of a Riemannian manifold (N, h) with non-positive sectional curvature, if $\int_M |\eta|^2 v_g < \infty$, then (M, g) is minimal in (N, h), i.e., $\eta \equiv 0$, where η is the mean curvature tensor field of (M, g) in (N, h). This gives a positive affirmative answer, under the condition $\int_M |\eta|^2 v_g < \infty$, to the *generalized B.Y. Chen's conjecture*: every biharmonic submanifold of a Riemannian manifold with non-positive sectional curvature must be minimal. This conjecture turned out false in case of an incomplete Riemannian manifold (M, g) by a counter example of Y-L. Ou and L. Tang [123]

1. Introduction and statement of results

This paper is an extension of our previous paper [108] to biharmonic submanifolds of any co-dimension of a Riemannian manifold of pon-positive curvature. Let us consider an isometric immersion $\varphi : (M, g) \rightarrow (N, h)$ of a Riemannian manifold (M, g) of dimension m into another Riemannian manifold (N, h) of dimension $n = m + p$ $(p \geq 1)$. We have

$$\nabla^N_{\varphi_* X} \varphi_* Y = \varphi_*(\nabla_X Y) + B(X, Y),$$

for vector fields X and Y on M, where ∇, ∇^N are the Levi-Civita connections of (M, g) and (N, h), and $B : \Gamma(TM) \times \Gamma(TM) \rightarrow \Gamma(TM)^\perp$ is the second fundamental form of the immersion φ corresponding to the decomposition:

$$T_{\varphi(x)} N = d\varphi(T_x M) \oplus d\varphi(T_x M)^\perp \quad (x \in M),$$

---

[1]This is due to [109]: N. Nakauchi and H. Urakawa, *Biharmonic submanifolds in a Riemannian manifold with non-positive curvature*, Results in Math., **63** (2013), 467–474.

respectively. Let η be the mean curvature vector field along φ defined by $\eta = \frac{1}{m} \sum_{i=1}^{m} B(e_i, e_i)$, where $\{e_i\}_{i=1}^{m}$ is a local orthonormal frame on (M, g). Then, the **generalized B.Y. Chen's conjecture** (cf. [17], [16], [21], [22], [121], [125], [123]) is that:

For an isometric immersion $\varphi : (M, g) \to (N, h)$, assume that the sectional curvature of (N, h) is non-positive. If φ is biharmonic (cf. See Sect. 2), then, it is minimal, i.e., $\eta \equiv 0$.

In this paper, we will show

THEOREM 1.1. *Assume that (M, g) is a complete Riemannian manifold of dimension m and (N, h) is a Riemannian manifold of dimension $m + p$ ($p \geq 1$) whose sectional curvature is non-positive. If $\varphi : (M, g) \to (N, h)$ is biharmonic and satisfies that $\int_M |\eta|^2 \, v_g < \infty$, then, φ is minimal.*

In our previous paper [**108**], we showed

THEOREM 1.2. *Assume that (M, g) is complete and the Ricci tensor Ric^N of (N, h) satisfies that*

$$\mathrm{Ric}^N(\xi, \xi) \leq |A|^2. \tag{1.1}$$

If $\varphi : (M, g) \to (N, h)$ is biharmonic and satisfies that

$$\int_M H^2 \, v_g < \infty, \tag{1.2}$$

then, φ has constant mean curvature, i.e., H is constant.

Notice that, in Theorem 1.2 in case of codimension one, we only need the weaker assumption, non-positivity of the Ricci curvature of (N, h) ([**123**]). On the other hand, in Theorem 1.1, we should treat with a complete submanifold of an arbitrary co-dimension $p \geq 1$, and we need the stronger assumption non-positivity of the sectional curvature of (N, h). In proving Theorem 1.1, the method of the proof of Theorem 1.2 ([**108**]) does not work anymore. We should turn our mind, and have a different and very simple proof. Finally, our Theorem 1.1 implies that *the generalized B.Y. Chen's conjecture holds true under the assumption that $\int_M |\eta|^2 \, v_g$ is finite and (M, g) is complete.*

2. Preliminaries

2.1. Harmonic maps and biharmonic maps. In this subsection, we prepare general materials about harmonic maps and biharmonic maps of a complete Riemannian manifold into another Riemannian manifold (cf. [**40**]).

Let (M, g) be an m-dimensional complete Riemannian manifold, and the target space (N, h) is an n-dimensional Riemannian manifold. For every C^∞ map φ of M into N. Let $\Gamma(\varphi^{-1}TN)$ be the space of C^∞ sections of the induced bundle $\varphi^{-1}TN$ of the tangent bundle TN by φ. The *tension field* $\tau(\varphi)$ is defined globally on M by

$$\tau(\varphi) = \sum_{i=1}^{m} B(\varphi)(e_i, e_i) \in \Gamma(\varphi^{-1}TN), \tag{2.1}$$

where the second fundamental form $B(\varphi)$ is defined by

$$B(\varphi)(X, Y) = \nabla^N_{\varphi_*(X)}\varphi_*(Y) - \varphi_*(\nabla_X Y)$$

for $X, Y \in \mathfrak{X}(M)$. Then, a C^∞ map $\varphi : (M, g) \to (N, h)$ is *harmonic* if $\tau(\varphi) = 0$. The *bitension field* $\tau_2(\varphi)$ is defined globally on M by

$$\tau_2(\varphi) = J(\tau(\varphi)) = \overline{\Delta}\tau(\varphi) - \mathcal{R}(\tau(\varphi)), \tag{2.2}$$

where

$$J(V) := \overline{\Delta}V - \mathcal{R}(V),$$

$$\overline{\Delta}V := \overline{\nabla}^* \overline{\nabla}V = -\sum_{i=1}^{m}\{\overline{\nabla}_{e_i}(\overline{\nabla}_{e_i}V) - \overline{\nabla}_{\nabla_{e_i}e_i}V\},$$

$$\mathcal{R}(V) := \sum_{i=1}^{m} R^N(V, \varphi_*(e_i))\varphi_*(e_i).$$

Here, $\overline{\nabla}$ is the induced connection on the induced bundle $\varphi^{-1}TN$, and R^N is the curvature tensor of (N, h) (cf. [**58**]) given by

$$R^N(U, V)W = [\nabla^N_U, \nabla^N_V]W - \nabla^N_{[U,V]}W \quad (U, V, W \in \mathfrak{X}(N)).$$

A C^∞ map $\varphi : (M, g) \to (N, h)$ is called to be *biharmonic* ([**16**], [**40**], [**74**]) if

$$\tau_2(\varphi) = 0. \tag{2.3}$$

2.2. Setting of isometric immersions. In this sebsection, we prepare fundamental materials of general facts on isometric immersions (cf. [**82**]). Let φ be an isometric immersion of an m-dimensional Riemannian into an $(m+p)$-dimensional Riemannian manifold (N, h).

Then, the induced bundle $\varphi^{-1}TN$ of the tangent bundle TN of N by φ is decomposed into the direct sum:

$$\varphi^{-1}TN = \tau M \oplus \nu M, \tag{2.4}$$

where $\varphi^{-1}TN = \cup_{x \in M}T_{\varphi(x)}N$, $\tau M = d\varphi(TM) = \cup_{x \in M}d\varphi(T_xM)$, and $\nu M = \cup_{x \in M}d\varphi(T_xM)^\perp$ is the normal bundle. For the induced connection $\overline{\nabla}$ on $\varphi^{-1}TN$ of the Levi-Civita connection ∇^N of (N,h) by φ, $\overline{\nabla}_X(d\varphi(Y))$ is decomposed corresponding to (2.4) as

$$\overline{\nabla}_X(d\varphi(Y)) = d\varphi(\nabla_X Y) + B(X,Y) \tag{2.5}$$

for all C^∞ vector fields X and Y on M. Here, ∇ is the Levi-Civita connection of (M,g) and $B(X,Y)$ is the second fundamental form of the immersion $\varphi : (M,g) \to (N,h)$.

Let $\{\xi_1, \cdots, \xi_p\}$ be a local unit normal vector fields along φ that are orthogonal at each point, and let us decompose $B(X,Y)$ as

$$B(X,Y) = \sum_{i=1}^{p} b^i(X,Y)\,\xi_i, \tag{2.6}$$

where $b^i(X,Y)$ $(i = 1, \cdots, p)$ are the p second fundamental forms of φ. For every $\xi \in \Gamma(\nu M)$, $\overline{\nabla}_X \xi$, denoted also by $\nabla_X^N \xi$ is decomposed correspondingly to (2.4) into

$$\nabla_X^N \xi = -A_\xi(X) + \nabla_X^\perp \xi, \tag{2.7}$$

where $\nabla^\perp$ is called the normal connection of νM. The linear operator A_ξ of $\Gamma(TM)$ into itself, called the shape operator with respect to ξ, satisfies that

$$\langle A_\xi(X), Y \rangle = \langle B(X,Y), \xi \rangle \tag{2.8}$$

for all C^∞ vector fields X and Y on M. Here, we denote the Riemannian metrics g and h simply by $\langle \cdot, \cdot \rangle$.

We denote the tension field $\tau(\varphi)$ of an isometric immersion $\varphi : (M,g) \to (N,h)$ as

$$
\begin{aligned}
\tau(\varphi) = \mathrm{Trace}_g(\widetilde{\nabla}d\varphi) &= \sum_{i=1}^{m} B(e_i, e_i) \\
&= \sum_{k=1}^{p} (\mathrm{Trace}_g b^k)\,\xi_k \\
&= m \sum_{k=1}^{p} H_k\,\xi_k \\
&= m\,\eta,
\end{aligned}
\tag{2.9}
$$

where $\widetilde{\nabla}$ is the induced connection on $TM \otimes \varphi^{-1}TN$, $H_k := \frac{1}{m}\mathrm{Trace}_g b^k = \frac{1}{m}\mathrm{Trace}_g(A_{\xi_k})$ $(k = 1, \cdots, p)$, and $\eta := \sum_{k=1}^{p} H_k\,\xi_k$ is the mean curvature vector field of φ. Let us recall that $\varphi : (M, g) \to (N, h)$ is *minimal* if $\eta \equiv 0$.

3. Proof of main theorem

Assume that $\varphi : (M, g) \to (N, h)$ is a biharmonic immersion. Then, since (2.9): $\tau(\varphi) = m\,\eta$, the biharmonic map equation

$$\tau_2(\varphi) = \overline{\Delta}(\tau(\varphi)) - \mathcal{R}(\tau(\varphi)) = 0 \tag{3.1}$$

is equivalent to that

$$\overline{\Delta}\eta - \sum_{i=1}^{m} R^N(\eta, d\varphi(e_i))d\varphi(e_i) = 0. \tag{3.2}$$

Take any point x_0 in M, and for every $r > 0$, let us consider the follwoing cut-off function λ on M:

$$\begin{cases} 0 \leq \lambda(x) \leq 1 & (x \in M), \\ \lambda(x) = 1 & (x \in B_r(x_0)), \\ \lambda(x) = 0 & (x \notin B_{2r}(x_0)) \\ |\nabla\lambda| \leq \dfrac{2}{r} & (\text{on } M), \end{cases}$$

where $B_r(x_0) := \{x \in M : d(x, x_0) < r\}$ and d is the distance of (M, g). In both sides of (3.2), taking inner product with $\lambda^2\,\eta$, and integrate them over M, we have

$$\int_M \langle \overline{\Delta}\eta, \lambda^2\,\eta \rangle\, v_g = \int_M \sum_{i=1}^{m} \langle R^N(\eta, d\varphi(e_i))d\varphi(e_i), \eta \rangle\, \lambda^2\, v_g. \tag{3.3}$$

Since the sectional curvature of (N, h) is non-positive, $h(R^N(u, v)v, u) \leq 0$ for all tangent vectors u and v at T_yN $(y \in N)$, the right hand side of (3.3) is non-positive, i.e.,

$$\int_M \langle \overline{\Delta}\eta, \lambda^2\,\eta \rangle\, v_g \leq 0. \tag{3.4}$$

On the other hand, the right hand side coincides with

$$\int_M \langle \nabla \eta, \nabla (\lambda^2 \eta) \rangle \, v_g = \int_M \sum_{i=1}^m \langle \overline{\nabla}_{e_i} \eta, \overline{\nabla}_{e_i} (\lambda^2 \eta) \rangle \, v_g$$

$$= \int_M \lambda^2 \sum_{i=1}^m |\overline{\nabla}_{e_i} \eta|^2 \, v_g$$

$$+ 2 \int_M \sum_{i=1}^m \lambda \, (e_i \lambda) \, \langle \overline{\nabla}_{e_i} \eta, \eta \rangle \, v_g, \qquad (3.5)$$

since $\overline{\nabla}_{e_i}(\lambda^2 \eta) = \lambda^2 \overline{\nabla}_{e_i} \eta + 2\lambda (e_i \lambda) \, \eta$. Therefore, we have

$$\int_M \lambda^2 \sum_{i=1}^m |\overline{\nabla}_{e_i} \eta|^2 \, v_g \leq -2 \int_M \sum_{i=1}^m \langle \lambda \, \overline{\nabla}_{e_i} \eta, (e_i \lambda) \, \eta \rangle \, v_g. \qquad (3.6)$$

Now apply with $V := \lambda \, \overline{\nabla}_{e_i} \eta$, and $W := (e_i \lambda) \, \eta$, to Young's inequality: for all $V, W \in \Gamma(\varphi^{-1} T N)$ and $\epsilon > 0$,

$$\pm 2 \, \langle V, W \rangle \leq \epsilon |V|^2 + \frac{1}{\epsilon} |W|^2,$$

the right hand side of (3.6) is smaller than or equal to

$$\epsilon \int_M \lambda^2 \sum_{i=1}^m |\overline{\nabla}_{e_i} \eta|^2 \, v_g + \frac{1}{\epsilon} \int_M |\eta|^2 \sum_{i-1}^m |e_i \lambda|^2 \, v_g. \qquad (3.7)$$

By taking $\epsilon = \frac{1}{2}$, we obtain

$$\int_M \lambda^2 \sum_{i=1}^m |\overline{\nabla}_{e_i} \eta|^2 \, v_g \leq \frac{1}{2} \int_M \lambda^2 \sum_{i=1}^m |\overline{\nabla}_{e_i} \eta|^2 \, v_g + 2 \int_M |\eta|^2 \sum_{i=1}^m |e_i \lambda|^2 \, v_g.$$

Thus, we have

$$\int_M \lambda^2 \sum_{i=1}^m |\overline{\nabla}_{e_i} \eta|^2 \, v_g \leq 4 \int_M |\eta|^2 \sum_{i=1}^m |e_i \lambda|^2 \, v_g$$

$$\leq \frac{16}{r^2} \int_M |\eta|^2 \, v_g < \infty. \qquad (3.8)$$

Since (M, g) is complete, we can tend r to infinity, and then the left hand side goes to $\int_M \sum_{i=1}^m |\overline{\nabla}_{e_i} \eta|^2 \, v_g$, we obtain

$$\int_M \sum_{i=1}^m |\overline{\nabla}_{e_i} \eta|^2 \, v_g \leq 0. \qquad (3.9)$$

Thus, we have $\overline{\nabla}_X \eta = 0$ for all vector field X on M.

Then, we can conclude that $\eta \equiv 0$. For, applying (2.7):

$$\overline{\nabla}_X \xi_k = -A_{\xi_k}(X) + \nabla_X^\perp \xi_k,$$

to $\eta = \sum_{k=1}^{p} H_k \xi_k$, we have

$$0 = \overline{\nabla}_X \eta = -A_\eta(X) + \nabla_X^\perp \eta, \tag{3.10}$$

which implies that, for all vector field X on M,

$$\begin{cases} A_\eta(X) = 0, \\ \nabla_X^\perp \eta = 0. \end{cases} \tag{3.11}$$

by comparing the tangential and normal components. Then, by the first equation of (3.11), we have

$$\langle B(X,Y), \eta \rangle = \langle A_\eta(X), Y \rangle = 0, \tag{3.12}$$

for all vector fields X and Y on M. This implies that $\eta \equiv 0$ since $\eta = \frac{1}{m} \sum_{i=1}^{m} B(e_i, e_i)$. $\qquad\square$

CHAPTER 5

Biharmonic Hypersurfaces in a Riemannian Manifold with Non-positive Ricci Curvature

1

ABSTRACT. We show that, for a biharmonic hypersurface (M, g) of a Riemannian manifold (N, h) of non-positive Ricci curvature, if $\int_M |H|^2 v_g < \infty$, where H is the mean curvature of (M, g) in (N, h), then (M, g) is minimal in (N, h). For a counter example (M, g) in the case of hypersurfaces to the generalized Chen's conjecture (cf. Sect.1), it holds that $\int_M |H|^2 v_g = \infty$.

1. Introduction and statement of results

In this paper, we consider an isometric immersion $\varphi : (M, h) \to (N, h)$, of a Riemannian manifold (M, g) of dimension m, into another Riemannian manifold (N, h) of dimension $n = m + 1$. We have

$$\nabla^N_{\varphi_* X} \varphi_* Y = \varphi_*(\nabla_X Y) + k(X, Y)\xi,$$

for vector fields X and Y on M, where ∇, ∇^N are the Levi-Civita connections of (M, g) and (N, h), respectively, ξ is the unit normal vector field along φ, and k is the second fundamental form. Let $A : T_x M \to T_x M$ ($x \in M$) be the shape operator defined by $g(AX, Y) = k(X, Y)$, $(X, Y \in T_x M)$, and H, the mean curvature defined by $H := \frac{1}{m} \mathrm{Tr}_g(A)$. Then, let us recall the following **B.Y. Chen's conjectrure** (cf. [21], [22]):

Let $\varphi : (M, g) \to (\mathbb{R}^n, g_0)$ be an isometric immersion into the standard Euclidean space. If φ is biharmonic (see Sect. 2), then, it is minimal.

This conjecture is still open up to now, and let us recall also the following **generalized B.Y. Chen's conjecture** (cf. [21], [16]):

---

[1] This chapter is due to [108]: N. Nakauchi and H. Urakawa, *Biharmonic hypersurfaces in a Riemannian manifold with non-positive Ricci curvature*, Ann. Global Anal. Geom., **40** (2011), 125–131.

Let $\varphi : (M,g) \to (N,h)$ be an isometric immersion, and the sectional curvature of (N,h) is non-positive. If φ is biharmonic, then, it is minimal.

Oniciuc ([**121**]) and Ou ([**125**]) showed this is true if H is constant.

In this paper, we show

THEOREM 1.1. *Assume that (M,g) is complete and the Ricci tensor Ric^N of (N,h) satisfies that*

$$\mathrm{Ric}^N(\xi,\xi) \le |A|^2. \tag{1.1}$$

If $\varphi : (M,g) \to (N,h)$ is biharmonic (cf. Sect. 2) and satisfies that

$$\int_M H^2 \, v_g < \infty, \tag{1.2}$$

then, φ has constant mean curvature, i.e., H is constant.

As a direct corollary, we have

COROLLARY 1.1. *Assume that (M,g) is a complete Riemannian manifold of dimension m and (N,h) is a Riemannian manifold of dimension $m+1$ whose Ricci curvature is non-positive. If an isometric immersion $\varphi : (M,g) \to (N,h)$ is biharmonic and satisfies that $\int_M H^2 \, v_g < \infty$, then, φ is minimal.*

By our Corollary 1.2, if there would exist a counter example (cf. [**123**]) in the case $\dim N = \dim M + 1$, then it must hold that

$$\int_M H^2 \, v_g = \infty, \tag{1.3}$$

which imposes the strong condition on the behaviour of the boundary of M at infinity. Indeed, (1.3) implies that either H is unbounded on M, or it holds that $H^2 \ge C$ on an open subset Ω of M with infinite volume, for some constant $C > 0$.

Acknowledgement. We express our thanks to the referee(s) who informed relevant references and gave useful comments to us.

2. Preliminaries

In this section, we prepare general materials about harmonic maps and biharmonic maps of a complete Riemannian manifold into another Riemannian manifold (cf. [40]).

Let (M, g) be an m-dimensional complete Riemannian manifold, and the target space (N, h) is an n-dimensional Riemannian manifold. For every C^∞ map φ of M into N, and relatively compact domain Ω in M, the *energy functional* on the space $C^\infty(M, N)$ of all C^∞ maps of M into N is defined by

$$E_\Omega(\varphi) = \frac{1}{2} \int_\Omega |d\varphi|^2 \, v_g,$$

and for a C^∞ one parameter deformation $\varphi_t \in C^\infty(M, N)$ $(-\epsilon < t < \epsilon)$ of φ with $\varphi_0 = \varphi$, the variation vector field V along φ is defined by $V = \frac{d}{dt}\big|_{t=0} \varphi_t$. Let $\Gamma_\Omega(\varphi^{-1}TN)$ be the space of C^∞ sections of the induced bundle $\varphi^{-1}TN$ of the tangent bundle TN by φ whose supports are contained in Ω. For $V \in \Gamma_\Omega(\varphi^{-1}TN)$ and its one-parameter deformation φ_t, the *first variation formula* is given by

$$\frac{d}{dt}\bigg|_{t=0} E_\Omega(\varphi_t) = -\int_\Omega \langle \tau(\varphi), V \rangle \, v_g.$$

The *tension field* $\tau(\varphi)$ is defined globally on M by

$$\tau(\varphi) = \sum_{i=1}^m B(\varphi)(e_i, e_i), \tag{2.1}$$

where

$$B(\varphi)(X, Y) = \nabla^N_{\varphi_*(X)} \varphi_*(Y) - \varphi_*(\nabla_X Y)$$

for $X, Y \in \mathfrak{X}(M)$. Then, a C^∞ map $\varphi : (M, g) \to (N, h)$ is *harmonic* if $\tau(\varphi) = 0$. For a harmonic map $\varphi : (M, g) \to (N, h)$, the *second variation formula* of the energy functional $E_\Omega(\varphi)$ is

$$\frac{d^2}{dt^2}\bigg|_{t=0} E_\Omega(\varphi_t) = \int_\Omega \langle J(V), V \rangle \, v_g$$

where

$$J(V) := \overline{\Delta} V - \mathcal{R}(V),$$

$$\overline{\Delta} V := \overline{\nabla}^* \overline{\nabla} V = -\sum_{i=1}^m \{\overline{\nabla}_{e_i}(\overline{\nabla}_{e_i} V) - \overline{\nabla}_{\nabla_{e_i} e_i} V\},$$

$$\mathcal{R}(V) := \sum_{i=1}^m R^N(V, \varphi_*(e_i))\varphi_*(e_i).$$

Here, $\overline{\nabla}$ is the induced connection on the induced bundle $\varphi^{-1}TN$, and R^N is the curvature tensor of (N, h) given by $R^N(U, V)W = [\nabla^N_U, \nabla^N_V]W - \nabla^N_{[U,V]}W$ $(U, V, W \in \mathfrak{X}(N))$.

The *bienergy functional* is defined by

$$E_{2,\Omega}(\varphi) = \frac{1}{2} \int_\Omega |\tau(\varphi)|^2 \, v_g,$$

and the *first variation formula* of the bienergy is given (cf. [**74**]) by

$$\frac{d}{dt}\bigg|_{t=0} E_{2,\Omega}(\varphi_t) = - \int_\Omega \langle \tau_2(\varphi), V \rangle \, v_g$$

where the *bitension field* $\tau_2(\varphi)$ is defined globally on M by

$$\tau_2(\varphi) = J(\tau(\varphi)) = \overline{\Delta}\tau(\varphi) - \mathcal{R}(\tau(\varphi)), \tag{2.2}$$

and a C^∞ map $\varphi : (M, g) \to (N, h)$ is called to be *biharmonic* if

$$\tau_2(\varphi) = 0. \tag{2.3}$$

3. Some lemma for the Schrödinger type equation

In this section, we prepare some simple lemma of the Schrödinger type equation of the Laplacian Δ_g on an m-dimensional non-compact complete Riemannian manifold (M, g) defined by

$$\Delta_g f := \sum_{i=1}^m e_i(e_i f) - \nabla_{e_i} e_i f \qquad (f \in C^\infty(M)), \tag{3.1}$$

where $\{e_i\}_{i=1}^m$ is a locally defined orthonormal frame field on (M, g).

LEMMA 3.1. *Assume that (M, g) is a complete non-compact Riemannian manifold, and L is a non-negative smooth function on M. Then, every smooth L^2 function f on M satisfying the Schrödinger type equation*

$$\Delta_g f = L f \qquad (on \ M) \tag{3.2}$$

must be a constant.

PROOF. Take any point x_0 in M, and for every $r > 0$, let us consider the following cut-off function η on M:

$$\begin{cases} 0 \le \eta(x) \le 1 & (x \in M), \\ \eta(x) = 1 & (x \in B_r(x_0)), \\ \eta(x) = 0 & (x \notin B_{2r}(x_0)), \\ |\nabla \eta| \le \dfrac{2}{r} & (on \ M), \end{cases} \tag{3.3}$$

where $B_r(x_0) = \{x \in M : d(x, x_0) < r\}$, and d is the distance of (M, g). Multiply $\eta^2 f$ on (3.2), and integrale it over M, we have

$$\int_M (\eta^2 f) \Delta_g f \, v_g = \int_M L \eta^2 f^2 \, v_g. \tag{3.4}$$

By the integration by part for the left hand side, we have

$$\int_M (\eta^2 f) \Delta_g f \, v_g = - \int_M g(\nabla(\eta^2 f), \nabla f) \, v_g. \tag{3.5}$$

Here, we have

$$g(\nabla(\eta^2 f), \nabla f) = 2\eta f \, g(\nabla \eta, \nabla f) + \eta^2 \, g(\nabla f, \nabla f)$$
$$= 2\eta f \langle \nabla \eta, \nabla f \rangle + \eta^2 \, |\nabla f|^2, \tag{3.6}$$

where we use $\langle \cdot, \cdot \rangle$ and $| \cdot |$ instead of $g(\cdot, \cdot)$ and $g(u, u) = |u|^2$ ($u \in T_x M$), for simplicity. Substitute (3.6) into (3.5), the right hand side of (3.5) is equal to

$$RHS \text{ of } (3.5) = - \int_M 2\eta f \langle \nabla \eta, \nabla f \rangle \, v_g - \int_M \eta^2 |\nabla f|^2 \, v_g$$
$$= -2 \int_M \langle f \nabla \eta, \eta \nabla f \rangle \, v_g - \int_M \eta^2 |\nabla f|^2 \, v_g. \tag{3.7}$$

Here, applying Young's inequality: for every $\epsilon > 0$, and every vectors X and Y at each point of M,

$$\pm 2 \langle X, Y \rangle \leq \epsilon |X|^2 + \frac{1}{\epsilon} |Y|^2, \tag{3.8}$$

to the first term of (3.7), we have

$$RHS \text{ of } (3.7) \leq \epsilon \int_M |\eta \nabla f|^2 \, v_g + \frac{1}{\epsilon} \int_M |f \nabla \eta|^2 \, v_g - \int_M \eta^2 |\nabla f|^2 \, v_g$$
$$= -(1 - \epsilon) \int_M \eta^2 |\nabla f|^2 \, v_g + \frac{1}{\epsilon} \int_M f^2 |\nabla \eta|^2 \, v_g. \tag{3.9}$$

Thus, by (3.5) and (3.9), we obtain

$$\int_M L \eta^2 f^2 \, v_g + (1 - \epsilon) \int_M \eta^2 |\nabla f|^2 \, v_g \leq \frac{1}{\epsilon} \int_M f^2 |\nabla \eta|^2 \, v_g. \tag{3.10}$$

Now, puttig $\epsilon = \frac{1}{2}$, (3.10) implies that

$$\int_M L \eta^2 f^2 \, v_g + \frac{1}{2} \int_M \eta^2 |\nabla f|^2 \, v_g \leq 2 \int_M f^2 |\nabla \eta|^2 \, v_g. \tag{3.11}$$

Since $\eta = 1$ on $B_r(x_0)$ and $|\nabla \eta| \leq \frac{2}{r}$, and $L \geq 0$ on M, we have

$$0 \leq \int_{B_r(x_0)} L f^2 \, v_g + \frac{1}{2} \int_{B_r(x_0)} |\nabla f|^2 \, v_g \leq \frac{8}{r^2} \int_M f^2 \, v_g. \tag{3.12}$$

Since (M, g) is non-compact and complete, r can tend to infinity, and $B_r(x_0)$ goes to M. Then we have

$$0 \leq \int_M L f^2 \, v_g + \frac{1}{2} \int_M |\nabla f|^2 \, v_g \leq 0 \tag{3.13}$$

since $\int_M f^2 \, v_g < \infty$. Thus, we have $L f^2 = 0$ and $|\nabla f| = 0$ (on M) which implies that f is a constant. □

4. Biharmonic isometric immersions

In this section, we consider a hypersurface M of an $(m + 1)$-dimensional Riemannian manifold (N, h). Recently, Y-L. Ou showed (cf. [**125**])

THEOREM 4.1. *Let $\varphi : (M, g) \to (N, h)$ be an isometric immersion of an m-dimensional Riemannian manifold (M, g) into another $(m + 1)$-dimensional Riemannian manifold (N, h) with the mean curvature vector field $\eta = H \xi$, where ξ is the unit normal vector field along φ. Then, φ is biharmonic if and only if the following equations hold:*

$$\begin{cases} \Delta_g H - H \, |A|^2 + H \, \mathrm{Ric}^N(\xi, \xi) = 0, \\ 2 \, A \, (\nabla H) + \dfrac{m}{2} \, \nabla(H^2) - 2 \, H \, (\mathrm{Ric}^N(\xi))^T = 0, \end{cases} \tag{4.1}$$

where $\mathrm{Ric}^N : T_y N \to T_y N$ is the Ricci transform which is defined by $h(\mathrm{Ric}^N(Z), W) = \mathrm{Ric}^N(Z, W) \, (Z, W \in T_y N)$, $(\cdot)^T$ is the tangential component corresponding to the decomposition of $T_{\varphi(x)} N = \varphi_*(T_x M) \oplus \mathbb{R}\xi_x \, (x \in M)$, and ∇f is the gradient vector field of $f \in C^\infty(M)$ on (M, g), respectively.

Due to Theorem 4.1 and Lemma 3.1, we can show immediately our Theorem 1.1.

(*Proof of Theorem 1.1.*)
Let us denote by $L := |A|^2 - \mathrm{Ric}^N(\xi, \xi)$ which is a smooth non-negative function on M due to our assumption. Then, the first equation is reduced to the following Schrödinger type equation:

$$\Delta_g f = L f, \tag{4.2}$$

where $f := H$ is a smooth L^2 function on M by the assumption (1.2).

Assume that M is compact. In this case, by (4.2) and the integration by part, we have

$$0 \leq \int_M L f^2 \, v_g = \int_M f \, (\Delta_g f) \, v_g = - \int_M g(\nabla f, \nabla f) \, v_g \leq 0, \tag{4.3}$$

which implies that $\int_M g(\nabla f, \nabla f) \, v_g = 0$, that is, f is constant.

Assume that M is non-compact. In this case, we can apply Lemma 3.1 to (4.2). Then, we have that $f = H$ is a constant. $\qquad\square$

(*Proof of Corollary 1.2.*)

Assume that Ric^N is non-positive. Since $L = |A|^2 - \mathrm{Ric}^N(\xi, \xi)$ is non-negative, H is constant due to Theorem 1.1. Then, due to (4.1), we have that $H \, L = 0$ and $H \, (\mathrm{Ric}^N(\xi))^T = 0$. If $H \neq 0$, then $L = 0$, i.e.,

$$\mathrm{Ric}^N(\xi, \xi) = |A|^2. \tag{4.4}$$

By our assumption, $\mathrm{Ric}^N(\xi, \xi) \leq 0$, and the right hand side of (4.4) is non-negative, so we have $|A|^2 = 0$, i.e., $A \equiv 0$. This contradicts $H \neq 0$. We have $H = 0$. $\qquad\square$

CHAPTER 6

Note on Biharmonic Map Equations

1

ABSTRACT. This chapter gives the biharmonic map equations for
a Riemannian submanifold of a Riemannian manifold (N, h) which
are related to due to [**109**], [**108**] and [**111**].

1. Preliminaries

1.1. Harmonic maps and biharmonic maps. We prepare here
general materials about harmonic maps and biharmonic maps of a
complete Riemannian manifold into another Riemannian manifold (cf.
[**40**]).

Let (M, g) be an m-dimensional complete Riemannian manifold,
and the target space (N, h) is an n-dimensional Riemannian manifold.
For every C^∞ map φ of M into N. Let $\Gamma(\varphi^{-1}TN)$ be the space of C^∞
sections of the induced bundle $\varphi^{-1}TN$ of the tangent bundle TN by
φ. The *tension field* $\tau(\varphi)$ is defined globally on M by

$$\tau(\varphi) = \sum_{i=1}^{m} B(\varphi)(e_i, e_i) \in \Gamma(\varphi^{-1}TN), \tag{1.1}$$

where the second fundamental form $B(\varphi)$ is defined by

$$B(\varphi)(X, Y) = \nabla^N_{\varphi_*(X)} \varphi_*(Y) - \varphi_*(\nabla_X Y)$$

for $X, Y \in \mathfrak{X}(M)$. Then, a C^∞ map $\varphi : (M, g) \to (N, h)$ is *harmonic*
if $\tau(\varphi) = 0$. The *bitension field* $\tau_2(\varphi)$ is defined globally on M by

$$\tau_2(\varphi) = J(\tau(\varphi)) = \overline{\Delta}\tau(\varphi) - \mathcal{R}(\tau(\varphi)), \tag{1.2}$$

---

[1]This chapter consists of an unpublished note of discussions between Professor
N. Nakauchi of Yamaguchi University and myself.

where

$$J(V) := \overline{\Delta} V - \mathcal{R}(V),$$

$$\overline{\Delta} V := \overline{\nabla}^* \overline{\nabla} V = -\sum_{i=1}^{m} \{ \overline{\nabla}_{e_i} (\overline{\nabla}_{e_i} V) - \overline{\nabla}_{\nabla_{e_i} e_i} V \},$$

$$\mathcal{R}(V) := \sum_{i=1}^{m} R^N (V, \varphi_*(e_i)) \varphi_*(e_i).$$

Here, $\overline{\nabla}$ is the induced connection on the induced bundle $\varphi^{-1} TN$, and R^N is the curvature tensor of (N, h) given by $R^N(U, V)W = [\nabla_U^N, \nabla_V^N]W - \nabla_{[U,V]}^N W$ $(U, V, W \in \mathfrak{X}(N))$.

A C^∞ map $\varphi : (M, g) \to (N, h)$ is called to be *biharmonic* if

$$\tau_2(\varphi) = 0. \tag{1.3}$$

1.2. Setting of isometric immersions. In this sebsection, we prepare fundamental materials of general facts on isometric immersions (cf. [82]). Let φ be an isometric immersion of an m-dimensional Riemannian into an $(m + p)$-dimensional Riemannian manifold (N, h). Then, the induced bundle $\varphi^{-1} TN$ of the tangent bundle TN of N by φ is decomposed into the direct sum:

$$\varphi^{-1} TN = \tau M \oplus \nu M, \tag{1.4}$$

where $\varphi^{-1} TN = \cup_{x \in M} T_{\varphi(x)} N$, $\tau M = d\varphi(TM) = \cup_{x \in M} d\varphi(T_x M)$, and $\nu M = \cup_{x \in M} d\varphi(T_x M)^\perp$ is the normal bundle. For the induced connection $\overline{\nabla}$ on $\varphi^{-1} TN$ of the Levi-Civita connection ∇^N of (N, h) by φ, $\overline{\nabla}_X(d\varphi(Y))$ is decomposed corresponding to (2.4) as

$$\overline{\nabla}_X(d\varphi(Y)) = d\varphi(\nabla_X Y) + B(X, Y) \tag{1.5}$$

for all C^∞ vector fields X and Y on M. Here, ∇ is the Levi-Civita connection of (M, g) and $B(X, Y)$ is the second fundamental form of the immersion $\varphi : (M, g) \to (N, h)$.

Let $\{\xi_1, \cdots, \xi_p\}$ be a local unit normal vector fields along φ that are orthogonal at each point, and let us decompose $B(X, Y)$ as

$$B(X, Y) = \sum_{i=1}^{p} b^i(X, Y) \xi_i, \tag{1.6}$$

where $b^i(X, Y)$ $(i = 1, \cdots, p)$ are the p second fundamental forms of φ. For every $\xi \in \Gamma(\nu M)$, $\overline{\nabla}_X \xi$, denoted also by $\nabla_X^N \xi$ is decomposed correspondingly to (2.4) into

$$\nabla_X^N \xi = -A_\xi(X) + \nabla_X^\perp \xi, \tag{1.7}$$

where $\nabla^\perp$ is called the normal connection of νM. The linear operator A_ξ of $\Gamma(TM)$ into itself, called the shape operator with respect to ξ, satisfies that

$$\langle A_\xi(X), Y \rangle = \langle B(X, Y), \xi \rangle \tag{1.8}$$

for all C^∞ vector fields X and Y on M. Here, we denote the Riemannian metrics g and h simply by $\langle \cdot, \cdot \rangle$.

We denote the tension field $\tau(\varphi)$ of an isometric immersion $\varphi : (M, g) \to (N, h)$ as

$$
\begin{aligned}
\tau(\varphi) = \mathrm{Trace}_g(\widetilde{\nabla} d\varphi) &= \sum_{i=1}^m B(e_i, e_i) \\
&= \sum_{k=1}^p (\mathrm{Trace}_g b^k)\, \xi_k \\
&= m \sum_{k=1}^p H_k\, \xi_k \\
&= m\, \eta,
\end{aligned}
\tag{1.9}
$$

where $\widetilde{\nabla}$ is the induced connection on $TM \otimes \varphi^{-1}TN$, $H_k := \frac{1}{m}\mathrm{Trace}_g b^k = \frac{1}{m} = \mathrm{Trace}_g(A_{\xi_k})$ $(k = 1, \cdots, p)$, and $\eta := \sum_{k=1}^p H_k\, \xi_k$ is the mean curvature vector field of φ. Let us recall that $\varphi : (M, g) \to (N, h)$ is *minimal* if $\eta \equiv 0$.

2. Biharmonic map equations of an isometric immersion

In this section, we will derive the biharmonic map equations of a general isometric immersion $\varphi : (M, g) \to (N, h)$ with $\dim M = m$ and $\dim N = m + p$. We will calculate the bitension field $\tau_2(\varphi)$. By (2.2),

we have

$$\tau_2(\varphi) = \overline{\Delta}\tau(\varphi) - \mathcal{R}(\tau(\varphi))$$

$$= -\sum_{i=1}^{m}\left\{\overline{\nabla}_{e_i}\overline{\nabla}_{e_i}\left(\sum_{k=1}^{p}\sum_{k=1}^{p} m\,H_k\,\xi_k\right) - \overline{\nabla}_{\nabla_{e_i}e_i}\left(\sum_{k=1}^{p} m\,H_k\,\xi_k\right)\right.$$

$$\left. -R^N\left(d\varphi(e_i),\sum_{k=1}^{p} m\,H_k\,\xi_k\right)d\varphi(e_i)\right\}$$

$$= -m\sum_{k=1}^{p}\sum_{i=1}^{m}\left\{\overline{\nabla}_{e_i}(e_i(H_k)\,\xi_k + H_k\,\nabla_{e_i}^N\xi_k)\right.$$

$$- (\nabla_{e_i}e_i)(H_k)\,\xi_k - H_k\nabla_{\nabla_{e_i}e_i}^N\xi_k$$

$$\left. -H_k\,R^N(d\varphi(e_i),\xi_k)d\varphi(e_i)\right\}$$

$$= -m\sum_{k=1}^{p}\sum_{i=1}^{m}\left\{e_i\,e_i(H_k)\,\xi_k + e_i(H_k)\,\nabla_{e_i}^N\xi_k + e_i(H_k)\,\nabla_{e_i}^N\xi_k\right.$$

$$+ H_k\,\nabla_{e_i}^N\nabla_{e_i}^N\xi_k - (\nabla_{e_i}e_i)(H_k)\,\xi_k - H_k\,\nabla_{\nabla_{e_i}e_i}^N\xi_k$$

$$\left. -H_k R^N(d\varphi(e_i),\xi_k)d\varphi(e_i)\right\}. \tag{2.1}$$

Recall here that

$$\Delta H_k = \sum_{i=1}^{m}\{e_i\,e_i(H_k) - (\nabla_{e_i}e_i)(H_k)\}, \tag{2.2}$$

$$\overline{\Delta}\xi_k = -\sum_{i=1}^{m}\left\{\nabla_{e_i}^N\nabla_{e_i}^N\xi_k - \nabla_{\nabla_{e_i}}^N\xi_k\right\}, \tag{2.3}$$

$$\sum_{i=1}^{m} e_i(H_k)\nabla_{e_i}^N\xi_k = -A_{\xi_k}(\operatorname{grad} H_k) + \nabla_{\operatorname{grad} H_k}^{\perp}\xi_k. \tag{2.4}$$

Indeed, for (3.4), by (2.7),

$$\sum_{i=1}^{m} e_i(H_k)\nabla_{e_i}^N\xi_k = \nabla_{\sum_{i=1}^{m} e_i(H_k)e_i}^N\xi_k$$

$$= \nabla_{\operatorname{grad} H_k}^N\xi_k$$

$$= -A_{\xi_k}(\operatorname{grad} H_k) + \nabla_{\operatorname{grad} H_k}^{\perp}\xi_k.$$

Therefore, by (3.1), (3.2), (3.3) and (3.4), we obtain

$$\tau_2(\varphi) = -\sum_{k=1}^{p}\left\{m\,(\Delta H_k)\,\xi_k - 2m\,A_{\xi_k}(\operatorname{grad} H_k) + 2m\,\nabla_{\operatorname{grad} H_k}^{\perp}\xi_k\right.$$

$$\left. -m\,H_k\,\overline{\Delta}\xi_k - m\,H_k\sum_{i=1}^{m} R^N(d\varphi(e_i),\xi_k)d\varphi(e_i)\right\}. \tag{2.5}$$

Now, we decompose and calculate the tangential and normal parts of (3.5).

The case of $\sum_{i=1}^{m} R^N(d\varphi(e_i), \xi_k)d\varphi(e_i)$ $(k = 1, \cdots, p)$. Since $\varphi : (M, g) \to (N, h)$ is an isometric immersion, we may assume that $\{d\varphi(e_1), \cdots, d\varphi(e_m), \xi_1, \cdots, \xi_p\}$ is an adapted orthonormal frame along φ in (N, h), denoted simply also by $\{e'_1, \cdots, e'_n\}$ $(n = m + p)$. Then, we first obtain the following equations:

$$\sum_{i,j=1}^{m} \langle R^N(d\varphi(e_i), \xi_k)d\varphi(e_i), e_j \rangle e_j + \sum_{\ell=1}^{p}\sum_{j=1}^{m} \langle R^N(\xi_\ell, \xi_k)\xi_\ell, e_j \rangle e_j$$

$$= -\sum_{j=1}^{m} \mathrm{Ric}^N(\xi_k, e_j) e_j$$

$$= -(\mathrm{Ric}^N(\xi_k))^T, \tag{2.6}$$

where $(\mathrm{Ric}^N(\xi))^T$ is the tangential part of the image of ξ by the Ricci transform Ric^N of (N, h), and in the first equation of (3.6), we used that for every tangent vectors u and v at a point along φ in N,

$$\mathrm{Ric}^N(u, v) := \sum_{i=1}^{n} \langle R^N(u, e'_i)e'_i, v \rangle$$

$$= -\sum_{i=1}^{m} \langle R^N(d\varphi(e_i), u)d\varphi(e_i), v \rangle - \sum_{\ell=1}^{p} \langle R^N(\xi_\ell, u)\xi_\ell, v \rangle.$$

Thus, by (3.6), the *tangential part* of $\sum_{i=1}^{m} R^N(d\varphi(e_i), \xi_k)d\varphi(e_i)$ is given by

$$\sum_{i,j=1}^{m} \langle R^N(d\varphi(e_i), \xi_k)d\varphi(e_i), e_j \rangle e_j = -(\mathrm{Ric}^N(\xi_k))^T$$

$$- \sum_{\ell=1}^{p}\sum_{j=1}^{m} \langle R^N(\xi_\ell, \xi_k)\xi_\ell, e_j \rangle e_j. \tag{2.7}$$

We calculate the *normal part*. Let us consider the normal part $(\mathrm{Ric}^N(\xi))^{\perp}$. We have

$$-(\mathrm{Ric}^N(\xi_k))^{\perp} = -\sum_{s=1}^{p} \mathrm{Ric}^N(\xi_k, \xi_s)\xi_s$$

$$= \sum_{s=1}^{p}\sum_{i=1}^{m} \langle R^N(d\varphi(e_i), \xi_k)d\varphi(e_i), \xi_s \rangle \xi_s$$

$$+ \sum_{s=1}^{p}\sum_{\ell=1}^{p} \langle R^N(\xi_\ell, \xi_k)\xi_\ell, \xi_s \rangle \xi_s. \tag{2.8}$$

Thus, we have

$$\sum_{s=1}^{p}\sum_{i=1}^{m}\langle R^N(d\varphi(e_i),\xi_k)d\varphi(e_i),\xi_s\rangle\xi_s = -(\text{Ric}^N(\xi_k))^\perp$$

$$-\sum_{s=1}^{p}\sum_{\ell=1}^{p}\langle R^N(\xi_\ell,\xi_k)\xi_\ell,\xi_s\rangle\xi_s. \tag{2.9}$$

The case of $\overline{\Delta}\xi_k$. Next, we decompose of $\overline{\Delta}\xi_k$ into tangential and normal components. By (2.5) and (2.7), we have

$$\nabla^N_{e_i}\xi_k = -A_{\xi_k}(e_i) + \nabla^\perp_{e_i}\xi_k, \tag{2.10}$$

and also

$$\nabla^N_{\nabla_{e_i}e_i}\xi_k = -A_{\xi_k}(\nabla_{e_i}e_i) + \nabla^\perp_{\nabla_{e_i}e_i}\xi_k. \tag{2.11}$$

Furthermore, we have

$$\nabla^N_{e_i}\nabla^N_{e_i}\xi_k = -\nabla^N_{e_i}A_{\xi_k}(e_i) + \nabla^N_{e_i}\nabla^\perp_{e_i}\xi_k$$

$$= -\nabla_{e_i}A_{\xi_k}(e_i) - B(e_i, A_{\xi_k}(e_i)) - A_{\nabla^\perp_{e_i}\xi_k}e_i + \nabla^\perp_{e_i}\nabla^\perp_{e_i}\xi_k$$

$$= -(\nabla_{e_i}A_{\xi_k}(e_i) + A_{\nabla^\perp_{e_i}\xi_k}e_i) - (B(e_i, A_{\xi_k}(e_i)) - \nabla^\perp_{e_i}\nabla^\perp_{e_i}\xi_k). \tag{2.12}$$

Thus, the *normal part* of $\overline{\Delta}\xi_k$ is as follows.

$$(\overline{\Delta}\xi_k)^\perp := \sum_{\ell=1}^{p}\langle\overline{\Delta}\xi_k,\xi_\ell\rangle\xi_\ell$$

$$= \sum_{\ell=1}^{p}\sum_{i=1}^{m}\langle -\nabla^N_{e_i}\nabla^N_{e_i}\xi_k + \nabla^N_{\nabla_{e_i}e_i}\xi_k,\xi_\ell\rangle\xi_\ell$$

$$= \sum_{\ell=1}^{p}\sum_{i=1}^{m}\langle B(e_i, A_{\xi_k}(e_i)) - \nabla^\perp_{e_i}\nabla^\perp_{e_i}\xi_k + \nabla^\perp_{\nabla_{e_i}e_i}\xi_k,\xi_\ell\rangle\xi_\ell$$

$$= \sum_{i=1}^{m}B(e_i, A_{\xi_k}(e_i)) - \Delta^\perp\xi_k, \tag{2.13}$$

where $\Delta^\perp$ is the *normal Laplacian* acting on $\Gamma(\nu M)$ defined by

$$\Delta^\perp\xi := \sum_{i=1}^{m}\{\nabla^\perp_{e_i}\nabla^\perp_{e_i}\xi - \nabla^\perp_{\nabla_{e_i}e_i}\xi\} \tag{2.14}$$

for $\xi \in \Gamma(\nu M)$. By (3.11) and (3.12), the *tangential part* of $\overline{\Delta}\xi_k$ is also given by

$$(\overline{\Delta}\xi_k)^T = \sum_{i=1}^{m}\{\nabla_{e_i}A_{\xi_k}(e_i) + A_{\nabla^\perp_{e_i}\xi_k}e_i - A_{\xi_k}(\nabla_{e_i}e_i)\}. \tag{2.15}$$

Therefore, by (3.5), (3.7), (3.9) and (3.15), the tangential and normal part of the bitension field $\tau_2(\varphi)$ are given as follows. The *tangential part* of $\tau_2(\varphi)$ is given by

$$
(\tau_2(\varphi))^T := \sum_{k=1}^{p} \Big\{ - 2m\, A_{\xi_k}(\operatorname{grad} H_k) - m\, H_k\, (\overline{\Delta}\xi_k)^T
$$
$$
- m\, H_k \left(\sum_{i=1}^{m} R^N(d\varphi(e_i), \xi_k) d\varphi(e_i) \right)^T \Big\}
$$
$$
= -m \sum_{k=1}^{p} \Big\{ 2 A_{\xi_k}(\operatorname{grad} H_k)
$$
$$
+ H_k \sum_{i=1}^{m} \Big(\nabla_{e_i} A_{\xi_k}(e_i) + A_{\nabla^{\perp}_{e_i}\xi_k} e_i - A_{\xi_k}(\nabla_{e_i} e_i) \Big)
$$
$$
- H_k(\operatorname{Ric}^N(\xi_k))^T - H_k \sum_{\ell=1}^{p}\sum_{j=1}^{m}\langle R^N(\xi_\ell, \xi_k)\xi_\ell, e_j\rangle\, e_j \Big\}.
$$
$$(2.16)$$

The *normal part* of $\tau_2(\varphi)$ is given by

$$
(\tau_2(\varphi))^{\perp} := \sum_{k=1}^{p} \Big\{ m\,(\Delta H_k)\,\xi_k - m\, H_k\,(\overline{\Delta}\xi_k)^{\perp}
$$
$$
- m\, H_k \left(\sum_{i=1}^{m} R^N(d\varphi(e_i), \xi_k) d\varphi(e_i) \right)^{\perp}
$$
$$
+ 2m\, \nabla^{\perp}_{\operatorname{grad} H_k}\xi_k \Big\}
$$
$$
= m \sum_{k=1}^{p} \Big\{ (\Delta H_k)\xi_k - H_k \sum_{i=1}^{m} B(e_i, A_{\xi_k}(e_i)) + H_k\, \Delta^{\perp}\xi_k
$$
$$
+ H_k\,(\operatorname{Ric}^N(\xi_k))^{\perp} + H_k \sum_{s,\ell=1}^{p} \langle R^N(\xi_\ell, \xi_k)\xi_\ell, \xi_s\rangle\, \xi_s
$$
$$
+ 2\nabla^{\perp}_{\operatorname{grad} H_k}\xi_k \Big\}.
$$
$$(2.17)$$

Thus, we obtain the following theorem:

THEOREM 2.1. *Let $\varphi : (M, g) \to (N, h)$ be an isometric immersion of an m-dimensional Riemannian into an $n = (m+p)$-dimensional Riemannian manifold. Then, φ is biharmonic if and only if the following* *two equations (3.18) and (3.19) hold:*

$$\sum_{k=1}^{p} \left\{ 2A_{\xi_k}(\operatorname{grad} H_k) \right.$$

$$+ H_k \sum_{i=1}^{m} \left(\nabla_{e_i} A_{\xi_k}(e_i) + A_{\nabla_{e_i}^{\perp} \xi_k} e_i - A_{\xi_k}(\nabla_{e_i} e_i) \right)$$

$$\left. - H_k(\operatorname{Ric}^N(\xi_k))^T - H_k \sum_{\ell=1}^{p} \sum_{j=1}^{m} \langle R^N(\xi_\ell, \xi_k)\xi_\ell, e_j \rangle \, e_j \right\} = 0, \tag{2.18}$$

and

$$\sum_{k=1}^{p} \left\{ (\Delta H_k)\xi_k - H_k \sum_{i=1}^{m} B(e_i, A_{\xi_k}(e_i)) + H_k \Delta^{\perp} \xi_k \right.$$

$$+ H_k \left(\operatorname{Ric}^N(\xi_k)\right)^{\perp} + H_k \sum_{s,\ell=1}^{p} \langle R^N(\xi_\ell, \xi_k)\xi_\ell, \xi_s \rangle \, \xi_s$$

$$\left. + 2\nabla_{\operatorname{grad} H_k}^{\perp} \xi_k \right\} = 0. \tag{2.19}$$

CHAPTER 7

Harmonic Maps into Compact Lie Groups and Integrable Systems

1

ABSTRACT. We obtain the formulation of the biharmonic map equation in terms of the Maurer-Cartan form for all smooth maps of a compact Riemannian manifold into a compact Lie group (G, h) with the bi-invariant Riemannian metric h. Using our formula, we determine exactly all biharmonic curves into compact Lie groups, and all the biharmonic maps of an open domain of $\mathbb{R}^2$ equipped with a Riemannian metric conformal to the standard Euclidean metric into (G, h).

1. Introduction and statement of results

The theory of harmonic maps of a Riemann surface into Lie groups, symmetric spaces or homogeneous spaces has been extensively studied in connection with the integrable systems ([13], [15], [33], [36], [42], [48], [55], [152], [153]). Let us recall the theory of harmonic maps of a Riemann surface M into a compact Lie group G, briefly. A harmonic map is a critical map of the energy functional defined by

$$E(\psi) := \frac{1}{2} \int_M |d\psi|^2 \, v_g.$$

For such a map ψ, let α be the pull back of the Maurer-Cartan form θ of G which is decomposed into the sum of the holomorphic part and the antiholomorphic one as $\alpha = \alpha' + \alpha''$. Then, it satisfies $d\alpha = \frac{1}{2}[\alpha \wedge \alpha] = 0$ (the integrability condition), and the harmonicity of ψ is equivalent to the condition $\delta\alpha = 0$. Introducing a parameter $\lambda \in \mathbb{C}^* = \mathbb{C}\backslash\{0\}$ as

$$\alpha_\lambda := \frac{1}{2} (1|\lambda) \, \alpha' + \frac{1}{2} (1|\lambda^{|1|}) \, \alpha'',$$

---

[1]This chapter is due to [**149**]: H. Urakawa, *Biharmonic maps into compact Lie groups and symmetric spaces*, "Alexandru Myller" Mathematical Seminar, 246–263, AIP Conf. Proc. **1329**, Amer. Snst. Phys., Melville, NY, 2011, and also [**152**]: H. Urakawa, *Biharmonic maps into compact Lie groups and integrable systems*, Hokkaido Math. J., **43** (2014), 73–103.

both the harmonicity and the integrability condition are equivalent to

$$d\alpha_\lambda + \frac{1}{2}[\alpha_\lambda \wedge \alpha_\lambda] = 0,$$

which implies that there exists an extended solution $\Phi_\lambda : M \to G$ satisfying $\Phi_\lambda{}^{-1} = \alpha_\lambda$ ([147]). Guest and Ohnita ([55]) showed that the loop group $\Lambda G^{\mathbb{C}}$ of G acts on the space of all harmonic maps of M into G, and Uhlenbeck ([147]) showed that every harmonic map from the two-sphere into G is a harmonic map of finite uniton number, and Wood ([163]) determined explicitly harmonic maps of finite uniton numbers. On the other hand, the theory of biharmonic maps was initiated by Eells and Lemaire ([41]) and Jiang ([74]). A biharmonic map is a natural extension of harmonic map, and is a critical map of the bienergy functional defined by

$$E_2(\psi) := \frac{1}{2}\int_M |\delta d\psi|^2\, v_g = \frac{1}{2}\int_M |\tau(\psi)|^2\, v_g,$$

where $\tau(\psi)$ is the tension field of ψ, and, by definition, ψ is harmonic if and only if $\tau(\psi) \equiv 0$.

In this paper, we study biharmonic maps of a compact Riemannian manifold (M, g) into a compact Lie group (G, h) with the bi-invariant Riemannian metric h. For every C^∞ map $\psi : (M, g) \to (G, h)$, let us consider again the pullback α of the Maurer-Cartan form θ. We first will show that the biharmonicity condition for ψ is that

$$\delta d\delta \alpha + \mathrm{Trace}_g([\alpha, d\delta\alpha]) = 0$$

(cf. Corollary 3.5) which is a natural extension of harmonicity. Due to this formula, we can determine all real analytic biharmonic curves into a compact Lie group (G, h) in terms of the initial data $F(0)$, $F'(0)$ and $F''(0)$, where $F(t) = \alpha\left(\frac{\partial}{\partial t}\right)$ (cf. Section 4). We give a characterization of biharmonic maps of $(\mathbb{R}^2, \mu^2 g_0)$, where g_0 is the standard Euclidean metric on $\mathbb{R}^2$ and μ is a positive real analytic function on $\mathbb{R}^2$ (cf. Sections 5, 6 and 7).

Acknowledgement: The author expresses his gratitude to Prof. J. Inoguchi who gave many useful suggestions and Prof. A. Kasue for his financial support during the preparation of this paper, and Dr. Y. Takenaka and the referees who read carefully and pointed out several mistakes in the first draft.

2. Preliminaries

In this section, we prepare general materials and facts on harmonic maps, biharmonic maps into Riemannian manifolds (cf. [40], [41], [42],

[74]). Let (M, g) be an m-dimensional compact Riemannian manifold, and (N, h), an n-dimensional Riemannian manifold.

The *energy functional* on the space $C^\infty(M, N)$ of all C^∞ maps of M into N is defined by

$$E(\psi) = \frac{1}{2} \int_M |d\psi|^2 \, v_g,$$

and for a compactly supported C^∞ one parameter deformation $\psi_t \in C^\infty(M, N)$ $(-\epsilon < t < \epsilon)$ of ψ with $\psi_0 = \psi$, the *first variation formula* is given by

$$\frac{d}{dt}\bigg|_{t=0} E(\psi_t) = -\int_M \langle \tau(\psi), V \rangle \, v_g,$$

where V is a variation vector field along ψ defined by $V = \frac{d}{dt}\big|_{t=0} \psi_t$ which belongs to the space $\Gamma(\psi^{-1}TN)$ of sections of the induced bundle of the tangent bundle TN by ψ. The *tension field* $\tau(\psi)$ is defined by

$$\tau(\psi) = -\delta(d\psi), \tag{2.1}$$

where recall the definition $\delta\alpha$ for a $\psi^{-1}TN$-valued 1-form α,

$$\delta\alpha = -\sum_{i=1}^m (\overline{\nabla}_{e_i}\alpha)(e_i) = -\sum_{i=1}^m \left\{ \overline{\nabla}(\alpha(e_i)) - \alpha(\nabla_{e_i}e_i) \right\}.$$

Here, ∇, ∇^h and $\overline{\nabla}$ are the Levi-Civita connections of (M, g), (N, h), and the induced connections on the induced bundle $\psi^{-1}TN$ from ∇^h, respectively. For a harmonic map $\psi : (M, g) \to (N, h)$, the *second variation formula* of the energy functional $E(\psi)$ is

$$\frac{d^2}{dt^2}\bigg|_{t=0} E(\psi_t) = \int_M \langle J(V), V \rangle \, v_g$$

where

$$J(V) = \overline{\Delta} V - \mathcal{R}(V),$$

$$\overline{\Delta} V = \overline{\nabla}^* \overline{\nabla} V = -\sum_{i=1}^m \{ \overline{\nabla}_{e_i}(\overline{\nabla}_{e_i} V) - \overline{\nabla}_{\nabla_{e_i} e_i} V \},$$

$$\mathcal{R}(V) = \sum_{i=1}^m R^h(V, d\psi(e_i)) d\psi(e_i).$$

Here, $\overline{\nabla}$ is the induced connection on the induced bundle $\psi^{-1}TN$, and R^h is the curvature tensor of (N, h) given by $R^h(U, V)W = [\nabla^h_U, \nabla^h V]W - \nabla^h_{[U,V]}W$ $(U, V, W \in \mathfrak{X}(N))$. The *bienergy functional* is defined by

$$E_2(\psi) = \frac{1}{2} \int_M |\delta d\psi|^2 \, v_g = \frac{1}{2} \int_M |\tau(\psi)|^2 \, v_g, \tag{2.2}$$

and the *first variation formula* of the bienergy is given ([**74**]) by

$$\frac{d}{dt}\bigg|_{t=0} E_2(\psi_t) = -\int_M \langle \tau_2(\psi), V \rangle \, v_g \tag{2.3}$$

where the *bitension field* $\tau_2(\psi)$ is defined by

$$\tau_2(\psi) = J(\tau(\psi)) = \overline{\Delta}\tau(\psi) - \mathcal{R}(\tau(\psi)), \tag{2.4}$$

and a C^∞ map $\psi : (M, g) \to (N, h)$ is called to be *biharmonic* if

$$\tau_2(\psi) = 0. \tag{2.5}$$

The biharmonic maps are real analytic when both (M, g) and (N, h) are real analytic. This is because the solutions of non-linear elliptic partial differential equations are real analytic.

3. Determination of the bitension field

Now, assume that (N, h) is an n-dimensional compact Lie group with Lie algebra $\mathfrak{g}$, and h, the bi-invariant Riemannian metric on G corresponding to the $\mathrm{Ad}(G)$-invariant inner product $\langle\,,\,\rangle$ on $\mathfrak{g}$. Let θ be the Maurer-Cartan form on G, i.e., a $\mathfrak{g}$-valued left invariant 1-form on G which is defined by $\theta_y(Z_y) = Z$, $(y \in G, Z \in \mathfrak{g})$. For every C^∞ map ψ of (M, g) into (G, h), let us consider a $\mathfrak{g}$-valued 1-form α on M given by $\alpha = \psi^*\theta$. Then it is well known (see for example, [**33**]) that

LEMMA 3.1. *For every C^∞ map $\psi : (M, g) \to (G, h)$,*

$$\theta(\tau(\psi)) = -\delta\alpha. \tag{3.1}$$

Thus, $\psi : (M, g) \to (G, h)$ is harmonic if and only if $\delta\alpha = 0$.

Let $\{X_s\}_{s=1}^n$ be an orthonormal basis of $\mathfrak{g}$ with respect to the inner product $\langle\,,\,\rangle$. Then, for every $V \in \Gamma(\psi^{-1}TG)$,

$$V(x) = \sum_{s=1}^n h_{\psi(x)}(V(x), X_{s\,\psi(x)}) \, X_{s\,\psi(x)} \in T_{\psi(x)}G,$$

$$\theta(V)(x) = \sum_{s=1}^n h_{\psi(x)}(V(x), X_{s\,\psi(x)}) \, X_s \in \mathfrak{g}, \tag{3.2}$$

for all $x \in M$. Then, for every $X \in \mathfrak{X}(M)$,

$$\theta(\overline{\nabla}_X V) = \sum_{s=1}^{n} h(\overline{\nabla}_X V, X_s) X_s$$

$$= \sum_{s=1}^{n} \{X\, h(V, X_s) - h(V, \overline{\nabla}_X X_s)\} X_s$$

$$= X(\theta(V)) - \sum_{s=1}^{n} h(V, \overline{\nabla}_X X_s) X_s, \qquad (3.3)$$

where we regarded a vector field $Y \in \mathfrak{X}(G)$ by $Y(x) = Y(\psi(x))\,(x \in M)$ to be an element in the space $\Gamma(\psi^{-1}TG)$ of smooth sections of $\psi^{-1}TG$. Here, let us recall that the Levi-Civita connection ∇^h of (G, h) is given (cf. [**82**] Vol. II, p. 201, Theorem 3.3) by

$$\nabla^h_{X_t} X_s = \frac{1}{2}[X_t, X_s] = \frac{1}{2} \sum_{\ell=1}^{n} C^\ell_{ts} X_\ell, \qquad (3.4)$$

where the structure constant C^ℓ_{ts} of $\mathfrak{g}$ is defined by $[X_t, X_s] = \sum_{\ell=1}^{n} C^\ell_{ts} X_\ell$, and satisfies

$$C^\ell_{ts} = \langle [X_t, X_s], X_\ell \rangle = -\langle X_s, [X_t, X_\ell] \rangle = -C^s_{t\ell}. \qquad (3.5)$$

Thus, we have by (3.4) and (3.5),

$$\sum_{s=1}^{n} h(V, \overline{\nabla}_X X_s) X_s = \frac{1}{2} \sum_{s,t=1}^{n} h\left(V, \sum_{\ell=1}^{n} h(\psi_* X, X_t) C^\ell_{ts} X_\ell\right) X_s$$

$$= -\frac{1}{2} \sum_{s,t,\ell=1}^{n} h(V, X_\ell)\, h(\psi_* X < X_t)\, C^s_{t\ell} X_s$$

$$= -\frac{1}{2} \sum_{t,\ell=1}^{n} h(V, X_\ell)\, h(\psi_* X, X_t)\, [X_t, X_\ell]$$

$$= -\frac{1}{2}\left[\sum_{t=1}^{n} h(\psi_* X, X_t) X_t, \sum_{\ell=1}^{n} h(V, X_\ell) X_\ell\right]$$

$$= -\frac{1}{2}[\alpha(X), \theta(V)], \qquad (3.6)$$

which is because we have

$$\alpha(X) = \theta(\psi_* X) = \sum_{t=1}^{n} h(\psi_* X, X_t) X_t, \qquad (3.7)$$

and

$$\theta(V) = \sum_{\ell=1}^{n} h(V, X_\ell)\, \theta(X_\ell) = \sum_{\ell=1}^{n} h(V, X_\ell\, X_\ell. \qquad (3.8)$$

Therefore, inserting (3.6) into (3.3), we obtain

LEMMA 3.2. *For every* C^∞ *map* $\psi : (M, g) \to (G, h)$,

$$\theta(\overline{\nabla}_X V) = X(\theta(V)) + \frac{1}{2}[\alpha(X), \theta(V)], \qquad (3.9)$$

where $V \in \Gamma(\psi^{-1}TG)$ *and* $X \in \mathfrak{X}(M)$.

We shall show

THEOREM 3.1. *For every* $\psi \in C^\infty(M, G)$, *we have*

$$\theta(\tau_2(\psi)) = \theta(J(\tau(\psi)))$$
$$= -\delta\, d\, \delta\alpha - \text{Trace}_g([\alpha, d\, \delta\alpha]), \qquad (3.10)$$

where $\alpha = \psi^*\theta$.

Here, let us recall the definition:

DEFINITION 3.1. *For two* $\mathfrak{g}$-*valued 1-formsff* α *and* β *on* M, *we define a* $\mathfrak{g}$-*valued symmetric 2-tensor* $[\alpha, \beta]$ *on* M *by*

$$[\alpha, \beta](X, Y) := \frac{1}{2}\{[\alpha(X), \beta(Y)] + [\alpha(Y), \beta(X)]\}, \quad (X, Y \in \mathfrak{X}(M)) \qquad (3.11)$$

and its trace $\text{Trace}_g([\alpha, \beta])$ *by*

$$\text{Trace}_g([\alpha, \beta]) := \sum_{i=1}^{m}[\alpha, \beta](e_i, e_i). \qquad (3.12)$$

Recall that the $\mathfrak{g}$-*valued 2-form* $[\alpha \wedge \beta]$ *on* M *is given by*

$$[\alpha \wedge \beta](X, Y) := \frac{1}{2}\{[\alpha(X), \beta(Y)] - [\alpha(Y), \beta(X)]\} \quad (X, Y \in \mathfrak{X}(\mathfrak{M})). \qquad (3.13)$$

Then, we have immediately by Theorem 3.3,

COROLLARY 3.1. *For every* $\psi \in C^\infty(M, G)$, *we have (1)* $\psi : (M, g) \to (G, h)$ *is harmonic if and only if*

$$\delta\alpha = 0. \qquad (3.14)$$

(2) $\psi : (M, g) \to (G, h)$ *is biharmonic if and only if*

$$\delta\, d\, \delta\alpha + \text{Trace}_g([\alpha, d\, \delta\alpha]) = 0. \qquad (3.15)$$

We give a proof of Theorem 3.3.

Proof. (*The first step*) We first show that, for all $V \in \Gamma(\psi^{-1}TG)$,

$$\theta(\overline{\Delta}V) = \Delta_g \theta(V) - \sum_{i=1}^{m} \left\{ \frac{1}{2} [e_i(\alpha(e_i)), \theta(V)] + [\alpha(e_i), e_i(\theta(V))] \right.$$

$$\left. + \frac{1}{4} [\alpha(e_i), [\alpha(e_i), \theta(V)]] - \frac{1}{2} [\alpha(\nabla_{e_i} e_i), \theta(V)] \right\},$$

$$(3.16)$$

where $\{e_i\}_{i=1}^{m}$ is a locally defined orthonormal frame field on (M, g), and Δ_g is the (positive) Laplacian of (M, g) acting on $C^\infty(M)$.

Indeed, we have by using Lemma 3.2 twice,

$$\theta(\overline{\Delta}V) = -\sum_{i=1}^{m} \left\{ \theta(\overline{\nabla}_{e_i}(\overline{\nabla}_{e_i} V)) - \theta(\overline{\nabla}_{\nabla_{e_i} e_i} V) \right\}$$

$$= -\sum_{i=1}^{m} \left\{ e_i(\theta(\overline{\nabla}_{e_i} V)) + \frac{1}{2} [\alpha(e_i), \theta(\overline{\nabla}_{e_i} V] \right.$$

$$\left. - \nabla_{e_i} e_i(\theta(V)) - \frac{1}{2} [\alpha(\nabla_{e_i} e_i), \theta(V)] \right\}$$

$$= -\sum_{i=1}^{m} \left\{ e_i \left(e_i(\theta(V) + \frac{1}{2} [\alpha(e_i), \theta(V)]) \right) \right.$$

$$+ \frac{1}{2} \left[\alpha(e_i), e_i(\theta(V)) + \frac{1}{2} [\alpha(e_i), \theta(V)] \right]$$

$$\left. - \nabla_{e_i} e_i(\theta(V)) - \frac{1}{2} [\alpha(\nabla_{e_i} e_i), \theta(V)] \right\}$$

$$= -\sum_{i=1}^{m} \{ e_i(e_i(\theta(V)) - \nabla_{e_i} e_i(\theta(V))) \}$$

$$- \sum_{i=1}^{m} \left\{ \frac{1}{2} e_i([\alpha(e_i), \theta(V)]) + \frac{1}{2} [\alpha(e_i), e_i(\theta(\theta(V)))] \right.$$

$$\left. + \frac{1}{4} [\alpha(e_i), [\alpha(e_i), \theta(V)]] - \frac{1}{2} [\alpha(\nabla_{e_i} e_i), \theta(V)] \right\}.$$

$$(3.17)$$

Here, we have

$$e_i([\alpha(e_i), \theta(V)] = [e_i(\alpha(e_i)), \theta(V)] + [\alpha(e_i), e_i(\theta(V))],$$

which we substitute into (3.17), and by definition of Δ_g, we have (3.16).

(*The second step*) On the other hand, we have to consider

$$-\sum_{i=1}^{m} R^h(V, \psi_* e_i)\psi_* e_i = -\sum_{i=1}^{m} R^h(L_{\psi(x)\,*}^{-1}V, L_{\psi(x)\,*}^{-1}\psi_* e_i)L_{\psi(x)\,*}^{-1}\psi_* e_i.$$

(3.18)

Under the identification $T_e G \ni Z_e \leftrightarrow Z \in \mathfrak{g}$, we have

$$T_e G \ni L_{\psi(x)\,*}^{-1}\psi_* e_i \leftrightarrow \alpha(e_i) \in \mathfrak{g}, \tag{3.19}$$

$$T_e G \ni L_{\psi(x)\,*}^{-1}V \leftrightarrow \theta(V) \in \mathfrak{g}, \tag{3.20}$$

respectively. Because, we have

$$L_{\psi(x)\,*}^{-1}\psi_* e_i = \sum_{s=1}^{n} h(\psi_* e_i, X_{s\,\psi(x)})\, X_{s\,e}$$

and

$$\alpha(e_i) = \psi^*\theta(e_i) = \theta(\psi_* e_i) = \sum_{s=1}^{n} h(\psi_* e_i, X_{s\,\psi(x)})\, \theta(X_{s\,\psi(x)})$$

$$= \sum_{s=1}^{n} h(\psi_* e_i, X_{s\,\psi(x)})\, X_s, \tag{3.21}$$

which implies that (3.19). Analogously, we obtain (3.20).

Under this identification, the curvature tensor of (G, h) is given as (see Kobayashi-Nomizu ([**82**], pp. 203-204)),

$$R^h(X, Y)_e = -\frac{1}{4}\,\mathrm{ad}([X, Y]) \qquad (X, Y \in \mathfrak{g}),$$

and then, we have

$$\theta\left(-\sum_{i=1}^{m} R^h(V, \psi_* e_i)\psi_* e_i\right) = \frac{1}{4}\sum_{i=1}^{m}[[\theta(V), \alpha(e_i)], \alpha(e_i)]$$

$$= \frac{1}{4}\sum_{i=1}^{m}[\alpha(e_i), [\alpha(e_i), \theta(V)]].$$

(3.22)

(*The third step*)   By (3.16) and (3.21), for $V \in \Gamma(\psi^{-1}TG)$, we have

$$\theta\left(\overline{\Delta}V - \sum_{i=1}^{m} R^h(V, \psi_* e_i)\psi_* e_i\right)$$

$$= \Delta_g \theta(V()$$

$$- \sum_{i=1}^{m}\left\{\frac{1}{2}[e_i(\alpha(e_i)), \theta(V)] + [\alpha(e_i), e_i(\theta(V))] + \frac{1}{4}[\alpha(e_i), [\alpha(e_i), \theta(V)]]\right.$$

$$\left. - \frac{1}{2}[\alpha(\nabla_{e_i} e_i), \theta((V)]\right\}$$

$$+ \frac{1}{4}\sum_{i=1}^{m}[\alpha(e_i), [\alpha(e_i), \theta(V)]]$$

$$= \Delta_g \theta(V) - \frac{1}{2}\sum_{i=1}^{m} e_i(\alpha(e_i)), \theta(V)] + \sum_{i=1}^{m}[\alpha(e_i), e_i(\theta(V))]$$

$$+ \frac{1}{2}\sum_{i=1}^{m}[\alpha(\nabla_{e_i} e_i), \theta(V)]$$

$$= \Delta_g \theta(V) - \frac{1}{2}\left[\sum_{i=1}^{m}(e_i(\alpha(e_i)) - \alpha(\nabla_{e_i} e_i)), \theta(V)\right] + \sum_{i=1}^{m}[\alpha(e_i), e_i(\theta(V))]$$

$$= \Delta_g \theta(V) + \frac{1}{2}[\delta\alpha, \theta(V)] + \sum_{i=1}^{m}[\alpha(e_i), e_i(\theta(V))]. \qquad (3.23)$$

(*The fourth step*)   For $V = \tau(\psi)$ in (3.22), since $\theta(\tau(\psi)) = -\delta\alpha$, we have

$$\theta(J(\tau(\psi))) = \Delta_g \theta(\tau(\psi)) + \frac{1}{2}[\delta\alpha, \theta(\tau(\psi))]$$

$$+ \sum_{i=1}^{m}[\alpha(e_i), e_i(\theta(\tau(\psi)))]$$

$$= -\Delta_g \delta\alpha - \frac{1}{2}[\delta\alpha, \delta\alpha] - \sum_{i=1}^{m}[\alpha(e_i), e_i(\delta\alpha)]$$

$$= -\Delta_g \delta\alpha - \sum_{i=1}^{m}[\alpha(e_i), e_i(\delta\alpha)]$$

$$= -\Delta_g \delta\alpha - \sum_{i=1}^{m}[\alpha(e_i), (d\delta\alpha)(e_i)]. \qquad (3.24)$$

Then, (3.23) implies the desired (3.10). $\square$

4. Biharmonic curves from $\mathbb{R}$ into compact Lie groups

In this section, we consider the simplest case: $(M, g) = (\mathbb{R}, g_0)$ is the standard 1-dimensional Euclidean space, and (G, h) is an n-dimensional compact Lie group with the bi-invariant Riemannian metric h.

4.1   First, let $\psi : \mathbb{R} \ni t \mapsto \psi(t) \in (G, h)$, a C^∞ curve in G. Then, $\alpha := \psi^* \theta$ is a $\mathfrak{g}$-valued 1-form on $\mathbb{R}$. So, α can be written at $t \in \mathbb{R}$ as

$$\alpha_t = F(t)\, dt, \tag{4.1}$$

where $F : \mathbb{R} \ni t \mapsto F(t) \in \mathfrak{g}$ is given by

$$F(t) = \alpha \left(\frac{\partial}{\partial t} \right) = \psi^* \theta \left(\frac{\partial}{\partial t} \right) = \theta \left(\psi_* \left(\frac{\partial}{\partial t} \right) \right). \tag{4.2}$$

Here, since

$$\psi'(t) := \psi_* \left(\frac{\partial}{\partial t} \right) = \sum_{s=1}^{n} h_{\psi(t)} \left(\psi_* \left(\frac{\partial}{\partial t} \right), X_{s\,\psi(t)} \right) X_{s\,\psi(t)}, \tag{4.3}$$

we have

$$F(t) = \sum_{s=1}^{n} h_{\psi(t)} \left(\psi_* \left(\frac{\partial}{\partial t} \right), X_{s\,\psi(t)} \right) X_s, \tag{4.4}$$

so that we have the following correspondence:

$$T_e G \ni L_{\psi(t)\,*}^{-1} \psi'(t) = \sum_{s=1}^{n} h_{\psi(t)} \left(\psi'(t), X_{s\,\psi(t)} \right) X_{s\,e}$$

$$\leftrightarrow F(t) = \theta \left(\psi_* \left(\frac{\partial}{\partial t} \right) \right) \in \mathfrak{g}. \tag{4.5}$$

4.2   We have that

$$\delta \alpha = -F'(t), \tag{4.6}$$

since we have $\delta \alpha = -e_1(\alpha(e_1)) = -e_1(F(t)) = -F'(t)$.

Therefore, we have $\psi : (\mathbb{R}, g_0) \to (G, h)$ is *harmonic* if and only if

$$\delta \alpha = 0 \quad \Longleftrightarrow \quad F' = 0$$
$$\Longleftrightarrow \quad \alpha = X \otimes dt \quad \text{(for some } X \in \mathfrak{g})$$
$$\Longleftrightarrow \quad \psi : \mathbb{R} \to (G, h), \text{is a } \textit{geodesic}, \tag{4.7}$$

since

$$F(t) = \theta(\psi'(t)) = L_{\psi(t)\,*}^{-1} \psi'(t), \tag{4.8}$$

we have

$$\psi'(t) = L_{\psi(t)*} X = X_{\psi(t)}, \tag{4.9}$$

for some $X \in \mathfrak{g}$ which yields that

$$\psi(t) = x \, \exp(tX).$$

Therefore, *any geodesic through $\psi(0) = x$ is given by*

$$\psi(t) = x \, \exp(tX), \ (t \in \mathbb{R}) \tag{4.10}$$

for some $X \in \mathfrak{g}$.

On the other hand, we want to determine a *biharmonic curve* $\psi :$ $(\mathbb{R}, g_0) \to (G, h)$. By (4.6), we have

$$\delta d \delta \alpha = -\frac{\partial^2}{\partial t^2}\left(-F'(t)\right) = F^{(3)}(t), \tag{4.11}$$

and

$$\mathrm{Trace}_g[\alpha, d\delta\alpha] = \left[\alpha\left(\frac{\partial}{\partial t}\right), d\delta\alpha\left(\frac{\partial}{\partial t}\right)\right] = [F(t), F''(t)], \tag{4.12}$$

so by (4.9), (4.10), and (3.16) in Corollary 3.5, $\psi : (\mathbb{R}, g_0) \to (G, h)$ is *biharmonic if and only if*

$$F^{(3)} - [F(t), F''(t)] = 0. \tag{4.13}$$

4.3   For a C^∞ curve $\psi : \mathbb{R} \to G$, let $\psi(t) := \exp X(t)$, where $X(t) \in \mathfrak{g}$. Then,

$$F(t) = \theta\left(\psi_*\left(\frac{\partial}{\partial t}\right)\right), \ \psi_*\left(\frac{\partial}{\partial t}\right) \in T_{\psi(t)}G, \tag{4.14}$$

and by the following formula (cf.[**58**], p.95)

$$\exp_{*X} = L_{\exp X *e} \circ \frac{1 - e^{-\mathrm{ad}\,X}}{\mathrm{ad}\,X} \quad (X \in \mathfrak{g}),$$

we have

$$\psi_*\left(\frac{\partial}{\partial t}\right) = \exp_{*X(t)} X'(t)$$

$$= L_{\exp X(t) *e}\left(\sum_{n=0}^{\infty} \frac{(-\mathrm{ad}\,X(t))^n}{(n+1)!}\,(X'(t))\right). \tag{4.15}$$

Since θ is a left invariant 1-form, we have

$$F(t) = \sum_{n=0}^{\infty} \frac{(-\mathrm{ad}\,X(t))^n}{(n+1)!}\,(X'(t)). \tag{4.16}$$

4.4   The initial value problem

$$\begin{cases} F^{(3)}(t) = [F(t), F''(t)], \\ F(0) = B_0, \, F'(0) = B_1, \, F''(0) = B_2, \end{cases} \tag{4.17}$$

for every $B_i \in \mathfrak{g}$ $(i = 0, 1, 2)$, has a unique solution $F(t)$. Assume that $X(t)$ is a real analytic curve in t, and $X(0) = 0$. Then, $F(t)$ is also real analytic in t, and we can write as

$$X(t) = \sum_{n=1}^{\infty} A_n \, t^n, \quad F(t) = \sum_{n=0}^{\infty} B_n \, t^n. \tag{4.18}$$

By (4.16), we have

$$F(t) = X'(t) + \frac{1}{2}[-X(t), X'(t)] + \frac{1}{6}[-X(t), [-X(t), X'(t)]]$$
$$+ \sum_{n=3}^{\infty} \frac{(-\mathrm{ad}\, X(t))^n}{(n+1)!} (X'(t)). \tag{4.19}$$

Since $X'(t) = \sum_{m=0}^{\infty} A_{m+1}(m+1)\, t^m$, we have

$$\frac{1}{2}[-X(t), X'(t)] = -\frac{1}{2}[A_1, A_2]\, t^2 + O(t^3),$$

and

$$\frac{1}{6}[-X(t), [-X(t), X'(t)]] = O(t^3),$$

so that we have

$$F(t) = A_1 + 2\, A_2\, t + \left(3\, A_3 - \frac{1}{2}[A_1, A_2]\right) t^2 + O(t^3).$$

Continuing this process, we have

$$\begin{cases} B_0 = A_1, \\ B_1 = 2\, A_2, \\ B_2 = 3\, A_3 - \frac{1}{2}[A_1, A_2], \\ \cdots\cdots\cdots\cdots\cdots\cdots\cdots\cdots \\ B_n = (n+1)\, A_{n+1} + G_n(A_1, \ldots, A_n), \end{cases} \tag{4.20}$$

where $G_n(x_1, \cdots, x_n)$ is a polynomial in $(x_1, \cdots, x_n)$. Notice that for arbitrary given data (B_0, B_1, B_2), all B_n $(n = 0, 1, \cdots)$ are determined, and by using (4.20), one can determine all A_n $(n = 1, 2, \cdots)$, uniquely. Therefore, by summarizing the above, we obtain

THEOREM 4.1. *For every C^∞ curve $\psi : \mathbb{R} \to G$, $\psi(t) = \exp X(t)$ $(X(t) \in \mathfrak{g})$, and*

$$\alpha\left(\frac{\partial}{\partial t}\right) = F(t) = \sum_{n=0}^{\infty} \frac{(-\mathrm{ad}\, X(t))^n}{(n+1)!}(X'(t)). \qquad (4.21)$$

(1) $\psi : (\mathbb{R}, g_0) \to (G, h)$ *is biharmonic if and only if*

$$F^{(3)}(t) = [F(t), F''(t)]. \qquad (4.22)$$

(2) *The initial value problem*

$$\begin{cases} F^{(3)}(t) = [F(t), F''(t)], \\ F(0) = B_0, \; F'(0) = B_1, \; F''(0) = B_2, \end{cases} \qquad (4.23)$$

has a unique solution $F(t)$ for arbitrary given data (B_0, B_1, B_2) in $\mathfrak{g}$.

(3) *Assume that $\psi : (\mathbb{R}, g_0) \to (G, h)$ is a real analytic biharmonic curve with $\psi(0) = e$. Then, $\psi(t)$ is uniquely determined by $F(0) = B_0$, $F'(0) = B_1$, and $F''(0) = B_2$.*

Example   If G is abelian, let us consider a C^∞ curve $\psi : \mathbb{R} \to G$ given by $\psi(t) = \exp X(t)$. Then, $F(t) = X'(t)$, and $\psi : (\mathbb{R}, g_0) \to (G, h)$ is biharmonic if and only if $F^{(3)}(t) = X^{(4)}(t) = 0$. Then, $X(t) = A_0 + A_1\, t + A_2\, t^2 + A_3\, t^3$. Thus, every biharmonic curve $\psi : (\mathbb{R}, g_0) \to (G, h)$ with $\psi(0) = e$ is given by

$$\psi(t) = \exp(A_1\, t + A_2\, t^2 + A_3\, t^3).$$

4.5   Now we will solve the ODE (4:22) for a biharmonic isometric immersion $\psi : (\mathbb{R}, g_0) \to G$ and a $\mathfrak{g}$-valued curve $F(t)$ in the case of $\mathfrak{g} = \mathfrak{su}(2)$. Let $G = SU(2)$ with the bi-invariant Riemannian metric h which corresponds to the following $\mathrm{Ad}(SU(2))$-invariant inner product $\langle\,,\,\rangle$ on

$$\mathfrak{g} = \mathfrak{su}(2) = \{X M(2, \mathbb{C}); X + {}^t\overline{X} = 0, \mathrm{Tr}(X) = 0\},$$
$$\langle X, Y \rangle = |2\mathrm{Tr}(XY) \quad (X; Y \in \mathfrak{su}(2)).$$

If we choose

$$X_1 = \begin{pmatrix} \frac{\sqrt{-1}}{2} & 0 \\ 0 & -\frac{\sqrt{-1}}{2} \end{pmatrix}, \; X_2 = \begin{pmatrix} 0 & \frac{1}{2} \\ -\frac{1}{2} & 0 \end{pmatrix}, \; X_3 = \begin{pmatrix} 0 & \frac{\sqrt{-1}}{2} \\ \frac{\sqrt{-1}}{2} & 0 \end{pmatrix},$$

then $\{X_1, X_2, X_3\}$ is an orthonormal basis of $(\mathfrak{su}(2), \langle\,,\,\rangle)$, and satisfies the Lie bracket relations:

$$[X_1, X_2] = X_3, \; [X_2, X_3] = X_1, \; [X_3, X_1] = X_2.$$

Thus, the ODE (4.22) becomes

$$\begin{cases} y_1^{(3)} = y_2\, y_3'' - y_3\, y_2'', \\ y_2^{(3)} = y_3\, y_1'' - y_1\, y_3'', \\ y_3^{(3)} = y_1\, y_2'' - y_2\, y_1'', \end{cases} \qquad (4.24)$$

which is equivalent to

$$\mathbf{y}^{(3)} = \mathbf{y} \times \mathbf{y}'', \qquad (4.25)$$

where $\mathbf{y} := {}^t(y_1, y_2, y_3) \in \mathbb{R}^3$, and $\mathbf{a} \times \mathbf{b}$ stands for the vector cross product in $\mathbb{R}^3$. Notice here that $\mathfrak{g}$ is non-abelian, but our equation (4.22) turns to the vector equation (4.26) depending on the time t of the Euclidean space $\mathbb{R}^3$ by identifying $\mathfrak{g} \ni \sum_{i=1}^3 y_i\, X_i \mapsto (y_1, y_2, y_3) \in \mathbb{R}^3$.

Then, the ODE (4.25) can be solved as follows:

Let $\mathbf{x}(s) = {}^t(x_1(s), x_2(s), x_3(s))$ be a C^∞ curve in $\mathbb{R}^3$ with arc length parameter s, and then

$$\mathbf{y}(s) = \mathbf{x}'(s) = \mathbf{e}_1(s).$$

Let $\{\mathbf{e}_1(s), \mathbf{e}_2(s), \mathbf{e}_3(s))\}$ be the Frenet frame field along $\mathbf{x}(s)$. Recall the Frenet-Serret formula:

$$\begin{cases} \mathbf{e}_1' = \qquad\quad \kappa\, \mathbf{e}_2 \\ \mathbf{e}_2' = -\kappa\, \mathbf{e}_1 \qquad\quad + \tau\, \mathbf{e}_3 \\ \mathbf{e}_3' = \qquad\quad -\tau\, \mathbf{e}_2 \end{cases}$$

where κ and τ are the curvature and torsion of $\mathbf{x}(s)$, respectively. Then, we have

$$\begin{cases} \mathbf{y}' = \kappa\, \mathbf{e}_2 \\ \mathbf{y}'' = -\kappa^2\, \mathbf{e}_1 + \kappa'\, \mathbf{e}_2 + \kappa\tau\, \mathbf{e}_3 \\ \mathbf{y}''' = -3\kappa\kappa'\, \mathbf{e}_1 + (\kappa'' - \kappa^3 - \kappa\tau^2)\, \mathbf{e}_2 + (2\kappa'\tau + \kappa\tau')\, \mathbf{e}_3. \end{cases} \qquad (4.26)$$

Thus, (4.24) is equivalent to

$$-3\kappa\kappa'\, \mathbf{e}_1 + (\kappa'' - \kappa^3 - \kappa\tau^2)\, \mathbf{e}_2 + (2\kappa'\tau + \kappa\tau')\, \mathbf{e}_3$$
$$= \mathbf{e}_1 \times (-\kappa^2\, \mathbf{e}_1 + \kappa'\, \mathbf{e}_2 + \kappa\tau\, \mathbf{e}_3)$$
$$= -\kappa\tau\, \mathbf{e}_2 + \kappa'\, \mathbf{e}_3 \qquad (4.27)$$

which is equivalent to

$$\begin{cases} -3\kappa\kappa' = 0 \\ \kappa'' - \kappa^3 - \kappa\tau^2 = -\kappa\tau \\ 2\kappa'\tau + \kappa\tau' = \kappa'. \end{cases} \qquad (4.28)$$

Then, the first equation of (4.28) turns out that $(\kappa^2)' = 0$, that is, κ^2 is constant, i.e., $\kappa \equiv 0$, or $\kappa \equiv \kappa_0 \neq 0$. In the case that $\kappa \equiv 0$, the solution of (4.28), $\mathbf{x}(s)$, is a line in $\mathbb{R}^3$.

For the case that $\kappa \equiv \kappa_0 \neq 0$, the only solution of (4.24) is

$$\begin{cases} \kappa \equiv \kappa_0 \neq 0, \\ \tau \equiv \tau_0, \text{ and} \\ \kappa_0{}^2 = \tau_0(1 - \tau_0), \end{cases} \tag{4.29}$$

and the unique solution of (4.25) is given by

$$\mathbf{x}(s) = \begin{pmatrix} x_1(s) \\ x_2(s) \\ x_3(s) \end{pmatrix} = \begin{pmatrix} a \cos \frac{s}{\sqrt{a^2+1}} + b \\ a \sin \frac{s}{a^2+1} + b \\ \frac{s}{\sqrt{a^2+1}} + b \end{pmatrix} \tag{4.30}$$

for some positive constant $a > 0$ and some constant b. Thus, $F(s)$ is given as follows:

$$F(s) = \mathbf{x}'(s) = \sum_{i=1}^{3} x_i{}'(s)\, X_i$$

$$= \left(-\frac{a}{\sqrt{a^2+1}} \sin \frac{s}{\sqrt{a^2+1}} \right) X_1 + \left(\frac{a}{\sqrt{a^2+1}} \cos \frac{s}{\sqrt{a^2+1}} \right) X_2$$

$$+ \left(\frac{1}{\sqrt{a^2+1}} \right) X_3, \tag{4.31}$$

for any constant $a > 0$. Conversely, it is easy to see that every such $F(s)$ in (4.31) is a solution of (4.22): $F^{(3)} = [F(s), F''(s)]$.

Remark   It is still difficult to determine $X(t)$ to satisfy (4.21):

$$F(t) = \sum_{n=0}^{\infty} \frac{(-\operatorname{ad} X(t))^n}{(n+1)!} (X'(t)),$$

in the case of $\mathfrak{su}(2)$.

5. Biharmonic maps from an open domain in $\mathbb{R}^2$

In this section, we consider a biharmonic map $\psi : (\mathbb{R}^2, g) \supset \Omega \to (G, h)$. Here, we assume that G is a linear compact Lie group, i.e., G is a subgroup of the unitary group $U(N)(\subset GL(N, \mathbb{C}))$ of degree N with a bi-invariant Riemannian metric h on G. Let $\mathfrak{g}$ be the Lie algebra of G which is a Lie subalgebra of the Lie algebra $\mathfrak{u}(N)$ of $U(N)$. The Riemannian metric g on $\mathbb{R}^2$ is a conformal metric which is given by $g = \mu^2 g_0$ with a C^∞ positive function μ on Ω and $g_0 = dx \cdot dx + dy \cdot dy$, where (x, y) is the standard coordinate on $\mathbb{R}^2$.

Let $\psi : \Omega \ni (x,y) \mapsto \psi(x,y) = \big(\psi_{ij}(x,y)\big) \in U(N)$ a C^∞ map. Let us consider

$$\frac{\partial \psi}{\partial x} := \left(\frac{\partial \psi_{ij}}{\partial x}\right), \qquad \frac{\partial \psi}{\partial y} := \left(\frac{\partial \psi_{ij}}{\partial y}\right).$$

Then,

$$A_x := \psi^{-1} \frac{\partial \psi}{\partial x}, \qquad A_y := \psi^{-1} \frac{\partial \psi}{\partial y} \tag{5.1}$$

are $\mathfrak{g}$-valued C^∞ functions on Ω. It is known that, for two given $\mathfrak{g}$-valued 1-forms A_x and A_y on Ω, there exists a C^∞ mapping $\psi : \Omega \to G$ satisfying the equations (5.1) if the *integrability condition* holds:

$$\frac{\partial A_y}{\partial x} - \frac{\partial A_x}{\partial y} + [A_x, A_y] = 0. \tag{5.2}$$

The pull back of the Maurer-Cartan form θ by ψ is given by

$$\begin{aligned}
\alpha := \psi^* \theta = \psi^{-1} d\psi &= \psi^{-1} \frac{\partial \psi}{\partial x} \, dx + \psi^{-1} \frac{\partial \psi}{\partial y} \, dy \\
&= A_x \, dx + A_y \, dy,
\end{aligned} \tag{5.3}$$

which is a $\mathfrak{g}$-valued 1-form on Ω.

Recall that the codifferential $\delta\alpha$ of a $\mathfrak{g}$-valued 1-form $\alpha = A_x \, dx + A_y \, dy$, where $A_x = \psi^{-1} \frac{\partial \psi}{\partial x}$ and $A_y = \psi^{-1} \frac{\partial \psi}{\partial y}$, is given by

$$\delta\alpha = -\mu^{-2} \left\{ \frac{\partial}{\partial x} A_x + \frac{\partial}{\partial y} A_y \right\}. \tag{5.4}$$

Then, we have the following well known facts:

LEMMA 5.1. *We have*

$$\delta\alpha = -\mu^{-2} \left\{ \frac{\partial}{\partial x} \left(\psi^{-1} \frac{\partial \psi}{\partial x}\right) + \frac{\partial}{\partial y} \left(\psi^{-1} \frac{\partial \psi}{\partial y}\right) \right\} \tag{5.5}$$

$$= -\mu^{-2} \left\{ \frac{\partial A_x}{\partial x} + \frac{\partial A_y}{\partial y} \right\}. \tag{5.6}$$

Therefore, the following three statements are equivalent:

$$(i) \qquad \psi : (\Omega, g) \to (G, h) \text{ is harmonic,}$$

$$(ii) \qquad \delta\alpha = 0, \tag{5.7}$$

$$(iii) \qquad \frac{\partial A_x}{\partial x} + \frac{\partial A_y}{\partial y} = 0. \tag{5.8}$$

Next, calculate the Laplacian Δ_g of $(\mathbb{R}^2, g)$ for $g = \mu^2 \, g_0$. We obtain

$$\Delta_g = -\sum_{i,j=1}^{2} g^{ij} \left(\frac{\partial^2}{\partial x^i \, \partial x^j} - \sum_{k=1}^{2} \Gamma_{ij}^k \frac{\partial}{\partial x^k} \right)$$

$$= -\mu^{-2} \left(\frac{\partial^2}{\partial x^2} + \frac{\partial^2}{\partial y^2} \right). \tag{5.9}$$

Thus we have

$$\delta d\delta\alpha = \Delta_g(\delta\alpha)$$

$$= \mu^{-2} \left(\frac{\partial^2}{\partial x^2} + \frac{\partial^2}{\partial y^2} \right) \left[\mu^{-2} \left\{ \frac{\partial}{\partial x} \left(\psi^{-1} \frac{\partial \psi}{\partial x} \right) + \frac{\partial}{\partial y} \left(\psi^{-1} \frac{\partial \psi}{\partial y} \right) \right\} \right]$$

$$= \mu^{-2} \left(\frac{\partial^2}{\partial x^2} + \frac{\partial^2}{\partial y^2} \right) \left[\mu^{-2} \left\{ \frac{\partial A_x}{\partial x} + \frac{\partial A_y}{\partial y} \right\} \right]$$

$$= -\mu^{-2} \left(\frac{\partial^2}{\partial x^2} + \frac{\partial^2}{\partial y^2} \right) (\delta\alpha). \tag{5.10}$$

On the other hand, by taking an orthonormal local frame field $\{e_1, e_2\}$ of $(\mathbb{R}^2, g)$, as $e_1 = \mu^{-1} \frac{\partial}{\partial x}$, $e_2 = \mu^{-1} \frac{\partial}{\partial y}$, we have

$$\mathrm{Trace}_g([\alpha, d\delta\alpha]) = [\alpha(e_1), d\delta\alpha(e_1)] + [\alpha(e_2), d\delta\alpha(e_2)]$$

$$= -\mu^{-2} \left[A_x, \frac{\partial}{\partial x} \left(\mu^{-2} \left\{ \frac{\partial A_x}{\partial x} + \frac{\partial A_y}{\partial y} \right\} \right) \right]$$

$$- \mu^{-2} \left[A_y, \frac{\partial}{\partial y} \left(\mu^{-2} \left\{ \frac{\partial A_x}{\partial x} + \frac{\partial A_y}{\partial y} \right\} \right) \right]$$

$$= \mu^{-2} [A_x, \frac{\partial}{\partial x} (\delta\alpha)] + \mu^{-2} [A_y, \frac{\partial}{\partial y} (\delta\alpha)]. \tag{5.11}$$

By (5.10) and (5.11), we obtain

$$\delta d\delta\alpha + \mathrm{Trace}_g([\alpha, d\delta\alpha])$$

$$= -\mu^{-2} \left(\frac{\partial^2}{\partial x^2} + \frac{\partial^2}{\partial y^2} \right) (\delta\alpha) + \mu^{-2} [A_x, \frac{\partial}{\partial x} (\delta\alpha)] + \mu^{-2} [A_y, \frac{\partial}{\partial y} (\delta\alpha)]$$

$$= -\mu^{-2} \left\{ \left(\frac{\partial^2}{\partial x^2} + \frac{\partial^2}{\partial y^2} \right) (\delta\alpha) - \frac{\partial}{\partial x} [A_x, \delta\alpha] - \frac{\partial}{\partial y} [A_y, \delta\alpha] \right\}, \tag{5.12}$$

where in the last equation in (5.11), we only notice that

$$\frac{\partial}{\partial x}[A_x, \delta\alpha] + \frac{\partial}{\partial y}[A_y, \delta\alpha]$$

$$= \left[\frac{\partial}{\partial x}A_x, \delta\alpha\right] + \left[A_x, \frac{\partial}{\partial x}(\delta\alpha)\right] + \left[\frac{\partial}{\partial y}A_y, \delta\alpha\right] + \left[A_y, \frac{\partial}{\partial y}(\delta\alpha)\right]$$

$$= \left[\frac{\partial}{\partial x}A_x + \frac{\partial}{\partial y}A_y, \delta\alpha\right] + \left[A_x, \frac{\partial}{\partial x}(\delta\alpha)\right] + \left[A_y, \frac{\partial}{\partial y}(\delta\alpha)\right]$$

$$= [-\mu^{-2}\delta\alpha, \delta\alpha] + \left[A_x, \frac{\partial}{\partial x}(\delta\alpha)\right] + \left[A_y, \frac{\partial}{\partial y}(\delta\alpha)\right]$$

$$= \left[A_x, \frac{\partial}{\partial x}(\delta\alpha)\right] + \left[A_y, \frac{\partial}{\partial y}(\delta\alpha)\right].$$

Thus, we have

THEOREM 5.1. *Let Ω be an open subset of $\mathbb{R}^2$, $g = \mu^2 g_0$, a Riemannian metric conformal to the standard metric g_0 on Ω with a C^∞ positive function μ on Ω, and $\psi : \Omega \to G$, a C^∞ map of Ω into a compact linear Lie group (G, h) with bi-invariant Riemannian metric h. Then,*

(1) *The 1-form α satisfies $d\alpha + \frac{1}{2}[\alpha \wedge \alpha] = 0$ which is equivalent to*

$$\frac{\partial A_y}{\partial x} - \frac{\partial A_x}{\partial y} + [A_x, A_y] = 0. \tag{5.13}$$

(2) *The following three are equivalent:*

(i) $\psi : (\Omega, g) \to (G, h)$ *is harmonic,*

(ii) $\delta\alpha = 0,$ \hfill (5.14)

(iii) $\dfrac{\partial}{\partial x}A_x + \dfrac{\partial}{\partial y}A_y = 0.$ \hfill (5.15)

(3) *The following three are equivalent:*

(i) $\psi : (\Omega, g) \to (G, h)$ *is biharmonic,*

(ii) $\delta d\delta\alpha + Trace_g([\alpha, d\delta\alpha]) = 0,$ \hfill (5.16)

(iii) $\left(\dfrac{\partial^2}{\partial x^2} + \dfrac{\partial^2}{\partial y^2}\right)(\delta\alpha) - \dfrac{\partial}{\partial x}[A_x, \delta\alpha] - \dfrac{\partial}{\partial y}[A_y, \delta\alpha] = 0.$
\hfill (5.17)

(4) *Let us consider two $\mathfrak{g}$-valued 1-forms β and Θ on Ω, defined by*

$$\beta := [A_x, \delta\alpha]\,dx + [A_y, \delta\alpha]\,dy, \tag{5.18}$$

$$\Theta := d\delta\alpha - \beta, \tag{5.19}$$

respectively. Then, $\psi : (\Omega, g) \to (G, h)$ is biharmonic if and only if

$$\delta\Theta = 0. \tag{5.20}$$

Proof   (1) is clear. We see already (2) and (3). For (4), we only have to see that (5.17) is equivalent to

$$0 = -\Delta_g(\delta\alpha) + \delta\beta = -\delta(d\delta\alpha - \beta) = -\delta\Theta \tag{5.21}$$

where

$$
\begin{aligned}
\Theta &:= d\delta\alpha - \beta \\
&= \frac{\partial}{\partial x}(\delta\alpha)\, dx + \frac{\partial}{\partial y}(\delta\alpha)\, dy - [A_x, \delta\alpha]\, dx - [A_y, \delta\alpha]\, dy \\
&= \left\{ \frac{\partial}{\partial x}(\delta\alpha) - [A_x, \delta\alpha] \right\} dx + \left\{ \frac{\partial}{\partial y}(\delta\alpha) - [A_y, \delta\alpha] \right\} dy.
\end{aligned}
\tag{5.22}
$$

$\square$

6. Complexification of the biharmonic map equation

We use the complex coordinate $z = x + iy$ ($i = \sqrt{-1}$) in Ω, and we put $A_z = \frac{1}{2}(A_x - i A_y)$ and $A_{\bar z} = \frac{1}{2}(A_x + i A_y)$ which are $\mathfrak{g}^{\mathbb{C}}$-valued functions with $A_{\bar z} = \overline{A_z}$. Then, it is well known that

$$
\begin{aligned}
&\frac{\partial}{\partial \bar z}A_z + \frac{\partial}{\partial z}A_{\bar z} = \frac{1}{2}\left\{ \frac{\partial}{\partial x}A_x + \frac{\partial}{\partial y}A_y \right\}, \\
&\frac{\partial}{\partial z}A_{\bar z} - \frac{\partial}{\partial \bar z}A_z + [A_z, A_{\bar z}] = \frac{i}{2}\left\{ \frac{\partial}{\partial x}A_y - \frac{\partial}{\partial y}A_x + [A_x, A_y] \right\},
\end{aligned}
\tag{6.1}
$$

and also

$$
\begin{aligned}
&\alpha = A_x\, dx + A_y\, dy = A_z\, dz + A_{\bar z}\, d\bar z, \\
&\frac{\partial^2}{\partial x^2} + \frac{\partial^2}{\partial y^2} = 4\frac{\partial^2}{\partial z \partial \bar z}, \\
&\delta\alpha = -\mu^{-2}\left(\frac{\partial}{\partial x}A_x + \frac{\partial}{\partial y}A_y \right) = -2\mu^{-2}\left(\frac{\partial}{\partial \bar z}A_z + \frac{\partial}{\partial z}A_{\bar z} \right).
\end{aligned}
\tag{6.2}
$$

Then, the condition (5.20) is equivalent to

$$\delta\widetilde{\Theta} = 0, \tag{6.3}$$

where

$$\tilde{\Theta} := \left\{ \frac{\partial}{\partial z}(\delta\alpha) - [A_z, \delta\alpha] \right\} dz + \left\{ \frac{\partial}{\partial \bar{z}}(\delta\alpha) - [A_{\bar{z}}, \delta\alpha] \right\} d\bar{z}. \tag{6.4}$$

The integrability condition (5.13) is equivalent to

$$\frac{\partial}{\partial z} A_{\bar{z}} - \frac{\partial}{\partial \bar{z}} A_z + [A_z, A_{\bar{z}}] = 0 \tag{6.5}$$

7. Determination of biharmonic maps

In this section, we want to show how to determine all the biharmonic maps of (Ω, g) into a compact Lie group (G, h) where $g = \mu^2 g_0$ with a positive C^∞ function on Ω and h is a bi-invariant Riemannian metric on G. Our method to obtain all the biharmonic maps can be divided into three steps:

(*The first step*) We first solve the equation:

$$\frac{\partial}{\partial \bar{z}} B_z + \frac{\partial}{\partial z} B_{\bar{z}} = 0 \tag{7.1}$$

Notice that, if these B_z and $B_{\bar{z}}$ satisfy furthermore, the integrability condition

$$\frac{\partial}{\partial z} B_{\bar{z}} - \frac{\partial}{\partial \bar{z}} B_z + [B_z, B_{\bar{z}}] = 0, \tag{7.2}$$

then, there exists a harmonic map $\Psi : (\Omega, g) \to (G, h)$ such that

$$\begin{cases} \Phi^{-1} \dfrac{\partial \Psi}{\partial z} = B_z, \\[2mm] \Phi^{-1} \dfrac{\partial \Phi}{\partial \bar{z}} = B_{\bar{z}}, \end{cases} \tag{7.3}$$

and the converse is true.

(*The second step*) For such two $\mathfrak{g}^{\mathbb{C}}$-valued functions B_z and $B_{\bar{z}}$ on Ω satisfying (7.1) not necessarily satisfying (7.2), we should detect two $\mathfrak{g}^{\mathbb{C}}$-valued functions A_z and $A_{\bar{z}}$ on Ω satisfying that

$$\begin{cases} \dfrac{\partial}{\partial z}\left(-2\mu^{-2}\left(\dfrac{\partial A_z}{\partial \bar{z}} + \dfrac{\partial A_{\bar{z}}}{\partial z} \right) \right) - \left[A_z, -2\mu^{-2}\left(\dfrac{\partial A_z}{\partial \bar{z}} + \dfrac{\partial A_{\bar{z}}}{\partial z} \right) \right] = B_z, \\[4mm] \dfrac{\partial}{\partial \bar{z}}\left(-2\mu^{-2}\left(\dfrac{\partial A_z}{\partial \bar{z}} + \dfrac{\partial A_{\bar{z}}}{\partial z} \right) \right) - \left[A_{\bar{z}}, -2\mu^{-2}\left(\dfrac{\partial A_z}{\partial \bar{z}} + \dfrac{\partial A_{\bar{z}}}{\partial z} \right) \right] = B_{\bar{z}}, \\[4mm] \dfrac{\partial}{\partial z} A_{\bar{z}} - \dfrac{\partial}{\partial \bar{z}} A_z + [A_z, A_{\bar{z}}] = 0. \end{cases}$$

$$\tag{7.4}$$

(*The third step*) Finally, for the above $\mathfrak{g}^{\mathbb{C}}$-valued functions A_z and $A_{\bar{z}}$ on Ω satisfying (7.4) and $a \in G$, there exists a C^{∞} mapping $\psi : \Omega \to G$ satisfying that

$$
\begin{cases}
\psi(x_0, y_0) = a, \\
\psi^{-1}\dfrac{\partial \psi}{\partial z} = A_z, \\
\psi^{-1}\dfrac{\partial \psi}{\partial \bar{z}} = A_{\bar{z}}.
\end{cases}
\tag{7.5}
$$

Then, $\psi : (\Omega, g) \to (G, h)$ is a *biharmonic map* due to (5.20), (6.1) and (7.4), and conversely, every biharmonic map $\psi : (\Omega, g) \to (G, h)$ could be obtained in this way. To do the these procedures rigorously, let us define

DEFINITION 7.1. (1) *Let us define the four sets* Λ, Λ_1, Λ_2, *and* Λ_0:
• *Let* Λ *be the set of all* $\mathfrak{g}$-*valued two functions* (A_x, A_y) *on* Ω, *(or all* $\mathfrak{g}^{\mathbb{C}}$-*valued two functions* $(A_z, A_{\bar{z}})$ *on* Ω *with* $A_{\bar{z}} = \overline{A_z}$,
• *let* Λ_1, *the set of* $(A_x, A_y) \in \Lambda$ *which satisfy the harmonic map equation (5.12) (or (7.1))*,
• *let* Λ_2, *the set of* $(A_x, A_y) \in \Lambda$ *which satisfy the biharmonic map equation (5.17) (or (6.1))*, *and*
• *let* Λ_0, *the set of* $(A_x, A_y) \in \Lambda$ *which satisfy the integrability condition (5.13), (or (6.3)), respectively.*

(2) *Let us define two sets* Ξ *and* Ξ_1:
• *Let* Ξ *be the set of all* $\mathfrak{g}$-*valued two real analytic functions* (B_x, B_y) *on* Ω *(or* $\mathfrak{g}^{\mathbb{C}}$-*valued two real analytic functions* $(B_z, B_{\bar{z}})$ *on* Ω *with* $B_{\bar{z}} = \overline{B_z}$), *and*
• *let* Ξ_1, *the set of all* $(B_x, B_y) = (B_z, B_{\bar{z}}) \in \Xi$ *satisfying the harmonic map equation (7.1), respectively.*

DEFINITION 7.2. *Let us define two* C^{∞} *mappings* Φ_i $(i = 1, 2)$ *of* Λ *into* Ξ *by*

$$
\begin{aligned}
\Phi_1(A_x, A_y) &:= \left(\frac{\partial}{\partial x}\left(-\mu^{-2}\left(\frac{\partial A_x}{\partial x} + \frac{\partial A_y}{\partial y} \right) \right) - \left[A_x, -\mu^{-2}\left(\frac{\partial A_x}{\partial x} + \frac{\partial A_y}{\partial y} \right) \right], \right. \\
&\left. \frac{\partial}{\partial y}\left(-\mu^{-2}\left(\frac{\partial A_x}{\partial x} + \frac{\partial A_y}{\partial y} \right) \right) - \left[A_y, -\mu^{-2}\left(\frac{\partial A_x}{\partial x} + \frac{\partial A_y}{\partial y} \right) \right] \right),
\end{aligned}
\tag{7.6}
$$

and also

$$\Phi_2(A_x, A_y) := \left(-\mu^{-2}\left(\frac{\partial^2 A_x}{\partial x^2} + \frac{\partial^2 A_x}{\partial y^2} - \frac{\partial}{\partial y}[A_x, A_y] \right) \right.$$

$$-\frac{\partial \mu^{-2}}{\partial x}\left(\frac{\partial A_x}{\partial x} + \frac{\partial A_y}{\partial y} \right) - \left[A_x, -\mu^{-2}\left(\frac{\partial A_x}{\partial x} + \frac{\partial A_y}{\partial y} \right) \right],$$

$$-\mu^{-2}\left(\frac{\partial^2 A_y}{\partial x^2} + \frac{\partial^2 A_y}{\partial y^2} - \frac{\partial}{\partial x}[A_x, A_y] \right)$$

$$\left. -\frac{\partial \mu^{-2}}{\partial y}\left(\frac{\partial A_x}{\partial x} + \frac{\partial A_y}{\partial y} \right) - \left[A_y, -\mu^{-2}\left(\frac{\partial A_x}{\partial x} + \frac{\partial A_y}{\partial y} \right) \right] \right),$$

(7.7)

respectively.

Then, we obtain

THEOREM 7.1. *Assume that Ω be a simply connected open domain in $\mathbb{R}^2$, and μ is a positive real analytic function on Ω. Then, we have:*

(1) *For every $(B_x, B_y) = (B_z, B_{\bar{z}}) \in \Xi$ there exists $(A_x, A_y) = (A_z, A_{\bar{z}}) \in \Lambda$ such that $\Phi_2(A_x, A_y) = (B_x, B_y)$ (or $\Phi_2(A_z, A_{\bar{z}}) = (B_z, B_{\bar{z}})$). The solution $(A_x, A_y) = (A_z, A_{\bar{z}})$ is uniquely determined by the initial data $A_x(x_0, y)$, $A_y(x_0, y)$, $\frac{\partial A_x}{\partial x}(x_0, y)$ and $\frac{\partial A_y}{\partial x}(x_0, y)$, $(x_0, y) \in \Omega$.*

(2) $\Phi_1 = \Phi_2$ *on Λ_0,*

(3) $\Phi_1^{-1}(\Xi_1) = \Lambda_2$, *and* $\Phi_1(\Lambda_2 \cap \Lambda_0) = \Phi_2(\Lambda_2 \cap \Lambda_0) = \Xi_1$.

Proof.   For (1), by definition of Φ_2, that $\Phi_2(A_x, A_y) = (B_x, B_y)$ is equivalent to the following two equations:

$$\frac{\partial^2 A_x}{\partial x^2} = -\frac{\partial^2 A_x}{\partial y^2} + \frac{\partial}{\partial y}[A_x, A_y]$$

$$- \mu^2 \frac{\partial \mu^{-2}}{\partial x}\left(\frac{\partial A_x}{\partial x} + \frac{\partial A_y}{\partial y} \right) - \mu^2\left[A_x, -\mu^{-2}\left(\frac{\partial A_x}{\partial x} + \frac{\partial A_y}{\partial y} \right) \right]$$

$$- \mu^2 B_x,$$

(7.8)

and also

$$\frac{\partial^2 A_y}{\partial x^2} = -\frac{\partial^2 A_y}{\partial y^2} + \frac{\partial}{\partial x}[A_x, A_y]$$

$$- \mu^2 \frac{\partial \mu^{-2}}{\partial y}\left(\frac{\partial A_x}{\partial x} + \frac{\partial A_y}{\partial y} \right) - \mu^2\left[A_y, -\mu^{-2}\left(\frac{\partial A_x}{\partial x} + \frac{\partial A_y}{\partial y} \right) \right]$$

$$- \mu^2 B_y.$$

(7.9)

Notice that the system of (7.8) and (7.9) satisfies all the conditions of the theorem of Cauchy-Kovalevskaya when $n_i = 2$ $(i = 1, 2)$ (cf. [46], p. 1305, 429 B; [101], p. 224; [65], p. 181)

THEOREM 7.2. *(Cauchy-Kovalevskaya) Let us consider the following Cauchy problem of unknown N functions $u_i(t, x)$ $(i = 1, \cdots, N)$ in t and $x = (x_1, \cdots, x_m)$,*

$$\begin{cases} \dfrac{\partial^{n_i} u_i}{\partial t^{n_i}} = F_i(t, x, D_t^k D_x^p u_j) & (i = 1, \cdots, N), \\[2mm] \dfrac{\partial^k u_i}{\partial t^k}(t_0, x) = \varphi_i^k(x) & (0 \le k \le n_i - 1; \ i = 1, \cdots, N), \end{cases} \quad (7.10)$$

where, for $p=(p_1, \cdots, p_m)$, $|p|=p_1 + \cdots + p_m$, $D_t^k D_x^p := \dfrac{\partial^k}{\partial t^k} \dfrac{\partial^{|p|}}{\partial x_1{}^{p_1} \cdots \partial x_m{}^{p_m}}$ and in the right hand side of the first equation of (7.10), k and p satisfy

$$k < n_j \quad \text{and} \quad k + |p| \le n_j \quad (j = 1, \cdots, N).$$

Assume that each F_i and φ_i^k are real analytic functions. Then, there exists a real analytic solution u_i $(i = 1, \cdots, N)$ of (7.10) and it is unique in the class of real analytic functions.

Then, for each $(B_x, B_y) \in \Xi$, there exists a real analytic solution (A_x, A_y) of the Cauchy problem (7.8) and (7.9) with the initial condition:

$$\begin{cases} \left(\dfrac{\partial A_x}{\partial x}\right)(x_0, y) = f_1(y), & A_x(x_0, y) = f_0(y), \\[3mm] \left(\dfrac{\partial A_y}{\partial x}\right)(x_0, y) = g_1(y), & A_y(x_0, y) = g_0(y), \end{cases} \quad (7.11)$$

and the real analytic solution (A_x, A_y) is unique for real analytic functions f_i and g_i $(i = 0, 1)$. By taking this process at each point (x_0, y_0) in Ω, we have a real analytic solution (A_x, A_y) of (7.8) and (7.9) in an open neighborhood of (x_0, y_0). Then, by the uniqueness theorem of the continuation of a real analytic function on a simply connected domain Ω, we have a solution (A_x, A_y) of (7.8) and (7.9) on Ω. We have (1).

For (2), we have to see $\Phi_1(A_x, A_y) = \Phi_2(A_x, A_y)$ for every $(A_x, A_y) \in \Lambda_0$, which follows from that

$$\frac{\partial}{\partial x}\left(\mu^{-2}\left(\frac{\partial A_x}{\partial x} + \frac{\partial A_y}{\partial y}\right)\right) = \mu^{-2}\left(\frac{\partial^2 A_x}{\partial x^2} + \frac{\partial^2 A_y}{\partial x \partial y}\right)$$

$$+ \frac{\partial \mu^{-2}}{\partial x}\left(\frac{\partial A_x}{\partial x} + \frac{\partial A_y}{\partial y}\right)$$

$$= \mu^{-2}\left(\frac{\partial^2 A_x}{\partial x^2} + \frac{\partial^2 A_x}{\partial y^2} - \frac{\partial}{\partial y}[A_x, A_y]\right)$$

$$+ \frac{\partial \mu^{-2}}{\partial x}\left(\frac{\partial A_x}{\partial x} + \frac{\partial A_y}{\partial y}\right), \qquad (7.12)$$

because of (5.13) and it is a similar for $\frac{\partial}{\partial y}\left(\mu^{-2}\left(\frac{\partial A_x}{\partial x} + \frac{\partial A_y}{\partial y}\right)\right)$, so that we have (2).

For (3), due to (2), we only have to see $\Phi_1^{-1}(\Xi_1) = \Lambda_2$ which is equivalent to that:

> for all $(B_x, B_y) \in \Xi$, exists a unique $(A_x, A_y) \in \Lambda_2$ such that $\Phi_1(A_x, A_y) = (B_x, B_y)$, and vice versa.

But, that $(B_x, B_y) = (B_z, B_{\bar{z}}) \in \Xi_1$ means that it satisfies the harmonic map equation (7.1). On the other hand, $\Phi_1(A_x, A_y) = (B_x, B_y)$ means that $\Phi_1(A_z, A_{\bar{z}}) = (B_z, B_{\bar{z}})$ which is equivalent to that the first two equations of (7.4) hold by definition of Φ_1, and notice here that $\Phi_1(A_x, A_y) = (B_x, B_y)$ is equivalent to the two following equations

$$\frac{\partial}{\partial x}\left(-\mu^{-2}\left(\frac{\partial A_x}{\partial x} + \frac{\partial A_y}{\partial y}\right)\right) - \left[A_x, -\mu^{-2}\left(\frac{\partial A_x}{\partial x} + \frac{\partial A_y}{\partial y}\right)\right] = B_x, \qquad (7.13)$$

$$\frac{\partial}{\partial y}\left(-\mu^{-2}\left(\frac{\partial A_x}{\partial x} + \frac{\partial A_y}{\partial y}\right)\right) - \left[A_y, -\mu^{-2}\left(\frac{\partial A_x}{\partial x} + \frac{\partial A_y}{\partial y}\right)\right] = B_y, \qquad (7.14)$$

which are also equivalent to

$$\frac{\partial}{\partial z}\left(-2\mu^{-2}\left(\frac{\partial A_z}{\partial \bar{z}} + \frac{\partial A_{\bar{z}}}{\partial z}\right)\right) - \left[A_z, -2\mu^{-2}\left(\frac{\partial A_z}{\partial \bar{z}} + \frac{\partial A_{\bar{z}}}{\partial z}\right)\right] = B_z, \qquad (7.15)$$

$$\frac{\partial}{\partial \bar{z}}\left(-2\mu^{-2}\left(\frac{\partial A_z}{\partial \bar{z}} + \frac{\partial A_{\bar{z}}}{\partial z}\right)\right) - \left[A_{\bar{z}}, -2\mu^{-2}\left(\frac{\partial A_z}{\partial \bar{z}} + \frac{\partial A_{\bar{z}}}{\partial z}\right)\right] = B_{\bar{z}}. \qquad (7.16)$$

But, by inserting both (7.14) and (7.15) into

$$\frac{\partial}{\partial \bar{z}}B_z + \frac{\partial}{\partial z}B_{\bar{z}} = 0, \qquad (7.17)$$

we obtain

$$
\frac{\partial^2}{\partial \bar{z} \partial z}\left(-2\mu^{-2}\left(\frac{\partial A_z}{\partial \bar{z}} + \frac{\partial A_{\bar{z}}}{\partial z}\right)\right) - \frac{\partial}{\partial \bar{z}}\left[A_{\bar{z}}, -2\mu^{-2}\left(\frac{\partial A_z}{\partial \bar{z}} + \frac{\partial A_{\bar{z}}}{\partial z}\right)\right]
$$
$$
+ \frac{\partial^2}{\partial z \partial \bar{z}}\left(-2\mu^{-2}\left(\frac{\partial A_z}{\partial \bar{z}} + \frac{\partial A_{\bar{z}}}{\partial z}\right)\right) - \frac{\partial}{\partial z}\left[A_{\bar{z}}, -2\mu^{-2}\left(\frac{\partial A_z}{\partial \bar{z}} + \frac{\partial A_{\bar{z}}}{\partial z}\right)\right]
$$
$$
= 0, \tag{7.18}
$$

which is just the biharmonic map equation for $(A_z, A_{\bar{z}})$: (6.1) $\delta\tilde{\Theta} - 0$. By the same way, one can see also immediately (A_x, A_y) satisfies the biharmonic map equation (5.20) if (B_x, B_y) satisfies the harmonic map equation (5.15) by using Theorem 5.2, (5.6) and (5.22). Thus, we obtain $\Phi_1^{-1}(\Xi_1) = \Lambda_2$ and (3). $\square$

Remark   The solution (A_x, A_y) in (1) of Theorem 7.3 can be chosen in such a way that they satisfy the integrability condition (5.13) at the initial value (x_0, y),

$$
\frac{\partial A_y}{\partial x}(x_0, y) - \frac{\partial A_x}{\partial y}(x_0, y) + [A_x(x_0, y), A_y(x_0, y)] = 0, \tag{7.19}
$$

for each y, i.e., the initial functions f_0, f_1 and g_1 may be chosen to satisfy that

$$
\frac{\partial A_x}{\partial y}(x_0, y) = g_1(y) + [f_0(y), f_1(y)]. \tag{7.20}
$$

Finally, we introduce a loop group formulation for biharmonic maps. We *first*, consider a $\mathfrak{g}^{\mathbb{C}}$-valued 1-forms

$$
\beta_\nu = \frac{1}{2}(1-\nu)\, B_z\, dz + \frac{1}{2}(1-\nu^{-1})\, B_{\bar{z}}\, d\bar{z} \tag{7.21}
$$

for a parameter $\nu \in S^1$, which satisfy that

$$
d\beta_\nu + [\beta_\nu \wedge \beta_\nu] = 0 \qquad (\forall\, \nu \in S^1), \tag{7.22}
$$

where for the definition of $[\beta_\nu \wedge \beta_\nu]$, see (3.13).
    Next, we consider $\mathfrak{g}^{\mathbb{C}}$-valued 1-forms

$$
\alpha_\nu = \frac{1}{2}(1-\nu)\, A_z\, dz + \frac{1}{2}(1-\nu^{-1})\, A_{\bar{z}}\, d\bar{z} \tag{7.23}
$$

which satisfy that

$$
\begin{cases}
\dfrac{\partial}{\partial z}(\delta\,\alpha_\nu) - \left[\dfrac{1}{2}\,(1-\nu)\,A_z, \delta\,\alpha_\nu\right] = B_z, \\[2mm]
\dfrac{\partial}{\partial\bar{z}}(\delta\,\alpha_\nu) - \left[\dfrac{1}{2}\,(1-\nu)\,A_{\bar{z}}, \delta\,\alpha_\nu\right] = B_{\bar{z}}, \\[2mm]
d\,\alpha_\nu + [\alpha_\nu \wedge \alpha_\nu] = 0,
\end{cases}
\tag{7.24}
$$

for each $\nu \in S^1$. Here, the co-differentiation $\delta\,\alpha_\nu$ of α_ν is given by

$$
\delta\,\alpha_\nu = -2\mu^{-2}\left(\frac{1}{2}\,(1-\nu)\,\frac{\partial}{\partial\bar{z}}A_z + \frac{1}{2}\,(1-\nu^{-1})\,\frac{\partial}{\partial z}A_{\bar{z}}\right).
\tag{7.25}
$$

Then, the mapping $\psi_\nu : \Omega \to G$ satisfying $\psi_\nu{}^*\theta = \alpha_\nu$ is a biharmonic map of (Ω, g) into (G, h) where $g = \mu^2\,g_0$ for a positive C^∞ function μ on Ω.

CHAPTER 8

Biharmonic Maps into Symmetric Spaces and Integrable Systems

1

ABSTRACT. We give the biharmonic map equations in terms of the Maurer-Cartan form for all smooth maps of a compact Riemannian manifold into a Riemannian symmetric space $(G/K, h)$ induced from the bi-invariant Riemannian metric h on G. By using this, we characterize exactly all the biharmonic curves into symmetric spaces are determined, and all the biharmonic maps of an open domain of $\mathbb{R}^2$ with the standard Riemannian metric into $(G/K, h)$.

1. Introduction and statement of results

This paper is a continuation of our previous one [148]. In our previous paper, we discussed the description of biharmonic maps into compact Lie groups in terms of the Maurer-Cartan form, and gave their exact constructions. In this paper, we consider biharmonic maps into Riemannian symmetric spaces.

The theory of harmonic maps into Lie groups, symmetric spaces or homogeneous spaces has been extensively studied in connection with the integrable systems by many authors (for instance, [13], [33], [36], [103], [104], [105], [121], [147], [163], [164]). Let us recall the loop group formulation of harmonic maps into symmetric spaces, briefly. Let φ be a smooth map of a Riemann surface M into a Riemannian symmetric space $(G/K, h)$ with a lift $\psi : M \to G$ so that $\pi \circ \psi = \varphi$. Let $\mathfrak{g} = \mathfrak{k} \oplus \mathfrak{m}$ be the corresponding Cartan decomposition of the Lie algebra $\mathfrak{g}$ of the Lie group G. Then, the pull back $\alpha = \psi^{-1} d\psi$ of the Maurer-Cartan form on G is decomposed as $\alpha = \alpha_{\mathfrak{k}} + \alpha_{\mathfrak{m}}$, correspondingly. Let us decompose $\alpha_{\mathfrak{m}}$ into the sum of the holomorphic part and the anti-holomorphic one: $\alpha_{\mathfrak{m}} = \alpha_{\mathfrak{m}}' + \alpha_{\mathfrak{m}}''$. Then, one can

---

[1] This chapter is due to [149]: H. Urakawa, *Biharmonic maps into compact Lie groups and symmetric spaces*, "Alexandru Myller" Mathematical Seminar, 246–263, AIP Conf. Proc. **1329**, Amer. Snst. Phys., Melville, NY, 2011, and also [153]: H. Urakawa, *Biharmonic maps into symmetric spaces and integrable systems*, Hokkaido Math. J., **43** (2014), 105–136.

obtain the extended solution $\widetilde{\psi}$ of M into a loop group ΛG satisfying $\widetilde{\psi}^{-1} d\widetilde{\psi} = \lambda \alpha_{\mathfrak{m}}' + \alpha_{\mathfrak{k}} + \lambda^{-1} \alpha_{\mathfrak{m}}''$ for all $\lambda \in U(1) = \{\lambda \in \mathbb{C} : |\lambda| = 1\}$. (cf. [36]). Then, $\varphi : M \to (G/K, h)$ is harmonic if and only if there exists a holomorphic and horizontal map $\widetilde{\psi}$ of M into the homogeneous $\Lambda G/K$ with $\widetilde{\psi}_1 = \psi$ (cf. [36], p. 648). Then, one can obtain a Weierstrass-type representation of harmonic maps (cf. [36], pp. 648–662).

On the other hand, the notion of harmonic map has been extended to the one of biharmonic map (cf. [41], [74]). In this paper, we will describe biharmonic maps into Riemannian symmetric spaces in terms of the pull back $\alpha = \alpha_{\mathfrak{k}} + \alpha_{\mathfrak{m}}$ of the Maurer-Cartan form (cf. Theorem 3.6), give some explicit solutions of the biharmonic map equation in Riemannian symmetric spaces, and construct several biharmonic maps into Riemannian symmetric spaces (Sections 4 and 5).

Acknowledgement: The author expresses his gratitude to Prof. J. Inoguchi and Prof. Y. Ohnita who gave many useful suggestions and discussions, and Prof. A. Kasue for his financial support during the preparation of this paper.

2. Preliminaries

In this section, we prepare general materials and facts on harmonic maps, biharmonic maps into Riemannian symmetric spaces (cf. [82]).

2.1.   Let (M, g) be an m-dimensional compact Riemannian manifold, and the target space (N, h), an n-dimensional Riemannian symmetric space $(G/K, h)$. Nemely, let $\mathfrak{g}$, $\mathfrak{k}$ be the Lie algebras of G, K, and $\mathfrak{g} = \mathfrak{k} \oplus \mathfrak{m}$ is the Cartan decomposition of $\mathfrak{g}$, and h, the G-invariant Riemannian metric on G/K corresponding to the $\mathrm{Ad}(K)$-invariant inner product $\langle \, , \, \rangle$ on $\mathfrak{m}$. Let k be a left invariant Riemannian metric on G such as the natural projection $\pi : G \to G/K$ is a Riemannian submersion of (G, k) onto $(G/K, h)$. For every C^∞ map φ of M into G/K, let us take its (local) *lift* $\psi : M \to G$ of φ, i.e., $\varphi = \pi \circ \psi$, $\varphi(x) = \psi(x) K \in G/K$ ($x \in U \subset M$), where U is an open subset of M.

The *energy functional* on the space $C^\infty(M, G/K)$ of all C^∞ maps of M into G/K is defined by

$$E(\varphi) = \frac{1}{2} \int_M |d\varphi|^2 \, v_g,$$

and for a C^∞ one parameter deformation $\varphi_t \in C^\infty(M, G/K)$ ($-\epsilon < t < \epsilon$) of φ with $\varphi_0 = \varphi$, the *first variation formula* is given by

$$\frac{d}{dt}\bigg|_{t=0} E(\varphi_t) = -\int_M \langle \tau(\varphi), V \rangle \, v_g,$$

where V is a variation vector field along φ defined by $V = \frac{d}{dt}\big|_{t=0}\varphi_t$ which belongs to the space $\Gamma(\varphi^{-1}T(G/K))$ of sections of the induced bundle of the tangent bundle $T(G/K)$ by φ. The *tension field* $\tau(\varphi)$ is defined by

$$\tau(\varphi) = \sum_{i=1}^m B(\varphi)(e_i, e_i), \tag{2.1}$$

where

$$B(\varphi)(X, Y) = \nabla^h_{d\varphi(X)} d\varphi(Y) - d\varphi(\nabla_X Y)$$

for $X, Y \in \mathfrak{X}(M)$. Here, ∇, and ∇^h, are the Levi-Civita connections of (M, g) and $(G/K, h)$, respectively. For a harmonic map $\varphi : (M, g) \to (G/K, h)$, the *second variation formula* of the energy functional $E(\varphi)$ is

$$\frac{d^2}{dt^2}\bigg|_{t=0} E(\varphi_t) = \int_M \langle J(V), V \rangle \, v_g$$

where

$$J(V) := \overline{\Delta} V - \mathcal{R}(V), \tag{2.2}$$

$$\overline{\Delta} V := \overline{\nabla}^* \overline{\nabla} V = -\sum_{i=1}^m \{\overline{\nabla}_{e_i}(\overline{\nabla}_{e_i} V) - \overline{\nabla}_{\nabla_{e_i} e_i} V\}, \tag{2.3}$$

$$\mathcal{R}(V) := \sum_{i=1}^m R^h(V, d\varphi(e_i)) d\varphi(e_i). \tag{2.4}$$

Here, $\overline{\nabla}$ is the induced connection on the induced bundle $\varphi^{-1}T(G/K)$, and is R^h is the curvature tensor of $(G/K, h)$ given by $R^h(U, V)W = [\nabla^h_U, \nabla^h_V]W - \nabla^h_{[U,V]}W$ ($U, V, W \in \mathfrak{X}(G/K)$).

The *bienergy functional* is defined by

$$E_2(\varphi) = \frac{1}{2}\int_M |(d + \delta)^2 \varphi|^2 \, v_g = \frac{1}{2}\int_M |\tau(\varphi)|^2 \, v_g, \tag{2.5}$$

and the *first variation formula* of the bienergy is given (cf. [**74**]) by

$$\frac{d}{dt}\bigg|_{t=0} E_2(\varphi_t) = -\int_M \langle \tau_2(\varphi), V \rangle \, v_g \tag{2.6}$$

where the *bitension field* $\tau_2(\varphi)$ is defined by

$$\tau_2(\varphi) = J(\tau(\varphi)) = \overline{\Delta}\tau(\varphi) - \mathcal{R}(\tau(\varphi)), \tag{2.7}$$

and a C^∞ map $\varphi : (M, g) \to (G/K, h)$ is said to be *biharmonic* if

$$\tau_2(\varphi) = 0. \tag{2.8}$$

2.2. Let k be a left invariant Riemannian metric on G corresponding to the inner product $\langle \cdot, \cdot \rangle$ on $\mathfrak{g}$ given by $\langle \cdot, \cdot \rangle = -B(\cdot, \cdot)$ if $(G/K, h)$ is of compact type, and by $\langle U + X, V + Y \rangle = -B(U, V) + B(X, Y)$ $(U, V \in \mathfrak{k}, X, Y \in \mathfrak{m})$ if $(G/K, h)$ is of non-compact type. Here, $B(\cdot, \cdot)$ is the Killing form of $\mathfrak{g}$. Then, the projection π of G onto G/K is a Riemannian submersion of (G, k) onto $(G/K, h)$, and we have also the orthogonal decomposition of the tangent space $T_{\psi(x)}G$ $(x \in M)$ with respect to the inner product $k_{\psi(x)}(\cdot, \cdot)$ $(x \in M)$ in such a way that

$$T_{\psi(x)}G = V_{\psi(x)} \oplus H_{\psi(x)}, \tag{2.9}$$

where the *vertical space* at $\psi(x) \in G$ is given by

$$V_{\psi(x)} = \mathrm{Ker}(\pi_{*\psi(x)}) = \{X_{\psi(x)} |\, X \in \mathfrak{k}\}, \tag{2.10}$$

and the *horizontal space* at $\psi(x)$ is given by

$$H_{\psi(x)} = \{Y_{\psi(x)} |\, Y \in \mathfrak{m}\}, \tag{2.11}$$

corresponding to the Cartan decomposition $\mathfrak{g} = \mathfrak{k} \oplus \mathfrak{m}$. Then, for every C^∞ section $W \in \Gamma(\psi^{-1}TG)$, we have the decomposition corresponding to (2.9),

$$W(x) = W^V(x) + W^H(x) \quad (x \in M), \tag{2.12}$$

where W^V, W^H, (denoted also by $\mathcal{V}W$, $\mathcal{H}W$, respectively) belong to $\Gamma(\psi^{-1}TG)$. We denote by $\Gamma(E)$, the space of all C^∞ sections of a vector bundle E. For $Y \in \mathfrak{m}$, define $\tilde{Y} \in \Gamma(\psi^{-1}TG)$ by $\tilde{Y}(x) := Y_{\psi(x)}$ $(x \in M)$. Let $\{X_i\}_{i=1}^n$ be an orthonormal basis of $\mathfrak{m}$ with respect to the inner product $\langle \cdot, \cdot \rangle$ of $\mathfrak{g}$ corresponding to the left invariant Riemannian metric k on G. Then, W^H can be written in terms of $\tilde{X}_i$ as

$$W^H = \sum_{i=1}^n f_i \tilde{X}_i$$

where $f_i \in C^\infty(M)$ $(i = 1, \cdots, n)$. Because, for every $x \in M$, $W^H(x) \in H_{\psi(x)}$, so that we have

$$W^H(x) = \sum_{i=1}^n f_i(x)\, X_{i\,\psi(x)} = \sum_{i=1}^n f_i(x)\tilde{X}_i(x).$$

We say $W \in \Gamma(\psi^{-1}TG)$ and $V \in \Gamma(\varphi^{-1}T(G/K))$ are *π-related*, denoted by $V = \pi_* W$, if it holds that

$$V(x) = \pi_* W(x) \quad (x \in M),$$

where $\pi_* : T_{\psi(x)}G \to T_{\varphi(x)}(G/K) = T_{\pi(\psi(x))}(G/K)$ is the differentiation of the projection π of G onto G/K at $\psi(x)$ for each $x \in M$.

Let be ∇, ∇^k, ∇^h, the Levi-Civita connections of (M, g), (G, k), $(G/K, h)$, and $\overline{\nabla}$, $\overline{\overline{\nabla}}$, the induced connection of ∇^k on the induced bundle $\psi^{-1}TG$ by $\psi : M \to G$, and the one of ∇^h on the induced bundle $\varphi^{-1}T(G/K)$ by $\varphi : M \to G/K$, respectively.

LEMMA 2.1. *Assume that* $W \in \Gamma(\psi^{-1}TG)$ *and* $V \in \Gamma(\varphi^{-1}T(G/K))$ *are* π-related, i.e., $V = \pi_*W$.
(1) *Then, we have*

$$\overline{\overline{\nabla}}_X V = \pi_* \nabla^k_{(\psi_*X)^H} W^H, \tag{2.13}$$

where $(\psi_*X)^H$ *is the horizontal component of* ψ_*X *for every* C^∞ *vector field* X *on* M.
(2) *If we express* $W^H = \sum_{i=1}^n f_i \widetilde{X}_i$ *and* $(\psi_*X)^H = \sum_{j=1}^n g_j \widetilde{X}_j$ *where* $f_i, g_j \in C^\infty(M)$ $(i, j = 1, \cdots, n)$, *then, it holds that*

$$\left(\nabla^k_{(\psi_*X)^H} W^H\right)_{\psi(x)} = \frac{1}{2}\sum_{i,j=1}^n f_i(x)\, g_j(x)\, [X_j, X_i]_{\psi(x)} + \sum_{i=1}^n X_x(f_i)\, \widetilde{X}_i(x)$$

$$\in V_{\psi(x)} \oplus H_{\psi(x)} \quad (x \in M), \tag{2.14}$$

correspondingly.
(3) *For every* $x \in M$, *we have*

$$\overline{\overline{\nabla}}_X V(x) = \sum_{i=1}^n X_x(\langle W, X_i{}_{\psi(x)}\rangle)\, \pi_*(X_i{}_{\psi(x)}). \tag{2.15}$$

Here, it holds that $\pi_*(X_{\psi(x)}) = t_{\psi(x)}{}_*\pi_*(X)$ $(X \in \mathfrak{m})$, *where* t_a *is the translation of* G/K *by* $a \in G$, i.e., $t_a(yK) := ayK$ $(y \in G)$.

PROOF. (1) Due to Lemmas 1 and 3 in [**121**], p.460, we have

$$\overline{\overline{\nabla}}_X V = \nabla^h_{\varphi_*X} V$$
$$= \nabla^h_{\pi_*(\psi_*X)} \pi_* W$$
$$= \pi_* \left(\mathcal{H}\, \nabla^k_{(\psi_*X)^H} W^H\right)$$
$$= \pi_* \nabla^k_{(\psi_*X)^H} W^H.$$

(2) Indeed, we have

$$
\begin{aligned}
\left(\nabla^k_{(\psi_* X)^H} W^H\right)_{\psi(x)} &= \sum_{j=1}^n g_j(x) \left(\nabla^k_{X_j} W^H\right)_{\psi(x)} \\
&= \sum_{j=1}^n g_j(x) \left(\sum_{i=1}^n \nabla^k_{X_j}(f_i\,\widetilde{X}_i)\right)_{\psi(x)} \\
&= \sum_{i,j=1}^n g_j(x) \left\{(X_j f_i)(x)\,\widetilde{X}_i(x) + f_i(x)\left(\nabla^k_{X_j}\widetilde{X}_i\right)_{\psi(x)}\right\} \\
&= \sum_{i,j=1}^n g_j(x) \left\{(X_j f_i)(x)\,\widetilde{X}_i(x) + \frac{1}{2} f_i(x)[X_j, X_i]_{\psi(x)}\right\} \\
&= \frac{1}{2}\sum_{i,j=1}^n f_i(x) g_j(x)[X_j, X_i]_{\psi(x)} + \sum_{i=1}^n X_x(f_i)\,\widetilde{X}_i(x),
\end{aligned}
$$

since it holds that

$$
\begin{aligned}
\left(\nabla^k_{X_j} X_i\right)_{\psi(x)} &= L_{\psi(x)\,*}\left(\nabla^k_{X_j} X_i\right)_e \\
&= L_{\psi(x)\,*}\left(\frac{1}{2}[X_j, X_i]_e\right) \\
&= \frac{1}{2}[X_j, X_i]_{\psi(x)}
\end{aligned}
$$

and $[\mathfrak{m}, \mathfrak{m}] \subset \mathfrak{k}$. For (3), notice that $W^H = \sum_{i=1}^n \langle W, X_{i\,\psi(\cdot)}\rangle\,\widetilde{X}_i$. Due to (1), (2), we have (3). $\qquad\square$

LEMMA 2.2. *Under the same assumption of Lemma 2.1, we have,*

$$
\overline{\overline{\nabla}}_X(\overline{\overline{\nabla}}_Y V) = \sum_{i=1}^n X_x(Y\langle W, X_{i\,\psi(\cdot)}\rangle)\,\pi_*(X_{i\,\psi(x)}) \in T_{\varphi(x)}(G/K),
\tag{2.16}
$$

at each $x \in M$, for every C^∞ vector fields X and Y on M.

PROOF. Let $Z := \overline{\overline{\nabla}}_Y V \in \Gamma(\varphi^{-1}T(G/K))$. Then, by Lemma 2.1 (1), we have

$$
\overline{\overline{\nabla}}_X(\overline{\overline{\nabla}}_Y V) = \overline{\overline{\nabla}}_X Z = \pi_* \nabla^k_{(\psi_* X)^H} Z^H
\tag{2.17}
$$

where by Lemma 2.1 (3), we have for every $y \in M$,

$$
Z^H(y) = \sum_{i=1}^n Y_y\langle W, X_{i\,\psi(\cdot)}\rangle\,X_{i\,\psi(y)},
$$

$Z(y) = \pi_* Z^H(y) \in T_{\varphi(y)}(G/K)$ and $Z \in \Gamma(\varphi^{-1}T(G/K))$. Then, at each $x \in M$, the right hand side of (2.17) which belong to $T_{\varphi(x)}(G/K)$, coincides with the following:

$$\sum_{j=1}^{n} X_x \langle Z^H, X_{j\,\psi(\cdot)} \rangle\, \pi_*(X_{j\,\psi(x)})$$

$$= \sum_{j=1}^{n} X_x \left\langle \sum_{i=1}^{n} Y_\bullet \langle W, X_{i\,\psi} \rangle\, X_{i\,\psi(\cdot)}, X_{j\,\psi(\cdot)} \right\rangle \pi_*(X_{j\,\psi(x)})$$

$$= \sum_{i,j=1}^{n} X_x\, (Y_\bullet \langle W, X_{i\,\psi} \rangle)\, \delta_{ij}\, \pi_*(X_{j\,\psi(x)})$$

$$= \sum_{i=1}^{n} X_x\, (Y_\bullet \langle W, X_{i\,\psi} \rangle)\, \pi_*(X_{j\,\psi(x)}).$$

Thus, we have (2.16).   □

PROPOSITION 2.1. *The rough Laplacian $\overline{\Delta}$ acting on $\Gamma(\varphi^{-1}T(G/K))$ can be calculated as follows: For $V \in \Gamma(\varphi^{-1}T(G/K))$ with $V = \pi_* W$ for $W \in \Gamma(\psi^{-1}TG)$,*

$$(\overline{\Delta} V)(x) = \sum_{i=1}^{n} \Delta_x\, \langle W, X_{i\,\psi(\cdot)} \rangle\, \pi_*(X_{i\,\psi(x)}) \in T_{\varphi(x)}(G/K),$$
(2.18)

for each $x \in M$. Here, since $f : M \ni x \mapsto \langle W(x), X_{i\,\psi(x)} \rangle_{\psi(x)} \in \mathbb{R}$ is a (local) C^∞ function on M, the Laplacian $\Delta_x = \delta\, d$ acting on $C^\infty(M)$ works well to this f.

Indeed, if we recall the definition (2.3) of the rough Laplacian $\overline{\Delta}$, and due to Lemmas 2.1 and 2.2, we have

$$\overline{\Delta} V = -\sum_{j=1}^{m} \{ \overline{\overline{\nabla}}_{e_j} (\overline{\overline{\nabla}}_{e_j} V) - \overline{\overline{\nabla}}_{\nabla_{e_j} e_j} V \},$$

$$= -\sum_{i=1}^{n} \sum_{j=1}^{n} ({e_j}^2 - \nabla_{e_j} e_j)\, \langle W, X_{i\,\psi(\cdot)} \rangle\, \pi_*(X_{i\,\psi(x)})$$

$$= \sum_{i=1}^{n} \Delta_x\, \langle W, X_{i\,\psi(\cdot)} \rangle\, \pi_*(X_{i\,\psi(x)}).$$

We have Proposition 2.3.   □

3. Determination of the bitension field

Now, let θ be the Maurer-Cartan form on G, i.e., a $\mathfrak{g}$-valued left invariant 1-form on G which is defined by $\theta_y(Z_y) = Z$ ($y \in G$, $Z \in \mathfrak{g}$). For every C^∞ map φ of (M, g) into $(G/K, h)$ with a lift $\psi : M \to G$, let us consider a $\mathfrak{g}$-valued 1-form α on M given by $\alpha = \psi^*\theta$ and the decomposition

$$\alpha = \alpha_{\mathfrak{k}} + \alpha_{\mathfrak{m}} \tag{3.1}$$

corresponding to the decomposition $\mathfrak{g} = \mathfrak{k} \oplus \mathfrak{m}$. Then, it is well known (see for example, [**33**]) that

LEMMA 3.1. *For every C^∞ map $\varphi : (M, g) \to (G/K, h)$,*

$$t_{\psi(x)^{-1}*}\tau(\varphi) = -\delta(\alpha_{\mathfrak{m}}j + \sum_{i=1}^{m}[\alpha_{\mathfrak{k}}(e_i), \alpha_{\mathfrak{m}}(e_i)], \quad (x \in M),$$
$$\tag{3.2}$$

where $\alpha = \varphi^*\theta$, and θ is the Maurer-Cartan form of G, $\delta(\alpha_{\mathfrak{m}})$ is the co-differentiation of $\mathfrak{m}$-valued 1-form $\alpha_{\mathfrak{m}}$ on (M, g).
Thus, $\varphi : (M, g) \to (G/K, h)$ is harmonic if and only if

$$\delta(\alpha_{\mathfrak{m}}) + \sum_{i=1}^{m}[\alpha_{\mathfrak{k}}(e_i), \alpha_{\mathfrak{m}}(e_i)] = 0. \tag{3.3}$$

Furthermore, we obtain

THEOREM 3.1. *We have*

$$t_{\psi(x)^{-1}*}\tau_2(\varphi) = \Delta_g \left(-\delta(\alpha_{\mathfrak{m}}) + \sum_{i=1}^{m}[\alpha_{\mathfrak{k}}(e_i), \alpha_{\mathfrak{m}}(e_i)] \right)$$
$$+ \sum_{s=1}^{m}\left[\left[-\delta(\alpha_{\mathfrak{m}}) + \sum_{i=1}^{m}[\alpha_{\mathfrak{k}}(e_i), \alpha_{\mathfrak{m}}(e_i)], \alpha_{\mathfrak{m}}(e_s) \right], \alpha_{\mathfrak{m}}(e_s) \right],$$
$$\tag{3.4}$$

where Δ_g is the (positive) Laplacian of (M, g) acting on C^∞ functions on M, and $\{e_i\}_{i=1}^{m}$ is a local orthonormal frame field on (M, g).

Therefore, we obtain immediately the following two corollaries.

COROLLARY 3.1. *Let $(G/K, h)$ be a Riemannian symmetric space, and $\varphi : (M, g) \to (G/K, h)$, a C^∞ mapping. Then, we have:*
(1)   the map $\varphi : (M, g) \to (G/K, h)$ is harmonic if and only if

$$-\delta(\alpha_{\mathfrak{m}}) + \sum_{i=1}^{m}[\alpha_{\mathfrak{k}}(e_i), \alpha_{\mathfrak{m}}(e_i)] = 0. \tag{3.5}$$

(2)   The map $\varphi : (M, g) \to (G/K, h)$ is biharmonic if and only if

$$\Delta_g \left(-\delta(\alpha_{\mathrm{m}}) + \sum_{i=1}^{m} [\alpha_{\mathfrak{k}}(e_i), \alpha_{\mathrm{m}}(e_i)] \right)$$
$$+ \sum_{s=1}^{m} \left[\left[-\delta(\alpha_{\mathrm{m}}) + \sum_{i=1}^{m} [\alpha_{\mathfrak{k}}(e_i), \alpha_{\mathrm{m}}(e_i)] , \alpha_{\mathrm{m}}(e_s) \right] , \alpha_{\mathrm{m}}(e_s) \right] = 0. \tag{3.6}$$

COROLLARY 3.2. *Let $(G/K, h)$ be a Riemannian symmetric space, and $\varphi : (M, g) \to (G/K, h)$, a C^∞ mapping with a horizontal lift $\psi : M \to G$, i.e., $\varphi = \pi \circ \psi$ and $\psi_x(T_x M) \subset H_{\psi(x)}$ which is equivalent to $\alpha_{\mathfrak{k}} \equiv 0$.*
Then, we have:
(1) the map $\varphi : (M, g) \to (G/K, h)$ is harmonic if and only if

$$\delta(\alpha_{\mathrm{m}}) = 0, \tag{3.7}$$

(2) and the map $\varphi : (M, g) \to (G/K, h)$ is biharmonic if and only if

$$\delta \, d \, \delta(\alpha_{\mathrm{m}}) + \sum_{s=1}^{m} [[\delta(\alpha_{\mathrm{m}}), \alpha_{\mathrm{m}}(e_s)], \alpha_{\mathrm{m}}(e_s)] = 0. \tag{3.8}$$

Proof of Theorem 3.2.
We need the following lemma:

LEMMA 3.2. *The tension field $\tau(\varphi)$ of a C^∞ map $\varphi : (M, g) \to (G/K, h)$ can be expressed as*

$$\tau(\varphi) = \pi_* W = \pi_*(W^H),$$

where $W \in \Gamma(\psi^{-1} TG)$, and W^H is the horizontal component of W in the decomposition $W(x) = W^V(x) + W^H(x) \in T_{\psi(x)} G = V_{\psi(x)} \oplus H_{\psi(x)}$ $(x \in M)$. If we define an $\mathfrak{m}$-valued function β on M by

$$\beta := \sum_{i=1}^{n} \langle W, \widetilde{X}_i \rangle X_i = \sum_{i=1}^{n} \langle W^H, \widetilde{X}_i \rangle X_i, \tag{3.9}$$

then, we have

$$t_{\psi(x)\,*}^{\;\;-1} \tau(\varphi) = \pi_* \beta. \tag{3.10}$$

If we define n $\mathfrak{m}$-valued functions β_i $(i = 1, \cdots, n)$ on M by

$$\beta_i := \sum_{j=1}^{n} \langle \psi_* e_i, X_j \,_{\psi(\cdot)} \rangle X_j \in \mathfrak{m}. \tag{3.11}$$

Then, it holds that

$$t_{\psi(x)\,*}^{\;\;-1}\varphi_*e_i = \pi_*\beta_i \text{ and } \beta_i = \alpha_{\mathfrak{m}}(e_i), \tag{3.12}$$

where $\alpha_{\mathfrak{m}}$ is the $\mathfrak{m}$-component of $\alpha := \psi^*\theta$, and the Maurer-Cartan form on G.

Indeed, (3.10) and the first part of (3.12) follow from the definition of β and the fact that

$$\begin{aligned}
\alpha(e_i) &= (\psi^*\theta)(e_i) \\
&= \theta(\psi_*e_i) \\
&= \theta\left(\sum_{j=1}^{n}\langle\psi_*e_i, X_{j\,\psi(x)}\rangle\, X_{j\,\psi(x)} + \sum_{j=n+1}^{\ell}\langle\psi_*e_i, X_{j\,\psi(x)}\rangle\, X_{j\,\psi(x)}\right) \\
&= \sum_{j=1}^{n}\langle\psi_*e_i, X_{j\,\psi(x)}\rangle\, X_j + \sum_{j=n+1}^{\ell}\langle\psi_*e_i, X_{j\,\psi(x)}\rangle\, X_j \\
&\in \mathfrak{m}\oplus\mathfrak{k},
\end{aligned}$$

since $\alpha = \psi^*\theta$. Thus, we have $\beta_i = \alpha_{\mathfrak{m}}(e_i)$. $\qquad\qquad\square$

(*Continued the proof of Theorem 3.2*)   We have

$$t_{\psi(x)\,*}^{\;\;-1}\varphi_*e_i = \sum_{j=1}^{n}\langle\psi_*e_i, X_{j\,\psi(x)}\rangle\, \pi_*(X_j) \in T_o(G/K), \tag{3.13}$$

where $o = \{K\} \in G/K$ is the origin of G/K. Because,

$$t_{\psi(x)\,*}^{\;\;-1}\varphi_*e_i = t_{\psi(x)\,*}^{\;\;-1}\pi_*\,\psi_*e_i = \pi_*\,L_{\psi(x)\,*}^{\;\;-1}\,\psi_*e_i = \pi_*(L_{\psi(x)\,*}^{\;\;-1}\psi_*\,e_i)_{\mathfrak{m}}$$

which coincides with

$$\sum_{j=1}^{n}\langle L_{\psi(x)\,*}^{\;\;-1}\psi_*e_i, X_j\rangle\, \pi_*(X_j) = \sum_{j=1}^{n}\langle\psi_*e_i, X_{j\,\psi(x)}\rangle\, \pi_*(X_j),$$

which imply (3.13).

Thus, we have

$$\varphi_*e_i = \pi_*W_i \quad (i = 1, \cdots, m), \tag{3.14}$$

where $W_i \in \Gamma(\psi^{-1}TG)$ and $\mathfrak{m}$-valued functions $\widetilde{W}_i$ on M ($i = 1, \cdots, m$) are given by

$$W_i(x) := \sum_{j=1}^{n}\langle\psi_*e_i, X_{j\,\psi(x)}\rangle\, X_{j\,\psi(x)}, \tag{3.15}$$

$$\widetilde{W}_i(x) := \sum_{j=1}^{n}\langle\psi_*e_i, X_{j\,\psi(x)}\rangle\, X_j \in \mathfrak{m}, \tag{3.16}$$

for each $x \in M$.

On the other hand, we have

$$\tau(\varphi) = \pi_* W, \tag{3.17}$$

where $W \in \Gamma(\psi^{-1}TG)$ and an $\mathfrak{m}$-valued function $\widetilde{W}$ on M are given by

$$W(x) := t_{\psi(x)\,*} \left(-\delta(\alpha_\mathfrak{m}) + \sum_{i=1}^{m} [\alpha_\mathfrak{k}(e_i), \alpha_\mathfrak{m}(e_i)] \right), \tag{3.18}$$

$$\widetilde{W} := -\delta(\alpha_\mathfrak{m}) + \sum_{i=1}^{m} [\alpha_\mathfrak{k}(e_i), \alpha_\mathfrak{m}(e_i)] \tag{3.19}$$

for each $x \in M$. And we also have

$$t_{\psi(x)\,*}^{-1} \overline{\Delta} \tau(\varphi)(x) = \Delta \left(-\delta(\alpha_\mathfrak{m}) + \sum_{i=1}^{m} [\alpha_\mathfrak{k}(e_i), \alpha_\mathfrak{m}(e_i)] \right)(x) \quad (x \in M), \tag{3.20}$$

where $\Delta = \delta\, d$ is the positive Laplacian acting on the space of all C^∞ $\mathfrak{m}$-valued functions on M.

We want to calculate $\mathcal{R}(\tau(\varphi)) = \sum_{i=1}^{m} R^h(\tau(\varphi), \varphi_* e_i)\varphi_* e_i$. Indeed, we have

$$t_{\psi(x)\,*}^{-1} \mathcal{R}(\tau(\varphi)) = -\sum_{i=1}^{m} [[\widetilde{W}, \widetilde{W_i}], \widetilde{W_i}]$$

$$= -\sum_{s=1}^{m} [[-\delta(\alpha_\mathfrak{m}) + \sum_{i=1}^{m} [\alpha_\mathfrak{k}(e_i), \alpha_\mathfrak{m}(e_i)],$$

$$\alpha_\mathfrak{m}(e_s)], \alpha_\mathfrak{m}(e_s)]. \tag{3.21}$$

Here, we used the formula of the curvature R^h of the Riemannian symmetric space $(G/K, h)$ ([82], p. 202, p.231, Theorem 3.2) :

$$(R^h(X, Y)Z)_o = -[[X, Y], Z]_o \quad (X, Y, Z \in \mathfrak{m}).$$

Thus, we obtain Theorem 3.2. $\qquad\qquad\qquad\qquad\qquad\qquad\qquad \square$

Let us recall the *integrability condition* for a C^∞ mapping $\varphi :$ $(M, g) \to (G/K, h)$. The Maurer-Cartan form θ on G satisfies

$$d\theta + \frac{1}{2}[\theta \wedge \theta] = 0, \tag{3.22}$$

so that the pull back $\alpha = \psi^*\theta$ of θ by the lift $\psi : M \to G$ of $\varphi : M \to$ G/K also satisfies that

$$d\alpha + \frac{1}{2}[\alpha \wedge \alpha] = 0, \tag{3.23}$$

which is equivalent to

$$\begin{cases} d\alpha_{\mathfrak{k}} + \dfrac{1}{2}[\alpha_{\mathfrak{k}} \wedge \alpha_{\mathfrak{k}}] + \dfrac{1}{2}[\alpha_{\mathfrak{m}} \wedge \alpha_{\mathfrak{m}}] = 0, \\ d\alpha_{\mathfrak{m}} + [\alpha_{\mathfrak{k}} \wedge \alpha_{\mathfrak{m}}] = 0. \end{cases} \qquad (3.24)$$

Summarizing the above, we have

THEOREM 3.2. *Let (M, g) be an m-dimensional compact Riemann-ian manifold, $(G/K, h)$, an n-dimensional Riemannian symmetric space, $\pi : G \to G/K$, the projection, and $\varphi : (M, g) \to (G/K, h)$, a C^∞ mapping with a local lift $\psi : M \to G$, $\varphi = \pi \circ \psi$. Let $\alpha = \psi^* \theta$ be the pull back of the Maurer-Cartan form θ, and $\alpha = \alpha_{\mathfrak{k}} + \alpha_{\mathfrak{m}}$, the de-composition of α corresponding to the Cartan decomposition $\mathfrak{g} = \mathfrak{k} \oplus \mathfrak{m}$.*
(I) The mapping $\varphi : (M, g) \to (G/K, h)$ is harmonic *if and only if*

$$-\delta(\alpha_{\mathfrak{m}}) + \sum_{i=1}^{m} [\alpha_{\mathfrak{k}}(e_i), \alpha_{\mathfrak{m}}(e_i)] = 0, \qquad (3.25)$$

where δ is the co-differentiation, and $\{e_i\}_{i=1}^{m}$ is a local orthonormal frame field on (M, g).
Furthermore, $\varphi : (M, g) \to (G/K, h)$ is biharmonic *if and only if*

$$\Delta\left(-\delta(\alpha_{\mathfrak{m}}) + \sum_{i=1}^{m} [\alpha_{\mathfrak{k}}(e_i), \alpha_{\mathfrak{m}}(e_i)]\right)$$
$$+ \sum_{s=1}^{m} \left[\left[-\delta(\alpha_{\mathfrak{m}}) + \sum_{i=1}^{m} [\alpha_{\mathfrak{k}}(e_i), \alpha_{\mathfrak{m}}(e_i)], \alpha_{\mathfrak{m}}(e_s)\right], \alpha_{\mathfrak{m}}(e_s)\right] = 0, \qquad (3.26)$$

where $\Delta = \delta d$ is the (positive) Laplacian of (M, g) acting on the space of $\mathfrak{g}$-valued C^∞ functions on (M, g).
(II) Conversely, let $\alpha = \alpha_{\mathfrak{k}} + \alpha_{\mathfrak{m}}$ be a $\mathfrak{g}$-valued 1-form on (M, g). If α satisfies (3.23) or (3.24), and satisfies (3.25) (resp. (3.26)), then, there exists a C^∞-mapping φ of M into G with a local lift $\psi : M \to G$, $\varphi = \pi \circ \psi$ and the initial value $\varphi(p) = a \in G$ at some $p \in M$ such that $\alpha = \psi^* \theta$ and φ is a harmonic (resp. biharmonic) map of (M, g) into $(G/K, h)$.

4. Biharmonic curves into Riemannian symmetric spaces

4.1.   Let $\varphi : (\mathbb{R}, g_0) \to (G/K, h)$ be a C^∞ curve, and $\psi : \mathbb{R} \to G$, a lift of φ, $(\varphi = \pi \circ \psi)$. Then, $\alpha = \psi^* \theta = \psi^{-1} d\psi = F(t) dt$ is a $\mathfrak{g}$-valued 1-form on $\mathbb{R}$ and F is a $\mathfrak{g}$-valued function on $\mathbb{R}$ satisfying $\psi(t)^{-1} \frac{d\psi}{dt} = F(t)$. Conversely, for a $\mathfrak{g}$-valued C^∞ function $F(t)$ on $\mathbb{R}$, there exists a unique C^∞-curve $\psi : \mathbb{R} \to G$ which satisfies

$$\begin{cases} \psi(t)^{-1}\dfrac{d\psi}{dt} = F(t), \\ \psi(0) = x \in G. \end{cases} \qquad (4.1)$$

To give an explicit solution ψ of (4.1) is very difficult, in general, since G is not abelian. However, corresponding to the decomposition $\mathfrak{g} = \mathfrak{k} \oplus \mathfrak{m}$, we decompose $F(t) = F_{\mathfrak{k}}(t) + F_{\mathfrak{m}}(t)$, $\alpha_{\mathfrak{k}} = F_{\mathfrak{k}}(t)dt$, and $\alpha_{\mathfrak{m}} = F_{\mathfrak{m}}(t)dt$, so we have

$$\delta\alpha = -(\overline{\nabla}_{e_1})(\alpha(e_1)) = -\nabla^h_{e_1}(\alpha(e_1)) = -e_1(F(t)) = -F'(t),$$

and

$$\delta\alpha_{\mathfrak{m}} = -F_{\mathfrak{m}}{}'(t).$$

Thus the harmonic map equation (3.25) is

$$F_{\mathfrak{m}}{}'(t) + [F_{\mathfrak{k}}(t), F_{\mathfrak{m}}(t)] = 0, \qquad (4.2)$$

and the biharmonic map equation (3.26) is

$$-\frac{d^2}{dt^2}\left(F_{\mathfrak{m}}{}'(t) + [F_{\mathfrak{k}}(t), F_{\mathfrak{m}}(t)]\right)$$
$$+ \left[[F_{\mathfrak{m}}{}'(t) + [F_{\mathfrak{k}}(t), F_{\mathfrak{m}}(t)], F_{\mathfrak{m}}], F_{\mathfrak{m}}\right] = 0. \qquad (4.3)$$

In these cases, the integrability condition (3.23) always holds, so that the existence of ψ of (4.1) is always true.

Let us recall that a lift $\psi(t)$ is *horizontal* if $\psi_*(T_xM) \subset L_{*\psi(x)}(\mathfrak{m})$ if and only if $F_{\mathfrak{k}} \equiv 0$. In this case, (4.2) is equivalent to

$$F_{\mathfrak{m}}{}'(t) = 0, \qquad (4.4)$$

which implies that $F_{\mathfrak{m}}(t) = X \in \mathfrak{m}$ (constant). So that $F(t) = X \in \mathfrak{m}$. Then, we have

$$\psi(t) = x \exp(tX), \quad \varphi(t) = x \exp(tX) K \in G/K. \qquad (4.5)$$

Furthermore, (4.3) is equivalent to

$$-F_{\mathfrak{m}}{}'''(t) + [[F_{\mathfrak{m}}{}'(t), F_{\mathfrak{m}}(t)], F_{\mathfrak{m}}(t)] = 0. \qquad (4.6)$$

Example 4.1. Assume that $(G/K, h)$ is *of the Euclidean type*. In this case, $\mathfrak{m}$ is an abelian ideal and $\mathfrak{k}$ acts on $\mathfrak{m}$ by $[T, X] = T \cdot X$ $(T \in \mathfrak{k}, X \in \mathfrak{m})$ regarding $\mathfrak{k}$ as a subalgebra of $\mathfrak{gl}(\mathfrak{m})$. Then, we have

(1) $\varphi : (\mathbb{R}, g_0) \to (G/K, h)$ is harmonic if and only if

$$F_{\mathfrak{m}}{}'(t) + F_{\mathfrak{k}}(t) \cdot F_{\mathfrak{m}}(t) = 0. \qquad (4.7)$$

(2) $\varphi : (\mathbb{R}, g_0) \to (G/K, h)$ is biharmonic if and only if

$$\frac{d^2}{dt^2} \left(F_{\mathfrak{m}}'(t) + F_{\mathfrak{k}}(t) \cdot F_{\mathfrak{m}}(t) \right) = 0 \tag{4.8}$$

which is equivalent to

$$F_{\mathfrak{m}}'(t) + F_{\mathfrak{k}}(t) \cdot F_{\mathfrak{m}}(t) = At + B \tag{4.9}$$

for some A and B in $\mathfrak{m}$. Thus, if $\psi : (\mathbb{R}, g_0) \to G$ is horizontal, i.e., $F_{\mathfrak{k}} \equiv 0$, then, $F_{\mathfrak{m}}(t) = C$ (a constant vector in $\mathfrak{m}$) for the case (1), and $F_{\mathfrak{m}}(t) = At^2 + Bt + C$ for the case (2). If $[A, B] = [B, C] = [C, A] = 0$, then $\psi(t) = \exp(t^2 A + t B + C)$ and $\varphi(t) = \psi(t) \cdot \{K\}$ is a biharmonic curve in a Riemannian symmetric space $(G/K, h)$ of the Euclidean type.

4.2. Biharmonic curves into rank one symmetric spaces.
In this subsection, we study biharmonic curves in a compact symmetric spaces $(G/K, h)$.

(1) *Case of the unit sphere (S^n, h).* Let $G = SO(n+1)$ act on $\mathbb{R}^{n+1}$ linearly, and $K = SO(n)$ be the isotropy subgroup of G at the origin $o = {}^t(1, 0, \cdots, 0)$. Their Lie algebras $\mathfrak{g} = \mathfrak{so}(n+1)$, $\mathfrak{k} = \mathfrak{so}(n)$ and the Cartan decomposition $\mathfrak{g} = \mathfrak{k} \oplus \mathfrak{m}$ are given by

$$\mathfrak{g} = \mathfrak{so}(n+1) = \{X \in \mathfrak{gl}(n+1) : X + {}^tX = O\},$$

$$\mathfrak{k} = \mathfrak{so}(n) = \left\{ \begin{pmatrix} \begin{array}{c|ccc} 0 & 0 & \cdots & 0 \\ \hline 0 & & & \\ \vdots & & X_1 & \\ 0 & & & \end{array} \end{pmatrix} : X_1 \in \mathfrak{gl}(n),\ X_1 + {}^tX_1 = O \right\},$$

$$\mathfrak{m} = \left\{ \begin{pmatrix} \begin{array}{c|c} 0 & -{}^tu \\ \hline u & O \end{array} \end{pmatrix} : u = {}^t(u_1, \cdots, u_n) \in \mathbb{R}^n \right\}.$$

For a $\mathfrak{m}$-valued C^∞ function $F_{\mathfrak{m}}(t)$ given by

$$F_{\mathfrak{m}}(t) = \begin{pmatrix} \begin{array}{c|ccc} 0 & -u_1(t) & \cdots & -u_n(t) \\ \hline u_1(t) & & & \\ \vdots & & O & \\ u_n(t) & & & \end{array} \end{pmatrix}, \tag{4.10}$$

and $F_{\mathfrak{k}} \equiv 0$, the biharmonic map equation (4.7) is equivalent to

$$-u_i''' + \sum_{j=1}^{n} (u_i\, u_j' - u_i'\, u_j) u_j = 0 \quad (i = 1, \cdots, n) \tag{4.11}$$

which is also equivalent to

$$-u''' + \langle u', u \rangle u - \langle u, u \rangle u' = 0, \tag{4.12}$$

where the inner product $\langle \, , \, \rangle$ on $\mathbb{R}^n$ is given by $\langle u, v \rangle = \sum_{i=1}^n u_i v_i$ for $u, v \in \mathbb{R}^n$.

Case of $n = 2$. Our problem is to find a C^∞ plane curve which satisfies (4.12). To do it, we assume that $u(t)$ is reparametrized in such a way that $u(s)$ is a tangent curve of a plane curve $\mathbf{p}(s)$: $u(s) = \mathbf{p}'(s) = \mathbf{e}_1(s)$. For the other cases, we have no idea to solve (4.12). Recall the Frenet-Serret formula for a plane curve $\mathbf{p}(s)$:

$$\begin{cases} \mathbf{p}'(s) = \mathbf{e}_1(s), \\ \mathbf{e}_1'(s) = \kappa(s) \, \mathbf{e}_2(s), \\ \mathbf{e}_2'(s) = -\kappa(s) \, \mathbf{e}_1(s). \end{cases} \tag{4.13}$$

Now we have

$$u = \mathbf{e}_1, \tag{4.14}$$
$$u' = \mathbf{e}_1' = \kappa \, \mathbf{e}_2, \tag{4.15}$$
$$u'' = \kappa' \, \mathbf{e}_2 + \kappa \, \mathbf{e}_2' = -\kappa^2 \, \mathbf{e}_1 + \kappa' \, \mathbf{e}_2, \tag{4.16}$$
$$u''' = -3\kappa \, \kappa' \, \mathbf{e}_1 + (\kappa'' - \kappa^3) \, \mathbf{e}_2. \tag{4.17}$$

Since $\langle u', u \rangle = 0$ and $\langle u, u \rangle = 1$, (4.12) is equivalent to

$$-3\kappa \, \kappa' = 0, \tag{4.18}$$
$$\kappa'' - \kappa^3 = -\kappa, \tag{4.19}$$

By (4.18), $\kappa = c$ (a constant), and by (4.19), $c = 0, 1, -1$. Thus, we have

(*i*) In the case of $c = 0$,

$$\mathbf{p}(s) = s \, \mathbf{a} + \mathbf{b}, \quad u(s) = \mathbf{a}, \quad (\mathbf{a}, \mathbf{b} \in \mathbf{R}^2), \tag{4.20}$$

(*ii*) in the case of $c = 1$,

$$\mathbf{p}(s) = (\cos s, \sin s), \quad u(s) = (-\sin s, \cos s), \tag{4.21}$$

(*iii*) in the case of $c = -1$,

$$\mathbf{p}(s) = (\cos s, -\sin s), \quad u(s) = (-\sin s, -\cos s). \tag{4.22}$$

Now it is easy to find $\psi : \mathbb{R} \to G$ and $\varphi(t) = \psi(t) \{K\} \in G/K$ satisfying $\psi(t)^{-1} \frac{d\psi}{dt} = F(t) = F_{\mathrm{m}}(t)$ for such $u(t)$ in (4.12).

Case (i):   If $\mathbf{a} = {}^t(a, b) \in \mathbb{R}^2$, we have due to (4.1),

$$\varphi(t) = \psi(t) \{K\} = x \begin{pmatrix} \cos(t\sqrt{a^2 + b^2}) \\ \frac{a}{\sqrt{a^2+b^2}} \sin(t\sqrt{a^2 + b^2}) \\ \frac{b}{\sqrt{a^2+b^2}} \sin(t\sqrt{a^2 + b^2}) \end{pmatrix}, \tag{4.23}$$

which is a great circle of the standard 2-sphere (S^2, h).

Cases (ii) and (iii): In these cases, if we assume $F_{\mathfrak{k}} \equiv 0$, we have

$$F_{\mathrm{m}}(t) = \left(\begin{array}{c|cc} 0 & \sin t & -\cos t \\ \hline -\sin t & 0 & 0 \\ \cos t & 0 & 0 \end{array} \right), \tag{4.24}$$

for Case (ii), and

$$F_{\mathrm{m}}(t) = \left(\begin{array}{c|cc} 0 & \sin t & \cos t \\ \hline -\sin t & 0 & 0 \\ -\cos t & 0 & 0 \end{array} \right), \tag{4.25}$$

for Case (iii). In these cases, because of $[F_{\mathrm{m}}(t), F_{\mathrm{m}}'(t)] \neq 0$, it is difficult for us to give explicitly a unique solution of the initial value problem of

$$\psi(t)^{-1} \frac{\psi(t)}{dt} = F(t) \quad \text{and } \psi(0) = a \in SO(3). \tag{4.26}$$

Case of $n = 3$. In this case, we have to solve for a C^∞ curve $u :$ $\mathbb{R} \to \mathbb{R}^3$, the equation (4.12) which is equivalent to

$$-u''' + u \times (u \times u') = 0. \tag{4.27}$$

To do it, we assume that $u(t)$ is parametrized in such a way that $u(s)$ is a tangent curve of a C^∞ curve in $\mathbb{R}^3$, $\mathbf{p}(s) : u(s) = \mathbf{p}'(s) = \mathbf{e}_1(s)$. Recall the Frene-Serret formula for a curve $\mathbf{p}(s)$:

$$\begin{cases} \mathbf{p} = \mathbf{e}_1 \\ \mathbf{e}_1' = \quad\quad \kappa \mathbf{e}_2 \\ \mathbf{e}_2' = -\kappa \mathbf{e}_1 \quad + \tau \mathbf{e}_3 \\ \mathbf{e}_3' = \quad\quad\quad -\tau \mathbf{e}_2 \end{cases} \tag{4.28}$$

where κ and τ are the curvature and torsion of $\mathbf{p}(s)$, respectively. By making use of (4.28), we have

$$\begin{cases} u' = \kappa \mathbf{e}_2 \\ u'' = -\kappa^2 \mathbf{e}_1 + \kappa' \mathbf{e}_2 + \kappa\tau \mathbf{e}_3 \\ u''' = -3\kappa\kappa' \mathbf{e}_1 + (\kappa'' - \kappa^3 - \kappa\tau^2)\mathbf{e}_2 + (2\kappa'\tau + \kappa\tau')\mathbf{e}_3. \end{cases} \tag{4.29}$$

Thus, (4.29) is equivalent to

$$\begin{cases} -3\kappa\kappa' = 0 \\ \kappa'' - \kappa^3 - \kappa\tau^2 = -\kappa \\ 2\kappa'\tau + \kappa\tau' = 0. \end{cases} \tag{4.30}$$

By the first equation of (4.30), $\kappa = \kappa_0$ (a constant). In the case $\kappa_0 = 0$, $u(t) = \mathbf{a} \in \mathbb{R}^3$ (a constant vector). In the case $\kappa_0 \neq 0$, by the third equation of (4.30), $\tau = \tau_0$ (a constant). By the second equation of (4.30), $\kappa_0^2 + \tau_0^2 = 1$. Then, $\mathbf{p}(s) = {}^t(a\cos t, a\sin t, bt)$, with $s = \sqrt{a^2 + b^2}\, t$. Here, $\kappa_0 = a/(a^2 + b^2)$, and $\tau_0 = b/(a^2 + b^2)$, and $1 = \kappa_0^2 + \tau_0^2 = 1/(a^2 + b^2)$, i.e., $a^2 + b^2 = 1$. Therefore, we have

$$\begin{cases} \mathbf{p}(t) = {}^t(a\cos t, a\sin t, bt), \\ u(t) = \mathbf{p}'(t) = {}^t(-a\sin t, a\cos t, b), \end{cases} \tag{4.31}$$

where a and b are constants with $a^2 + b^2 = 1$. Thus, $F_{\mathrm{m}}(t)$ with $F_{\mathfrak{k}} \equiv 0$, is given by

$$F_{\mathrm{m}}(t) = \begin{pmatrix} 0 & a\sin t & -a\cos t & -b \\ -a\sin t & & & \\ a\cos t & & O & \\ b & & & \end{pmatrix}. \tag{4.32}$$

Case of $n \geq 2$.   In this case, the other-type solutions exist:

Let $u = (u_1, \cdots, u_n) = (0, \cdots, 0, \overset{i\text{ th}}{v}, 0, \cdots, 0)$ $(i = 1, \cdots, n)$. Then, for such u, the equation (4.12) is reduced to $v''' = 0$. Thus, we have $v(t) = D_t := at^2 + bt + c$ for some constants a, b and c. Thus, $F_{\mathrm{m}}(t)$ is given by

$$F_{\mathrm{m}}(t) = D_t \begin{pmatrix} 0 & 0 & \cdots & -1 & \cdots & 0 \\ 0 & & & & & \\ \vdots & & & & & \\ 1 & & & O & & \\ \vdots & & & & & \\ 0 & & & & & \end{pmatrix}. \tag{4.33}$$

Thus,

$$\psi(t) = x \, \exp\left(\int_0^t F(s)ds\right) = x \begin{pmatrix} \cos d_t & 0 & \cdots & -\sin d_t & \cdots & 0 \\ 0 & 0 & \cdots & 0 & \cdots & 0 \\ \vdots & \vdots & & \vdots & & \vdots \\ \sin d_t & 0 & \cdots & \cos d_t & \cdots & 0 \\ \vdots & \vdots & & \vdots & & \vdots \\ 0 & 0 & \cdots & 0 & \cdots & 0 \end{pmatrix},$$

where $d_t := \frac{a}{3}t^3 + \frac{b}{2}t^2 + ct$. So, we have a *biharmonic curve* into (S^n, h):

$$\varphi(t) = \psi(t)\{K\} = x \, {}^t(\cos d_t, 0, \cdots, 0, \sin d_t, 0, \cdots, 0),$$

$$(4.34)$$

for $x \in SO(n+1)$, where $d_t := \frac{a}{3}t^3 + \frac{b}{2}t^2 + ct$. Furthermore, $\varphi(t)$ is harmonic if and only if $a = b = 0$.

(2) *Case of the complex projective space* $(\mathbb{C}P^n, h)$.
Let $G = SU(n+1)$ act on the projective space linearly on $\mathbb{C}P^n = \{[z] : z \in \mathbb{C}^{n+1}\backslash\{0\}\}$, and K, the isotropy subgroup of G at $o = {}^t[1, 0, \cdots, 0]$. The Cartan decomposition $\mathfrak{g} = \mathfrak{k} \oplus \mathfrak{m}$ is given by

$$\mathfrak{g} = \{X \in \mathfrak{gl}(n+1, \mathbb{C}) : X + {}^t\overline{X} = O, \ \mathrm{tr}X = 0\},$$

$$\mathfrak{k} = \left\{ \begin{pmatrix} \sqrt{-1}a & 0 \\ 0 & X \end{pmatrix} : a \in \mathbb{R}, X \in \mathfrak{gl}(n, \mathbb{C}), \ {}^t\overline{X} + X = O, \right.$$

$$\left. \sqrt{-1}a + \mathrm{tr}X = 0 \right\},$$

$$\mathfrak{m} = \left\{ \begin{pmatrix} 0 & -{}^t\overline{z} \\ z & O \end{pmatrix} : z \in \mathbb{C}^n \right\}.$$

For a C^∞ $\mathfrak{m}$-valued function $F_\mathfrak{m}(t)$ given by

$$F_\mathfrak{m}(t) = \begin{pmatrix} 0 & -\overline{z_1(t)} & \cdots & -\overline{z_n(t)} \\ z_1(t) & & & \\ \vdots & & O & \\ z_n(t) & & & \end{pmatrix},$$

$$(4.35)$$

where $z_i(t) = u_i(t) + \sqrt{-1}v_i(t)$ $u_i(t)$ and $v_i(t)$ are real valued C^∞ functions ($i = 1, \cdots, n$), and $F_\mathfrak{k} \equiv 0$, the biharmonic map equation (4.6) is equivalent to

$$-z_i''' + \sum_{j=1}^n \{(z_i \overline{z_j}' - z_i' \overline{z_j}) z_j - z_i(\overline{z_j}z_j' - \overline{z_j}' z_j)\} = 0$$

$$(4.36)$$

for all $i = 1, \cdots, n$. Notice here that this (4.36) can be written as

$$-z''' + 2\langle z, z'\rangle\, z - \langle z', z\rangle\, z - \langle z, z\rangle\, z' = 0, \tag{4.37}$$

where $\langle z, w\rangle = \sum_{i=1}^{n} z_i \overline{w_i}$ for two $\mathbb{C}^n$-valued functions z and w in t. If we write $z = u + \sqrt{-1}v$, where u and v are $\mathbb{R}^n$-valued functions, then (4.37) is equivalent to

$$\begin{cases} -u''' + 4n\left(-v^2\, u' + u\, v\, v'\right) = 0 \\ -v''' + 4n\left(u\, v\, u' - u^2\, v'\right) = 0. \end{cases} \tag{4.38}$$

One can find the following solutions of (4.38):

(i) $u = D_t = a\, t^2 + b\, t + c$ and $v \equiv 0$,
(ii) $u \equiv 0$ and $v = D_t = a\, t^2 + b\, t + c$, or
(iii) $u = v = D_t = a\, t^2 + b\, t + c$,

where a, b and c are constant vectors in $\mathbb{R}^n$. Corresponding to these, we can find $F_m(t)$ of (4.35) as follows:

$$F_m(t) = D_t \begin{pmatrix} 0 & -\overline{z_1(t)} & \cdots & -\overline{z_n(t)} \\ z_1(t) & & & \\ \vdots & & O & \\ z_n(t) & & & \end{pmatrix}, \tag{4.39}$$

where $z_1(t), \cdots, z_n(t)$ are
   Case (i): $z_1(t) = \cdots = z_n(t) = 1$,
   Case (ii): $z_1(t) = \cdots = z_n(t) = \sqrt{-1}$,
   Case (iii): $z_1(t) = \cdots = z_n(t) = 1 + \sqrt{-1}$,
correspondingly. In each cases, we can find $\psi(t)$ by the same way as the case of (S^n, h), and a *biharmonic curve* in $(\mathbb{C}P^n, h)$:
   Case (i): $\varphi(t) = x\,{}^t[\cos(\sqrt{n}\, d_t), \frac{1}{\sqrt{n}}\sin(\sqrt{n}\, d_t), \cdots, \frac{1}{\sqrt{n}}\sin(\sqrt{n}\, d_t)]$,
   Case (ii):

$$\varphi(t) = x\,{}^t[\cos(\sqrt{n}\, d_t), \tfrac{\sqrt{-1}}{\sqrt{n}}\sin(\sqrt{n}\, d_t), \cdots, \tfrac{\sqrt{-1}}{\sqrt{n}}\sin(\sqrt{n}\, d_t)],$$

   Case (iii):

$$\varphi(t) = x\,{}^t[\cos(\sqrt{2n}\, d_t), \tfrac{1+\sqrt{-1}}{\sqrt{2n}}\sin(\sqrt{2n}\, d_t), \cdots, \tfrac{1+\sqrt{-1}}{\sqrt{2n}}\sin(\sqrt{2n}\, d_t)],$$

where $d_t := \frac{a}{3}t^3 + \frac{b}{2}t^2 + c\, t$, a, b and c are constant real numbers, and $x \in SU(n+1)$. Each $\varphi : (\mathbb{R}, g_0) \to (\mathbb{C}P^n, h)$ is *harmonic* if and only if $a = b = 0$.

(3) *Case of the quaternion projective space* $(\mathbb{H}P^n, h)$.
   Let $G = Sp(n+1) = \{x \in U(2n+2) |\, {}^t x\, J_{n+1} x = J_{n+1}\}$, where
$J_{n+1} = \begin{pmatrix} O & I_{n+1} \\ -I_{n+1} & O \end{pmatrix}$, and I_{n+1} is the identity matrix of order $n+1$.
G acts on the quaternion projective space linearly on $\mathbb{H}P^n = \{[z] : z \in$

$\mathbb{H}^{n+1}\backslash\{0\}\}$, and $K = Sp(1) \times Sp(n)$ is the isotropy subgroup K of G at $o = {}^t[1, 0, \cdots, 0]$. The Cartan decomposition $\mathfrak{g} = \mathfrak{k} \oplus \mathfrak{m}$ is given by

$$\mathfrak{g} = \mathfrak{sp}(n+1) = \left\{ \begin{pmatrix} A & B \\ -\overline{B} & \overline{A} \end{pmatrix} \mid A, B \in M_{n+1}(\mathbb{C}), \, {}^t\overline{A} + A = O, {}^tB = B \right\},$$

$$\mathfrak{k} = \mathfrak{sp}(1) \times \mathfrak{sp}(n) = \left\{ \begin{pmatrix} x & 0 & y & 0 \\ 0 & X & 0 & Y \\ -\overline{y} & 0 & \overline{x} & 0 \\ 0 & -\overline{Y} & 0 & \overline{X} \end{pmatrix} \mid x \in \sqrt{-1}\mathbb{R}, \, y \in \mathbb{C}, \right.$$

$$\left. X, Y \in M_n(\mathbb{C}), {}^t\overline{X} + X = 0, {}^tY = Y \right\},$$

$$\mathfrak{m} = \left\{ \begin{pmatrix} 0 & Z & 0 & W \\ -{}^t\overline{Z} & O & {}^tW & O \\ 0 & -\overline{W} & 0 & \overline{Z} \\ -{}^t\overline{W} & O & -{}^tZ & O \end{pmatrix} \mid Z, W \in M(1, n, \mathbb{C}) \right\}.$$

For a C^∞ $\mathfrak{m}$-valued function $F_\mathfrak{m}(t)$ given by

$$F_\mathfrak{m}(t) = \begin{pmatrix} 0 & Z & 0 & W \\ -{}^t\overline{Z} & O & {}^tW & O \\ 0 & -\overline{W} & 0 & \overline{Z} \\ -{}^t\overline{W} & O & -{}^tZ & O \end{pmatrix}, \tag{4.40}$$

where $Z = Z(t) = (z_1(t), \cdots, z_n(t))$, $W = W(t) = (w_1(t), \cdots, w_n(t))$, and for $F_\mathfrak{m}$ in (4.40) with $F_\mathfrak{k} \equiv 0$, the biharmonic map equation (4.6) is equivalent to

$$\begin{cases} -Z''' - (|Z|^2 + |W|^2)Z \\ \qquad + (2\langle Z, Z'\rangle + 2\langle W, W'\rangle - \langle Z', Z\rangle - \langle W', W\rangle)Z \\ \qquad + (\langle Z', \overline{W}\rangle - \langle W', \overline{Z}\rangle)\overline{W} = 0, \\ -W''' - (|Z|^2 + |W|^2)W \\ \qquad + (2\langle Z, Z'\rangle + 2\langle W, W'\rangle - \langle Z', Z\rangle - \langle W', W\rangle)W \\ \qquad + 3(\langle Z', \overline{W}\rangle - \langle W', \overline{Z}\rangle)\overline{Z} = 0, \end{cases} \tag{4.41}$$

where $Z' = (z_1'(t), \cdots, z_n'(t))$ and $\langle Z, W\rangle := \sum_{i=1}^n z_i(t) \overline{w_i(t)}$.

We find the following solutions of (4.41):

Case (i): $z_1(t) = \cdots = z_n(t) = D_t$ and $w_1(t) = \cdots = w_n(t) = 0$.

Case (ii): $z_1(t) = \cdots = z_n(t) = \sqrt{-1}D_t$ and $w_1(t) = \cdots = w_n(t) = 0$.

Case (iii): $z_1(t) = \cdots = z_n(t) = 0$ and $w_1(t) = \cdots = w_n(t) = D_t$.

Case (iv): $z_1(t) = \cdots = z_n(t) = 0$ and $w_1(t) = \cdots = w_n(t) = \sqrt{-1}D_t$.

The corresponding biharmonic curves into the quaternion projective spaces $\mathbb{H}P^n$ are given as follows:

Case (i):

$$\varphi(t) = x \left[\cos(\sqrt{n}\, d_t), -\frac{1}{\sqrt{n}} \sin(\sqrt{n}\, d_t), \cdots, -\frac{1}{\sqrt{n}} \sin(\sqrt{n}\, d_t) \right].$$

Case (ii):

$$\varphi(t) = x \left[\cos(\sqrt{n}\, d_t), i\frac{1}{\sqrt{n}} \sin(\sqrt{n}\, d_t), \cdots, i\frac{1}{\sqrt{n}} \sin(\sqrt{n}\, d_t) \right].$$

Case (iii):

$$\varphi(t) = x \left[\cos(\sqrt{n}\, d_t), -j\frac{1}{\sqrt{n}} \sin(\sqrt{n}\, d_t), \cdots, -j\frac{1}{\sqrt{n}} \sin(\sqrt{n}\, d_t) \right].$$

Case (iv):

$$\varphi(t) = x \left[\cos(\sqrt{n}\, d_t), k\frac{1}{\sqrt{n}} \sin(\sqrt{n}\, d_t), \cdots, k\frac{1}{\sqrt{n}} \sin(\sqrt{n}\, d_t) \right].$$

Here, $x \in Sp(n+1)$, i, j and k are the quaternions satisfying $i^2 = j^2 = k^2 = -1$ and $ij = k$, and $d_t = \frac{a}{3}t^3 + \frac{b}{2}t^2 + ct$, a, b and c are constant real numbers. In each case, φ is harmonic if and only if $a = b = 0$.

5. Biharmonic maps from plane domains

5.1. Setting and deriving the equations. In this section, we will treat biharmonic maps of (M, g) into a Riennian symmetric space $(G/K, h)$, with dim $M = 2$. We assume that $(M, g) = (\Omega, g)$ is an open domain in the 2-dimensional Euclidean space $\mathbb{R}^2$ with $g = \mu^2 g_0$, where μ is a positive C^∞ function on Ω, $g_0 = (dx)^2 + (dy)^2$ is the standard Euclidean metric and (x, y) is the standard coordinate on $\mathbb{R}^2$.

Let φ be a C^∞ map from Ω into a symmetric space $N = G/K$ with a local lift $\psi : \Omega \to G$ satisfying $\varphi = \pi \circ \psi$, where $\pi : G \to G/K$ is the standard projection. The pull back of the Maurer-Cartan form θ on G by ψ is given by

$$\alpha = \psi^{-1} d\psi = \psi^{-1} \frac{\partial \psi}{\partial x} dx + \psi^{-1} \frac{\partial \psi}{\partial y} dy$$
$$= A_x \, dx + A_y \, dy,$$

where we decompose two $\mathfrak{g}$-valued functions $A_x := \psi^{-1}\frac{\partial\psi}{\partial x}$ and $A_y := \psi^{-1}\frac{\partial\psi}{\partial y}$ on Ω according to the Cartan decomposition $\mathfrak{g} = \mathfrak{k}\oplus\mathfrak{m}$ as follows:

$$A_x = A_{x,\mathfrak{k}} + A_{x,\mathfrak{m}}, \quad A_y = A_{y,\mathfrak{k}} + A_{y,\mathfrak{m}},$$

which yield the decomposition of α: $\alpha = \alpha_{\mathfrak{k}} + \alpha_{\mathfrak{m}}$, where

$$\alpha_{\mathfrak{k}} = A_{x,\mathfrak{k}}\,dx + A_{y,\mathfrak{k}}\,dy, \quad \alpha_{\mathfrak{m}} = A_{x,\mathfrak{m}}\,dx + A_{y,\mathfrak{m}}\,dy.$$

Then, we have by a direct computation,

$$\delta(\alpha_{\mathfrak{m}}) = -\mu^{-2}\left\{\frac{\partial A_{x,\mathfrak{m}}}{\partial x} + \frac{\partial A_{y,\mathfrak{m}}}{\partial y}\right\}. \tag{5.1}$$

Indeed, if we take, as an orthonormal frame field with respect to g, $e_1 = \frac{1}{\mu}\frac{\partial}{\partial x}$ and $e_2 = \frac{1}{\mu}\frac{\partial}{\partial y}$. Then, we have

$$\alpha_{\mathfrak{m}}(\nabla_{e_1}e_1) = -\mu^{-3}\frac{\partial\mu}{\partial y}A_{y,\mathfrak{m}}, \quad \alpha_{\mathfrak{m}}(\nabla_{e_2}e_2) = -\mu^{-3}\frac{\partial\mu}{\partial x}A_{x,\mathfrak{m}},$$

and

$$\delta(\alpha_{\mathfrak{m}}) = -\sum_{i=1}^{2}\left\{\nabla_{e_i}(\alpha_{\mathfrak{m}}(e_i)) - \alpha_{\mathfrak{m}}(\nabla_{e_i}e_i)\right\},$$

we have (5.1). □

Next, we have to calculate the harmonic map equation (3.25), and the biharmonic map equation (3.26) in this case.

First, for the left hand side of (3.25), we have

$$-\delta(\alpha_{\mathfrak{m}}) + \sum_{i=1}^{m}[\alpha_{\mathfrak{k}}(e_i), \alpha_{\mathfrak{m}}(e_i)]$$

$$= \mu^{-2}\left\{\frac{\partial A_{x,\mathfrak{m}}}{\partial x} + \frac{\partial A_{y,\mathfrak{m}}}{\partial y}\right\} + [\mu^{-1}A_{x,\mathfrak{k}}, \mu^{-1}A_{x,\mathfrak{m}}] + [\mu^{-1}A_{y,\mathfrak{k}}, \mu^{-1}A_{y,\mathfrak{m}}]$$

$$= \mu^{-2}\left\{\frac{\partial A_{x,\mathfrak{m}}}{\partial x} + \frac{\partial A_{y,\mathfrak{m}}}{\partial y} + [A_{x,\mathfrak{k}}, A_{x,\mathfrak{m}}] + [A_{y,\mathfrak{k}}, A_{y,\mathfrak{m}}]\right\}. \tag{5.2}$$

For the left hand side of (3.26), since $\Delta_g = -\mu^{-2}\left\{\frac{\partial^2}{\partial x^2} + \frac{\partial^2}{\partial y^2}\right\}$, we have

$$\Delta_g\left(-\delta(\alpha_{\mathrm{m}}) + \sum_{i=1}^{m}[\alpha_{\mathfrak{k}}(e_i), \alpha_{\mathrm{m}}(e_i)]\right)$$

$$+ \sum_{s=1}^{m}\left[\left[-\delta(\alpha_{\mathrm{m}}) + \sum_{i=1}^{m}[\alpha_{\mathfrak{k}}(e_i), \alpha_{\mathrm{m}}(e_i)], \alpha_{\mathrm{m}}(e_s)\right], \alpha_{\mathrm{m}}(e_s)\right]$$

$$= -\mu^{-2}\left\{\frac{\partial^2}{\partial x^2} + \frac{\partial^2}{\partial y^2}\right\}\left(\mu^{-2}\left\{\frac{\partial A_{x,\mathrm{m}}}{\partial x} + \frac{\partial A_{y,\mathrm{m}}}{\partial y}\right.\right.$$

$$\left.\left. + [A_{x,\mathfrak{k}}, A_{x,\mathrm{m}}] + [A_{y,\mathfrak{k}}, A_{y,\mathrm{m}}]\right\}\right)$$

$$+ \mu^{-4}\left[\left[\frac{\partial A_{x,\mathrm{m}}}{\partial x} + \frac{\partial A_{y,\mathrm{m}}}{\partial y} + [A_{x,\mathfrak{k}}, A_{x,\mathrm{m}}] + [A_{y,\mathfrak{k}}, A_{y,\mathrm{m}}], A_{x,\mathrm{m}}\right], A_{x,\mathrm{m}}\right]$$

$$+ \mu^{-4}\left[\left[\frac{\partial A_{x,\mathrm{m}}}{\partial x} + \frac{\partial A_{y,\mathrm{m}}}{\partial y} + [A_{x,\mathfrak{k}}, A_{x,\mathrm{m}}] + [A_{y,\mathfrak{k}}, A_{y,\mathrm{m}}], A_{y,\mathrm{m}}\right], A_{y,\mathrm{m}}\right].$$
$$(5.3)$$

Therefore, we have that $\varphi : (\Omega, g) \to (G/K, h)$ is biharmonic if and only if the right hand side of (5.3) vanishes.

Second, we have to examine the integrability condition (3.23) or (3.24). We have

$$d\alpha_{\mathfrak{k}} + \frac{1}{2}[\alpha_{\mathfrak{k}} \wedge \alpha_{\mathfrak{k}}] + \frac{1}{2}[\alpha_{\mathrm{m}} \wedge \alpha_{\mathrm{m}}]$$

$$= \left\{-\frac{\partial A_{x,\mathfrak{k}}}{\partial y} + \frac{\partial A_{y,\mathfrak{k}}}{\partial x} + [A_{x,\mathfrak{k}}, A_{y,\mathfrak{k}}] + [A_{x,\mathrm{m}}, A_{y,\mathrm{m}}]\right\} dx \wedge dy$$

$$= 0,$$

so that we have

$$-\frac{\partial A_{x,\mathfrak{k}}}{\partial y} + \frac{\partial A_{y,\mathfrak{k}}}{\partial x} + [A_{x,\mathfrak{k}}, A_{y,\mathfrak{k}}] + [A_{x,\mathrm{m}}, A_{y,\mathrm{m}}] = 0. \qquad (5.4)$$

For the second equation of (3.24), $d\alpha_{\mathrm{m}} + [\alpha_{\mathfrak{k}} \wedge \alpha_{\mathrm{m}}] = 0$, we have

$$-\frac{\partial A_{x,\mathrm{m}}}{\partial y} + \frac{\partial A_{y,\mathrm{m}}}{\partial x} + [A_{x,\mathfrak{k}}, A_{y,\mathrm{m}}] + [A_{x,\mathrm{m}}, A_{y,\mathfrak{k}}] = 0. \qquad (5.5)$$

Summing up the above, we obtain

THEOREM 5.1. *Let $\Omega \subset \mathbb{R}^2$ an open domian, $g = \mu^2 g_0$, $\mu > 0$, a positive C^∞ function on Ω, and $g_0 = (dx)^2 + (dy)^2$ is standard Riemannian metric on $\mathbb{R}^2$. on which (x, y) is the standard coordinate. Let $(G/K, h)$ a Riemannian symmetric space, with $\pi : G \to G/K$, the projection. For every C^∞ map from Ω into G/K with a local lift $\psi : \Omega \to G$ such that $\varphi = \pi \circ \psi$, let $\alpha = \psi^*\theta$, the pull back of the Maurer-Cartan form θ on G by ψ and decompose it in such a way that*

$\alpha = \alpha_{\mathfrak{k}} + \alpha_{\mathfrak{m}}$ *corresponding to the Cartan decomposition* $\mathfrak{g} = \mathfrak{k} \oplus \mathfrak{m}$.
Then,

(1) $\varphi : (\Omega, g) \to (G/K, h)$ *is harmonic if and only if*

$$\frac{\partial A_{x,\mathfrak{m}}}{\partial x} + \frac{\partial A_{y,\mathfrak{m}}}{\partial y} + [A_{x,\mathfrak{k}}, A_{x,\mathfrak{m}}] + [A_{y,\mathfrak{k}}, A_{y,\mathfrak{m}}] = 0. \qquad (5.6)$$

(2) $\varphi : (\Omega, g) \to (G/K, h)$ *is biharmonic if and only if* (5.3) *vanishes.*

(3) *For the integrability condition,* (5.4) *and* (5.5) *must hold.*

(4) *In particular, for a horizontal lift* ψ, *i.e.,* $\alpha_{\mathfrak{k}} \equiv 0$, *we have*

$$-\left\{\frac{\partial^2}{\partial x^2} + \frac{\partial^2}{\partial y^2}\right\}\left(\mu^{-2}\left\{\frac{\partial P}{\partial x} + \frac{\partial Q}{\partial y}\right\}\right) + \left[\left[\mu^{-2}\left\{\frac{\partial P}{\partial x} + \frac{\partial Q}{\partial y}\right\}, P\right], P\right]$$
$$+ \left[\left[\mu^{-2}\left\{\frac{\partial P}{\partial x} + \frac{\partial Q}{\partial y}\right\}, Q\right], Q\right] = 0, \qquad (5.7)$$

$$[P, Q] = 0, \qquad (5.8)$$

$$-\frac{\partial P}{\partial y} + \frac{\partial Q}{\partial x} = 0, \qquad (5.9)$$

where we put $P := \alpha_{x,\mathfrak{m}}$ *and* $Q := \alpha_{y,\mathfrak{m}}$. *In the case* $\mu = 1$, *the following three equations must hold for the biharmonic map* φ:

$$- P_{xxx} - P_{xyy} - Q_{xxy} - Q_{yyy}$$
$$+ [[P_x + Q_y, P], P] + [[P_x + Q_y, Q], Q] = 0,$$
$$\qquad (5.10)$$

$$[P, Q] = 0, \qquad (5.11)$$

$$P_y - Q_x = 0, \qquad (5.12)$$

where we denote $P_x = \frac{\partial P}{\partial x}$, *etc.*

5.2. Solving the biharmonic map equations.
In this subsection, we want to give the solutions of the equations (5.10), (5.11) and (5.12).

To do it, let us consider the special case that $P_y \equiv 0$ and $Q_x \equiv 0$, i.e., $P(x, y) = P(x)$ and $Q(x, y) = Q(y)$. Then, (5.12) holds clearly. The left hand side of (5.10) coincides with

$$\left\{ - P_{xxx} + [[P_x, P], P]\right\} + \left\{ - Q_{yyy} + [[Q_y, Q], Q]\right\}$$
$$+ [[Q_y, P], P] + [[P_x, Q], Q] = 0. \qquad (5.13)$$

Here, we have that $[[Q_y, P], P] = 0$ and $[[P_x, Q], Q] = 0$. Because, we have due to $Q_x = 0$

$$\frac{\partial}{\partial x}[[P, Q], Q] = [[P_x, Q], Q]. \tag{5.14}$$

But, due to (5.11) the left hand side of (5.14) must vanish. By the same way, we have $[[Q_y, P], P] = 0$.

Thus, (5.13) turns out that

$$-\big\{ - P_{xxx} + \big[[P_x, P], P\big]\big\} = -Q_{yyy} + \big[[Q_y, Q], Q\big] \tag{5.15}$$

But, notice that the left hand side of (5.15) is an $\mathfrak{m}$-valued function only in x and the right hand side of (5.15) is the one only in y, so we have

$$\begin{cases} -P_{xxx} + \big[[P_x, P], P\big] = c, \\ -Q_{yyy} + \big[[Q_y, Q], Q\big] = -c, \end{cases} \tag{5.16}$$

where $c \in \mathfrak{m}$ is a constant vector.

Notice here that both two equations of (5.16) are the same as (4.6) in the case $c = 0$. So, we can obtain the following two theorems by carrying out the similar calculations as in 4.2.

Thus, we have

THEOREM 5.2. *Let $(G/K, h)$ be a Riemannian symmetric space whose rank is bigger than or equal to two, $\mathfrak{g} = \mathfrak{k} \oplus \mathfrak{m}$, the Cartan decomposition, $\mathfrak{a}$, a maximal abelian subalgebra of $\mathfrak{g}$ contained in $\mathfrak{m}$. Let $X, Y \in \mathfrak{a}$ be two elements in $\mathfrak{a}$ which are linearly independent.*

(1) Let us take two $\mathfrak{m}$-valued functions $P(x, y) = (a_1 x^2 + b_1 x + c_1) X$ and $Q(x, y) = (a_2 y^2 + b_2 y + c_2) Y$, where a_i, b_i and c_i $(i = 1, 2)$ are constant real numbers. Then, P and Q are solutions of (5.10), (5.11) and (5.12). For such P and Q, there exists a unique C^∞ map ψ from Ω into G such that $\varphi = \pi \circ \psi$ is a biharmonic mapping form (Ω, g_0) into $(G/K, h)$ with $\varphi(0, 0) = x_0 \in G$ for a fixed point $x_0 \in G/K$. $\varphi : (\Omega, g) \to (G/K, h)$ is harmonic if and only if $a_i = b_i = 0$ $(i = 1, 2)$.

(2) Assume that G is a matrix Lie group, i.e., a subgroup of $GL(N, \mathbb{C})$. Then, the above C^∞ maps $\psi : \Omega \to G$ and $\varphi = \pi \circ \psi$ are given by

$$\begin{cases} \psi(x, y) = x_0 \exp(d_x X + d_y Y) \in G, \\ \varphi(x, y) = x_0 \exp(d_x X + d_y Y) \cdot o \in G/K, \end{cases} \tag{5.17}$$

where $o = \{K\} \in G/K$, $d_x = \frac{a_1}{3} x^3 + \frac{b_1}{2} x^2 + c_1 x$ and $d_y = \frac{a_2}{3} y^3 + \frac{b_2}{2} y^2 + c_2 y$, respectively.

PROOF. We only have to verify the statement (2). By the assumption that $\{X, Y\}$ is abelian, we have for the $\psi(x, y)$ of the form (5.17), as a matrix of degree N,

$$\frac{\partial \psi}{\partial x} = x_0 \exp(d_x X + d_y Y) \cdot \frac{\partial}{\partial x}(d_x X + d_y Y)$$
$$= \psi \cdot (a_1 x^2 + b_1 x + c_1) X$$
$$= \psi P,$$

so we have $\psi^{-1}\frac{\partial \psi}{\partial x} = P$. By the same way, $\psi^{-1}\frac{\partial \psi}{\partial y} = Q$, so we have $\psi^{-1} d\psi = P\,dx + Q\,dy = \alpha$. The mapping ψ is the desired C^∞ mapping of Ω into G, and due to Theorem 3.6, we obtain a biharmonic mapping of (Ω, g_0) into $(G/K, h)$. □

Remark. When $a_i = b_i = 0$ $(i = 1, 2)$, the mapping $\varphi : \mathbb{R}^2 \to (G/K, h)$ is a well known totally geodesic immersion into a Riemannian symmetric space $(G/K, h)$.

By a calculation similar to that in the subsection 4.2, we obtain

THEOREM 5.3. *For the cases of the standard unit sphere (S^n, h), the complex projective space $(\mathbb{C}P^n, h)$, the quaternion one $(\mathbb{H}P^n, h)$, we obtain the following biharmonic mappings of $(\mathbb{R}^2, g_0)$ into them, respectively.*
(1) *Case of (S^n, h):*

$$\varphi_1(t) = x_0\,{}^t(\cos(\sqrt{n}(d_x + d_y)),$$
$$\frac{1}{\sqrt{n}}\sin(\sqrt{n}(d_x + d_y)), \cdots, \frac{1}{\sqrt{n}}\sin(\sqrt{n}(d_x + d_y))),$$

$$(5.18)$$

is a biharmonic mapping of $(\mathbb{R}^2, g_0)$ into (S^n, h), where $x_0 \in G = SO(n+1)$.
(2) *Case of $(\mathbb{C}P^n, h)$:*

$$\varphi_2(t) = x_0\,{}^t(\cos(\sqrt{n}(d_x + d_y)),$$
$$\frac{\sqrt{-1}}{\sqrt{n}}\sin(\sqrt{n}(d_x + d_y)), \cdots, \frac{\sqrt{-1}}{\sqrt{n}}\sin(\sqrt{n}(d_x + d_y))),$$

$$(5.19)$$

is a biharmonic mapping of $(\mathbb{R}^2, g_0)$ into $(\mathbb{C}P^n, h)$, where $x_0 \in G = SU(n+1)$.

(3) *Case of* $(\mathbb{H}P^n, h)$:

$$\varphi_3(t) = x_0{}^t(\cos(\sqrt{n}(d_x + d_y)),$$

$$\frac{k}{\sqrt{n}}\sin(\sqrt{n}(d_x + d_y)), \cdots, \frac{k}{\sqrt{n}}\sin(\sqrt{n}(d_x + d_y))),$$

$$(5.20)$$

is a biharmonic mapping of $(\mathbb{R}^2, g_0)$ *into* $(\mathbb{H}P^n, h)$, *where* $x_0 \in G = Sp(n+1)$, *and* i, j *and* k *are the quaternions satisfying* $i^2 = j^2 = k^2 = -1$ *and* $ij = k$.

Here, *in all the cases,* $d_t = \frac{a}{3}t^3 + \frac{b}{2}t^2 + ct$ *for* $t = x$ *or* $t = y$.

Furthermore, φ_i $(i = 1, 2, 3)$ *are harmonic maps of* $(\mathbb{R}^2, g_0)$ *into* (S^n, h), $(\mathbb{C}P^n, h)$ *or* $(\mathbb{H}P^n, h)$ *if and only if* $a = b = 0$, *respectively.*

CHAPTER 9

Bubbling of Harmonic Maps and Biharmonic Maps

[1]

ABSTRACT. In this chapter, we show the bubbling phenomena of harmonic maps and biharmonic maps. The bubbling phenomena means a kind of compactness, namely, how small the totality of harmonic maps and the one of biharmonic maps are small.

1. Introduction

The theory of bubbling phenomena of harmonic maps was first studied by Sacks and Uhlenbeck [138] and extended to several variational problems including Yang-Mills theory (see Freed and Uhlenbeck [51])). For the bubbling phenomena of biharmonic maps, we will show

THEOREM 1.1. (*cf.* **Theorem 5.1**) *Let (M, g) and (N, h) be two compact Riemannian manifolds. Assume that $m = \dim M \geq 3$. For every positive constant $C > 0$, let us consider a family of smooth biharmonic maps of (M, g) into (N, h),*

$$\mathcal{F} = \Big\{ \varphi : (M, g) \to (N, h), \text{ smooth biharmonic } |$$

$$\int_M | \, d\varphi \, |^m \, v_g \leq C \text{ and } \int_M |\tau(\varphi)|^2 \, v_g \leq C \Big\}. \tag{1.1}$$

Assume that $m = \dim M \geq 3$. Then, any sequence in $\mathcal{F}$ causes a **bubbling***: Namely, for any sequence $\{\varphi_i\} \in \mathcal{F}$, there exist a finite set $\mathcal{S}$ in M, say, $\mathcal{S} = \{x_1, \cdots, x_\ell\}$, and a smooth biharmonic map $\varphi_\infty : (M \backslash \mathcal{S}, g) \to (N, h)$ suth that,*
(1) a subsequence φ_{i_j} converges φ_∞ in the C^∞-topology on $M \backslash \mathcal{S}$, as $j \to \infty$, and

---

[1]This chapter is due to [110]: N. Nakauchi and H. Urakawa, *Bubbling phenomena of biharmonic maps*, J. Geom. Phys. **98** (2015), 355–375.

(2) *the Radon measures* $|d\varphi_{i_j}|^m v_g$ *converges to a measure*

$$|d\varphi_\infty|^m v_g + \sum_{i=1}^{\ell} a_k \, \delta_{x_k}, \qquad (1.2)$$

as $j \to \infty$. *Here* a_k *is a constant, and* δ_{x_k} *is the Dirac measure whose support is* $\{x_k\}$ $(k = 1 \cdots, \ell)$.

As an application, we have the bubbling theorem for harmonic maps (cf. Theorem 5.2).

Acknowledgement.   We would like to express our gratitude to Professor H. Naito for his useful suggestions and comments during the preparations of this paper.

2. Preliminaries

In this section, we prepare materials for the first and second variational formulas for the bienergy functional and biharmonic maps. Let us recall the definition of a harmonic map $\varphi : (M, g) \to (N, h)$, of a compact Riemannian manifold (M, g) into another Riemannian manifold (N, h), which is an extremal of the *energy functional* defined by

$$E(\varphi) = \int_M e(\varphi)\, v_g,$$

where $e(\varphi) := \frac{1}{2}|d\varphi|^2$ is called the energy density of φ. That is, for any variation $\{\varphi_t\}$ of φ with $\varphi_0 = \varphi$,

$$\frac{d}{dt}\Big|_{t=0} E(\varphi_t) = -\int_M h(\tau(\varphi), V) v_g = 0, \qquad (2.1)$$

where $V \in \Gamma(\varphi^{-1}TN)$ is a variation vector field along φ which is given by $V(x) = \frac{d}{dt}|_{t=0}\varphi_t(x) \in T_{\varphi(x)}N$, $(x \in M)$, and the *tension field* is given by $\tau(\varphi) = \sum_{i=1}^{m} B(\varphi)(e_i, e_i) \in \Gamma(\varphi^{-1}TN)$, where $\{e_i\}_{i=1}^{m}$ is a locally defined frame field on (M, g), and $B(\varphi)$ is the second fundamental form of φ defined by

$$\begin{aligned}
B(\varphi)(X, Y) &= (\widetilde{\nabla}d\varphi)(X, Y) \\
&= (\widetilde{\nabla}_X d\varphi)(Y) \\
&= \overline{\nabla}_X (d\varphi(Y)) - d\varphi(\nabla_X Y) \\
&= \nabla^N_{d\varphi(X)} d\varphi(Y) - d\varphi(\nabla_X Y), \qquad (2.2)
\end{aligned}$$

for all vector fields $X, Y \in \mathfrak{X}(M)$. Furthermore, ∇, and ∇^N, are connections on TM, TN of (M, g), (N, h), respectively, and $\overline{\nabla}$, and $\widetilde{\nabla}$

are the induced ones on $\varphi^{-1}TN$, and $T^*M \otimes \varphi^{-1}TN$, respectively. By (2.1), φ is harmonic if and only if $\tau(\varphi) = 0$.

The second variation formula is given as follows. Assume that φ is harmonic. Then,

$$\left.\frac{d^2}{dt^2}\right|_{t=0} E(\varphi_t) = \int_M h(J(V), V)v_g, \tag{2.3}$$

where J is an elliptic differential operator, called *Jacobi operator* acting on $\Gamma(\varphi^{-1}TN)$ given by

$$J(V) = \overline{\Delta}V - \mathcal{R}(V), \tag{2.4}$$

where $\overline{\Delta}V = \overline{\nabla}^*\overline{\nabla}V = -\sum_{i=1}^m \{\overline{\nabla}_{e_i}\overline{\nabla}_{e_i}V - \overline{\nabla}_{\nabla_{e_i}e_i}V\}$ is the *rough Laplacian* and $\mathcal{R}$ is a linear operator on $\Gamma(\varphi^{-1}TN)$ given by $\mathcal{R}V = \sum_{i=1}^m R^N(V, d\varphi(e_i))d\varphi(e_i)$, and R^N is the curvature tensor of (N, h) given by $R^N(U, V) = \nabla^N_U \nabla^N_V - \nabla^N_V \nabla^N_U - \nabla^N_{[U,V]}$ for $U, V \in \mathfrak{X}(N)$.

J. Eells and L. Lemaire [**40**] proposed polyharmonic (k-harmonic) maps and Jiang [**74**] studied the first and second variation formulas of biharmonic maps. Let us consider the *bienergy functional* defined by

$$E_2(\varphi) = \frac{1}{2} \int_M |\tau(\varphi)|^2 v_g, \tag{2.5}$$

where $|V|^2 = h(V, V)$, $V \in \Gamma(\varphi^{-1}TN)$.

Then, the first variation formula of the bienergy functional is given as follows.

THEOREM 2.1.   (*the first variation formula*)

$$\left.\frac{d}{dt}\right|_{t=0} E_2(\varphi_t) = -\int_M h(\tau_2(\varphi), V)v_g. \tag{2.6}$$

Here,

$$\tau_2(\varphi) := J(\tau(\varphi)) = \overline{\Delta}\tau(\varphi) - \mathcal{R}(\tau(\varphi)), \tag{2.7}$$

which is called the bitension field *of φ, and J is given in (2.4).*

DEFINITION 2.1. *A smooth map φ of M into N is said to be* biharmonic *if $\tau_2(\varphi) = 0$.*

3. The Bochner-type estimation for the tension field of a biharmonic map

In this section, we give the Bochner-type estimations for the tension fields of biharmonic maps into a Riemannian manifold (N, h) of nonpositive curvature.

LEMMA 3.1. *Assume that the sectional curvature of (N, h) is non-positive, and $\varphi : (M \backslash \mathcal{S}, g) \to (N, h)$ is a biharmonic mapping for some closed set $\mathcal{S}$ of M. Then, it holds that*

$$\Delta \, |\tau(\varphi)|^2 \geq 2 \, |\overline{\nabla} \tau(\varphi)|^2 \qquad (3.1)$$

at each point in $M \backslash \mathcal{S}$. Here, Δ is the Laplace-Beltrami operator of (M, g).

Proof   Let us take a local orthonormal frame field $\{e_i\}_{i=1}^m$ on $M \backslash \mathcal{S}$, and $\varphi : (M \backslash \mathcal{S}, g) \to (N, h)$, a biharmonic map. Then, for $V := \tau(\varphi) \in \Gamma(\varphi^{-1} TN)$, we have

$$\begin{aligned}
\frac{1}{2} \Delta \, |V|^2 &= \frac{1}{2} \sum_{i=1}^m \left\{ e_i^2 \, |V|^2 - \nabla_{e_i} e_i \, |V|^2 \right\} \\
&= \sum_{i=1}^m \left\{ e_i \, h(\overline{\nabla}_{e_i} V, V) - h(\overline{\nabla}_{\nabla_{e_i} e_i} V, V) \right\} \\
&= \sum_{i=1}^m \left\{ h(\overline{\nabla}_{e_i} \overline{\nabla}_{e_i} V, V) - h(\overline{\nabla}_{\nabla_{e_i} e_i} V, V) \right\} \\
&\quad + \sum_{i=1}^m h(\overline{\nabla}_{e_i} V, \overline{\nabla}_{e_i} V) \\
&= h(-\overline{\Delta} V, V) + |\overline{\nabla} V|^2 \\
&= h(-\mathcal{R}(V), V) + |\overline{\nabla} V|^2 \\
&\geq |\overline{\nabla} V|^2, \qquad (3.2)
\end{aligned}$$

because for the second last equality, we used $\overline{\Delta} V - \mathcal{R}(V) = J(V) = 0$ for $V = \tau(\varphi)$, due to the biharmonicity of $\varphi : (M \backslash \mathcal{S}, g) \to (N, h)$, and for the last inequality of (3.2), we used

$$h(\mathcal{R}(V), V) = \sum_{i=1}^m h(R^N(V, \varphi_* e_i) \varphi_* e_i, V) \leq 0 \qquad (3.3)$$

since the sectional curvature of (N, h) is non-positive.   □

By Lemma 3.1, we have

LEMMA 3.2. *Under the same assumptions as Lemma 3.1, we have*

$$|\tau(\varphi)| \, \Delta \, |\tau(\varphi)| \geq 0. \qquad (3.4)$$

Proof Due to Lemm 3.1, we have

$$\begin{aligned}
2 \, |\overline{\nabla} \tau(\varphi)|^2 &\leq \Delta \, |\tau(\varphi)|^2 \\
&= 2 \, |\tau(\varphi)| \, \Delta \, |\tau(\varphi)| + 2 \, | \nabla \, |\tau(\varphi)| \, |^2. \qquad (3.5)
\end{aligned}$$

Thus, we have

$$|\tau(\varphi)| \, \Delta \, |(\tau(\varphi)| \geq |\overline{\nabla}\tau(\varphi)|^2 - |\, \nabla \, |\tau(\varphi)| \, |^2$$
$$\geq 0. \tag{3.6}$$

Here, to see the last inequality of (3.6), it suffices to notice that for all $V \in \Gamma(\varphi^{-1}TN)$,

$$|\overline{\nabla}V| \geq |\, \nabla \, |V| \, | \tag{3.7}$$

which follows from that

$$|V| \, |\, \nabla \, |V| \, | = \frac{1}{2} |\, \nabla \, |V|^2 \, |$$
$$= \frac{1}{2} |\nabla \, h(V,V)|$$
$$= |\, h(\overline{\nabla}V, V) \, |$$
$$\leq |\overline{\nabla}V| \, |V|. \tag{3.8}$$

This proves Lemma 3.2. □

Then, by using Moser's iteration technique due to this Lemma 3.2, we have the following theorem.

THEOREM 3.1. *Assume that* (M, g) *is a compact Riemannian manifold of* $\dim M = m \geq 3$, *and the sectional curvature of* (N, h) *is nonpositive. Then, there exists a positive constant* $C > 0$ *depending only on* $\dim M$ *such that for every biharmonic mapping* $\varphi : (M \backslash S, g) \to (N, h)$ *with* $S = \{x_1, \cdots, x_\ell\}$, *every positive number* $r > 0$ *and each point* $x_i \in S$,

$$\sup_{B_r(x_i)} |\tau(\varphi)| \leq \frac{C}{r^{m/2}} \left(\int_{B_{2r}(x_i)} |\tau(\varphi)|^2 v_g \right)^{1/2}, \tag{3.9}$$

where $B_r(x_i) = \{x \in M; \, r(x, x_i) < r\}$ *is the metric ball in* (M, g) *around* x_i *of radius* r, *for every sufficiently small* $r > 0$ *in such a way that* $B_r(x_i) \cap B_r(x_j) = \emptyset \, (i \neq j)$.

4. Moser's iteration technique and proof of Theorem 3.3

(*The first step*)   For a fixed point $x_i \in S$, and for every $0 < \rho_1 < \rho_2 < \infty$, we first take a cutoff C^∞ function η on M (for instance, see

[**83**]) satisfying that

$$
\begin{cases}
0 \le \eta(x) \le 1 & (x \in M), \\
1 & (x \in B_{\rho_1}(x_i)), \\
0 & (x \notin B_{\rho_2}(x_i)), \\
|\nabla \eta| \le \dfrac{2}{\rho_2 - \rho_1} & (x \in M).
\end{cases}
\tag{4.1}
$$

For $2 \le p < \infty$, multiply $|\tau(\varphi)|^{p-2}\eta^2$ to both hand sides of the inequality (3.4) in Lemma 3.2, and integrate over M, we have

$$
\begin{aligned}
0 &\le \int_M |\tau(\varphi)|^{p-1}\eta^2 \Delta \left(|\tau(\varphi)|\right) v_g \\
&= -\int_M g(\nabla(|\tau(\varphi)|^{p-1}\eta^2), \nabla |\tau(\varphi)|) v_g \\
&= -(p-1)\int_M |\tau(\varphi)|^{p-2}\eta^2 \, | \nabla(|\tau(\varphi)|) |^2 v_g \\
&\quad -2 \int_M |\tau(\varphi)|^{p-1} \eta \, g(\nabla(|\tau(\varphi)|), \nabla \eta) v_g \\
&= -\frac{4(p-1)}{p^2} \int_M |\nabla(|\tau(\varphi)|^{p/2})|^2 \, \eta^2 \, v_g \\
&\quad -\frac{4}{p} \int_M g(\eta \, \nabla(|\tau(\varphi)|^{p/2}), |\tau(\varphi)|^{p/2} \, \nabla \eta) v_g.
\end{aligned}
\tag{4.2}
$$

Therefore, by using Young's inequality, we have, for every positive real number $\epsilon > 0$,

$$
\begin{aligned}
\int_M |\nabla(\,|\tau(\varphi)|^{p/2}\,)|^2 \, \eta^2 \, v_g &\le \frac{p}{p-1} \int_M g(\eta \, \nabla(|\tau(\varphi)|^{p/2}), |\tau(\varphi)|^{p/2} \, \nabla \eta) v_g \\
&\le \frac{p}{2(p-1)} \Bigg\{ \epsilon \int_M \eta^2 \, | \nabla(|\tau(\varphi)|^{p/2})|^2 \, v_g \\
&\quad + \frac{1}{\epsilon} \int_M |\tau(\varphi)|^p \, |\nabla \eta|^2 \, v_g \Bigg\}.
\end{aligned}
\tag{4.3}
$$

By (4.3), we have

$$
\begin{aligned}
\left(1 - \frac{p}{2(p-1)}\epsilon\right) \int_M \eta^2 \, | \nabla(|\tau(\varphi)|^{p/2})|^2 \, v_g \\
\le \frac{p}{2(p-1)} \frac{1}{\epsilon} \int_M |\tau(\varphi)|^p \, |\nabla \eta|^2 \, v_g.
\end{aligned}
\tag{4.4}
$$

By choosing $\epsilon = \frac{p-1}{p}$ in (4.4), we have

$$\int_M \eta^2 \, |\nabla(|\tau(\varphi)|^{p/2})|^2 \, v_g \leq \frac{p^2}{(p-1)^2} \int_M |\tau(\varphi)|^p \, |\nabla\eta|^2 \, v_g.$$
(4.5)

Here, by using

$$\nabla(|\tau(\varphi)|^{p/2} \, \eta) = \eta \, \nabla(|\tau(\varphi)|^{p/2}) + |\tau(\varphi)|^{p/2} \, \nabla\eta,$$

$|A+B|^2 \leq 2\,|A|^2 + 2\,|B|^2$ and (4.5), and then, by (4.1), we have

$$\int_M |\nabla(|\tau(\varphi)|^{p/2} \, \eta)|^2 \, v_g \leq 2 \int_M \eta^2 \, |\nabla(|\tau(\varphi)|^{p/2})|^2 \, v_g$$

$$+ \, 2 \int_M |\tau(\varphi)|^p \, |\nabla\eta|^2 \, v_g$$

$$\leq 4 \, \frac{p^2}{(p-1)^2} \int_M |\tau(\varphi)|^p \, |\nabla\eta|^2 \, v_g$$

$$\leq \frac{p^2}{(p-1)^2} \, \frac{16}{(\rho_2-\rho_1)^2} \int_{B_{\rho_2}(x_i)} |\tau(\varphi)|^p \, v_g.$$
(4.6)

For the left hand side of (4.6), let us recall the Sobolev embedding theorem (cf. [3], p. 55; [51], p. 95):

$$H_1^2(M) \subset L^\gamma(M),$$
(4.7)

where $\gamma := \frac{2m}{m-2}$, i.e., there exists a pocitive constant $C > 0$ such that

$$\left(\int_M |f|^\gamma \, v_g \right)^{1/\gamma} \leq C \left(\int_M |\nabla f|^2 \, v_g \right)^{1/2} \qquad (\forall f \in H_1^2(M)).$$
(4.8)

Therefore, we have

$$\int_M |\nabla(|\tau(\varphi)|^{p/2} \, \eta)|^2 \, v_g \geq \frac{1}{C} \left(\int_M \left\{ |\tau(\varphi)|^{p/2} \eta \right\}^\gamma v_g \right)^{2/\gamma}$$

$$\geq \frac{1}{C} \left(\int_{B_{\rho_1}(x_i)} \left(|\tau(\varphi)|^{p/2} \right)^\gamma v_g \right)^{2/\gamma},$$
(4.9)

where we used (4.1).

Thus, together with (4.6) and (4.9), we have

LEMMA 4.1. *Assume that (M,g) is a compact Riemannian manifold, the sectional curvature of (N,h) is non-positive, and $\varphi\colon (M\backslash\mathcal{S}, g) \to (N,h)$ is a biharmonic mapping, where $\mathcal{S} = \{x_1, \cdots, x_\ell\} \subset$*

M. Then, for each $0 < \rho_1 < \rho_2 < \infty$, and $2 \le p < \infty$, it holds that for each $i = 1, \cdots, \ell$,

$$
\left(\int_{B_{\rho_1}(x_i)} \left(|\tau(\varphi)|^{p/2} \right)^\gamma v_g \right)^{1/\gamma} \le \frac{p}{p-1} \frac{C'}{\rho_2 - \rho_1} \times
$$

$$
\times \left(\int_{B_{\rho_2}(x_i)} \left(|\tau(\varphi)|^{p/2} \right)^2 v_g \right)^{1/2}, \tag{4.10}
$$

where $C' = 4\sqrt{C}$, and $C > 0$ is the Sobolev constant in (4.8) and $\gamma := \frac{2m}{m-2}$, $m = \dim M \ge 3$.

(*The second step*)   Here, let us define

$$
\begin{cases}
\overline{\gamma} := \dfrac{m}{m-2} = \dfrac{1}{2}\gamma, \\
p_k := 2\,\overline{\gamma}^{k-1} \to \infty \quad (k \to \infty), \\
r_k := \left(1 + \dfrac{1}{2^{k-1}}\right) r \to r \quad (k \to \infty),
\end{cases} \tag{4.11}
$$

and in (4.10), let us put

$$
\begin{cases}
p := p_k, \\
\rho_1 := r_{k+1}, \\
\rho_2 := r_k.
\end{cases}
$$

Then, we have

$$
\begin{cases}
\dfrac{p\,\gamma}{2} = p_k\,\overline{\gamma} = 2\,\overline{\gamma}^k = p_{k+1}, \\
\rho_2 - \rho_1 = r_k - r_{k+1} = \left(\dfrac{1}{2^{k-1}} - \dfrac{1}{2^k}\right) r = \dfrac{1}{2^k}\, r, \tag{4.12}
\end{cases}
$$

so that (4.10) can be rewritten as follows.

$$
\left(\int_{B_{r_{k+1}}(x_i)} |\tau(\varphi)|^{p_{k+1}} v_g \right)^{1/\gamma} \le \frac{2\,\overline{\gamma}^{k-1}}{2\,\overline{\gamma}^{k-1} - 1} \frac{2^k}{r} \times
$$

$$
\times \left(\int_{B_{r_k}(x_i)} |\tau(\varphi)|^{p_k} v_g \right)^{1/2}. \tag{4.13}
$$

By taking $\frac{1}{\overline{\gamma}^{k-1}}$ power of (4.13), we have

$$
\|\tau(\varphi)\|_{L^{p_{k+1}}(B_{r_{k+1}}(x_i))} \le \left(\frac{2\,\overline{\gamma}^{k-1}}{2\,\overline{\gamma}^{k-1} - 1} \right)^{2/p_k} \frac{2^{(k/\overline{\gamma}^{k-1})}}{r^{(1/\overline{\gamma}^{k-1})}} \times
$$

$$
\times \|\tau(\varphi)\|_{L^{p_k}(B_{r_k}(x_i))} \tag{4.14}
$$

since, for the power of the left hand side of (4.13), we calculated as

$$\frac{1}{\gamma}\frac{1}{\overline{\gamma}^{k-1}} = \frac{1}{2\overline{\gamma}}\frac{1}{\overline{\gamma}^{k-1}} = \frac{1}{2\overline{\gamma}^k} = \frac{1}{p_{k+1}}.$$

(*The third step*)   Now iterate (4.14), then we have

$$\|\tau(\varphi)\|_{L^{p_{k+1}}(B_{r_{k+1}}(x_i))} \le \prod_{k=1}^{\infty}\left(\frac{2\,\overline{\gamma}^{k-1}}{2\,\overline{\gamma}^{k-1}-1}\right)^{2/p_k} \frac{2^{(k/\overline{\gamma}^{k-1})}}{r^{(1/\overline{\gamma}^{k-1})}} \times$$
$$\times \|\tau(\varphi)\|_{L^2(B_{2r}(x_i))} \qquad (4.15)$$

since $p_1 = 2$ and $r_1 = 2r$. Here, we notice that

$$\prod_{k=1}^{\infty}\frac{1}{r^{(1/\overline{\gamma}^{k-1})}} = \frac{1}{r^{(\sum_{k=1}^{\infty}1/\overline{\gamma}^{k-1})}} = \frac{1}{r^{m/2}} \qquad (4.16)$$

since

$$\sum_{k=1}^{\infty}\frac{1}{\overline{\gamma}^{k-1}} = \frac{1}{1-\frac{1}{\overline{\gamma}}} = \frac{1}{1-\frac{m-2}{m}} = \frac{m}{2}.$$

Notice also that

$$\prod_{k=1}^{\infty}\frac{1}{(2\,\overline{\gamma}^{k-1}-1)^{2/p_k}} \le 1 \qquad (4.17)$$

since $2\overline{\gamma}^{k-1} - 1 > 2 - 1 = 1$ when $\overline{\gamma} = m/(m-2) > 1$ $(m \ge 3)$. And also notice that

$$\prod_{k=1}^{\infty}2^{(k/\overline{\gamma}^{k-1})} = 2^{\sum_{k=1}^{\infty}\frac{k}{\overline{\gamma}^{k-1}}} < \infty, \qquad (4.18)$$

$$\prod_{k=1}^{\infty}(2\,\overline{\gamma})^{2(k-1)/p_k} = \gamma^{2\sum_{k=1}^{\infty}\frac{k-1}{p_k}} = \gamma^{\sum_{k=1}^{\infty}\frac{k-1}{\overline{\gamma}^{k-1}}} < \infty.$$
$$\qquad (4.19)$$

Therefore, (4.15) turns out that

$$\|\tau(\varphi)\|_{L^{p_{k+1}}(B_{r_{k+1}}(x_i))} \le C''\frac{1}{r^{m/2}}\|\tau(\varphi)\|_{L^2(B_{2r}(x_i))}$$
$$\qquad (4.20)$$

for some positive constant C'' depending only on $m = \dim M$.

(*The fourth step*)   Now, let k tend to infinity. Then, by (4.11), the norm $\|\tau(\varphi)\|_{L^{p_{k+1}}(B_{r_{k+1}}(x_i))}$ tends to

$$\|\tau(\varphi)\|_{L^{\infty}(B_r(x_i))} = \sup_{B_r(x_i)}|\tau(\varphi)|.$$

Thus, we obtain

$$\sup_{B_r(x_i)} |\tau(\varphi)| \le \frac{C''}{r^{m/2}} \|\tau(\varphi)\|_{L^2(B_{2r}(x_i))}, \tag{4.21}$$

which is the desired inequality (3.9). We have Theorem 3.3.   □

Due to Theorem 3.3 , we have immediately

THEOREM 4.1. *Assume that* (M, g) *is a compact Riemannian manifold of* $\dim M = m \ge 3$, *and the sectional curvature of* (N, h) *is non-positive, and there exists a finite set* $\mathcal{S}$ *of points in* M, *say* $\mathcal{S} = \{x_1, \cdots, x_\ell\}$, *such that* $\varphi : (M \backslash \mathcal{S}, g) \to (N, h)$ *is a biharmonic map and have the finite bienergy:*

$$E_2(\varphi) = \frac{1}{2} \int_M |\tau(\varphi)|^2 \, v_g < \infty. \tag{4.22}$$

Then, the norm $|\tau(\varphi)|$ *of the tension field* $\tau(\varphi)$ *is bounded on* M. *So,* $|\tau(\varphi)|$ *has a unique continuous extension on* (M, g).

REMARK 4.1. *In the case of* $\dim M = m = 2$, *Theorem 4.2 does not hold. Indeed, consider a real valued function* φ *on* $\mathbf{B}^2 \backslash \{o\}$ *defined by*

$$\varphi(x) = \frac{r^2}{4}(\log r - 1),$$

where $r = r(x) = |x|$ *for* $x \in \mathbf{B}^2 \backslash \{o\}$. *Since* $\Delta \varphi = \log r$, $E_2(\varphi) < \infty$, *but* $|\tau(\varphi)| = |\Delta \varphi|$ *is singular at the origin* o.

REMARK 4.2. *If we assume the boundedness of* $|d\varphi|$ *on* M *in addition the assumptions of Theorem 4.2, then* φ *can be uniquely extended to a biharmonic map of* (M, g) *into* (N, h). *However, notice here that the function* $\varphi(z) := \frac{1}{z}$ *on* $(\mathbb{C} \cup \{\infty\}) \backslash \{0\}$ *cannot be extended to* $\mathbb{C} \cup \{\infty\}$. *Indeed, it is holomorphic and harmonic, but* $|d\varphi|$ *is not bounded on* $(\mathbb{C} \cup \{\infty\}) \backslash \{0\}$.

5. Bubbling theorem of biharmonic maps

We have the following bubbling theorem for biharmonic maps.

THEOREM 5.1. (*Bubbling for Biharmonic Maps*) *Let* (M, g) *and* (N, h) *be two compact Riemannian manifolds. Assume that* $\dim M = m \ge 3$. *For every positive constant* $C > 0$, *consider a family of smooth* *biharmonic maps of (M, g) into (N, h),*

$$\mathcal{F} = \left\{ \varphi : (M, g) \to (N, h), \text{ smooth biharmonic } \Big| \right.$$

$$\left. \int_M |d\varphi|^m v_g \leq C \text{ and } \int_M |\tau(\varphi)|^2 v_g \leq C \right\}. \tag{5.1}$$

Then, any sequence in $\mathcal{F}$ causes a **bubbling***: Namely, for any sequence $\{\varphi_i\} \in \mathcal{F}$, there exist a finite set $\mathcal{S}$ in M, say, $\mathcal{S} = \{x_1, \cdots, x_\ell\}$, and a smooth biharmonic map $\varphi_\infty : (M\backslash\mathcal{S}, g) \to (N, h)$ suth that,*

(1) a subsequence φ_{i_j} converges φ_∞ in the C^∞-topology on $M\backslash\mathcal{S}$, as $j \to \infty$, and

(2) the Radon measures $|d\varphi_{i_j}|^m v_g$ converges to a measure

$$|d\varphi_\infty|^m v_g + \sum_{i=1}^\ell a_k \, \delta_{x_k}, \tag{5.2}$$

as $j \to \infty$. Here a_k is a constant, and δ_{x_k} is the Dirac measure center at $\{x_k\}$ $(k = 1 \cdots, \ell)$.

As a corollary, we have immediately

THEOREM 5.2. (*Bubbling for Harmonic Maps*) *Let (M, g) and (N, h) be two compact Riemannian manifolds. Assume that $\dim M = m \geq 3$. For every positive constant $C > 0$, let us consider a family of smooth harmonic maps of (M, g) into (N, h),*

$$\mathcal{F}^h = \left\{ \varphi : (M, g) \to (N, h), \text{ smooth harmonic } \Big| \int_M |d\varphi|^m v_g \leq C \right\}. \tag{5.3}$$

Then, any sequence in $\mathcal{F}^h$ causes a **bubbling***: Namely, for any sequence $\{\varphi_i\} \in \mathcal{F}^h$, there exist a finite set $\mathcal{S}$ in M, say, $\mathcal{S} = \{x_1, \cdots, x_\ell\}$, and a smooth harmonic map $\varphi_\infty : (M\backslash\mathcal{S}, g) \to (N, h)$ suth that,*

(1) a subsequence φ_{i_j} converges φ_∞ in the C^∞-topology on $M\backslash\mathcal{S}$, as $j \to \infty$, and

(2) the Radon measures $|d\varphi_{i_j}|^m v_g$ converges to a measure

$$|d\varphi_\infty|^m v_g + \sum_{i=1}^\ell a_k \, \delta_{x_k}, \tag{5.4}$$

as $j \to \infty$. Here a_k is a constant, and δ_{x_k} is the Dirac measure center at $\{x_k\}$ $(k = 1 \cdots, \ell)$.

PROOF. For any sequence in $\{\varphi_i\} \in \mathcal{F}^h$, the limit φ_∞ in Theorem 5.1 is a smooth biharmonic map of $(M\backslash\mathcal{S}, g)$ into (N, h). Due to (1) of Theorem 5.1, φ_{i_j} converges to φ_∞ in the C^∞-topology on $M\backslash\mathcal{S}$, so that $\tau(\varphi_{i_j})$ converges to $\tau(\varphi_\infty)$ pointwise on $M\backslash\mathcal{S}$. Since $\tau(\varphi_{i_j}) \equiv 0$, we have $\tau(\varphi_\infty) \equiv 0$ on $M\backslash\mathcal{S}$, i.e., φ_∞ is harmonic on $M\backslash\mathcal{S}$. And (1) and (2) hold also due to Theorem 5.1.   □

6. Basic inequalities

To prove Theorem 5.1, it is necessary to prepare the following two basic inequalities.

LEMMA 6.1. *Assume that the sectional curvature of (N, h) is bounded above by a constant C. Then, we have*

$$\frac{1}{2} \Delta |V|^2 + C |d\varphi|^2 |V|^2 \geq |\overline{\nabla} V|^2 \tag{6.1}$$

for all $V \in \Gamma(\varphi^{-1}TN)$.

PROOF. Indeed, let us recall (3.2)

$$\frac{1}{2} \Delta |V|^2 = h(-\mathcal{R}(V), V) + |\overline{\nabla} V|^2, \tag{6.2}$$

for all $V \in \Gamma(\varphi^{-1}TN)$. Since

$$h(\mathcal{R}(V), V) = \sum_{i=1}^{m} h(R^N(V, d\varphi(e_i))d\varphi(e_i), V),$$

the right hand side of (6.2) is bigger than or equal to

$$-C\sum_{i=1}^{m} |d\varphi(e_i)|^2 |V|^2 + |\overline{\nabla} V|^2 = -C |d\varphi|^2 |V|^2 + |\overline{\nabla} V|^2.$$

We have (6.1).   □

LEMMA 6.2. *Under the same assumption as Lemma 6.1, we have*

$$|\tau(\varphi)| \Delta |\tau(\varphi)| + C |d\varphi|^2 |\tau(\varphi)|^2 \geq 0 \tag{6.3}$$

for all $\varphi \in C^\infty(M, N)$.

PROOF. The proof goes in the same way as Lemma 3.2. Indeed, by the equality of (3.5) in the proof of Lemma 3.2 and also Lemma 6.1,

we have

$$|\tau(\varphi)| \, \Delta \, |\tau(\varphi)| + |\nabla \, |\tau(\varphi)| \, |^2 + C \, |d\varphi|^2 \, |\tau(\varphi)|^2$$

$$\geq \frac{1}{2} \Delta \, |\tau(\varphi)| + C \, |d\varphi|^2 \, |\tau(\varphi)|^2$$

$$\geq |\overline{\nabla} \tau(\varphi)|^2. \tag{6.4}$$

So that we have

$$|\tau(\varphi)| \, \Delta \, |\tau(\varphi)| + C \, |d\varphi|^2 \, |\tau(\varphi)|^2$$

$$\geq |\overline{\nabla} \tau(\varphi)|^2 - |\nabla \, |\tau(\varphi)| \, |^2, \tag{6.5}$$

due to (3.6) in the proof of Lemma 3.2. Notice that the right hand side of (6.5) is non-negative for $V = \tau(\varphi)$ by (3.7). This proves Lemma 6.2. $\qquad\square$

PROPOSITION 6.1. *Assume that the sectional curvature of (N, h) is bounded above by a positive constant $C > 0$. Then, there exists a positive number $\epsilon_0 > 0$ depending only on the Sobolev constant of (M, g) and C such that for every smooth biharmonic map φ of (M, g) into (N, h), if*

$$\int_{B_r(x_0)} |d\varphi|^m \, v_g \leq \epsilon_0, \tag{6.6}$$

then

$$\sup_{B_{r/2}(x_0)} |\tau(\varphi)|^2 \leq \frac{C'}{r^{m/2}} \int_{B_r(x_0)} |\tau(\varphi)|^2 \, v_g. \tag{6.7}$$

for some positive constant $C' > 0$ depending only on C and $m = \dim M$.

PROOF. The proof of Proposition 6.3 goes in the same line of the one of Theorem 3.3. We retain the situation in Section Three. Multiply $|\tau(\varphi)|^{p-2} \eta^2$ to both hand sides of (6.3) and integrate it over M. Then, we have

$$0 \leq \int_M |\tau(\varphi)|^{p-1} \eta^2 \, \Delta(|\tau(\varphi)|) \, v_g + C \int_M |d\varphi|^2 \, |\tau(\varphi)|^p \, \eta^2 \, v_g. \tag{6.8}$$

In order to estimate the second term of the right hand side of (6.8), we need the following lemma.

LEMMA 6.3. *We have*

$$\int_M |d\varphi|^2 \, |\tau(\varphi)|^p \, \eta^2 \, v_g$$

$$\leq C'' \left\{ \int_{B_r(x_0)} |d\varphi|^m \, v_g \right\}^{2/m} \int_M |\nabla(\, |\tau(\varphi)|^{p/2} \, \eta)\,|^2 \, v_g, \tag{6.9}$$

where $C'' > 0$ is a positive constant independent on φ.

We postpone the proof of Lemma 6.4, and we continue the proof of Proposition 6.3.

In the first step of the proof of Theorem 3.3, we have instead of (4.2), by (6.8),

$$0 \leq \int_M |\tau(\varphi)|^{p-1} \, \eta^2 \, \Delta(|\tau(\varphi)|) \, v_g + C \int_M |d\varphi|^2 \, |\tau(\varphi)|^p \, \eta^2 \, v_g$$

$$= -\frac{4(p-1)}{p^2} \int_M |\nabla(|\tau(\varphi)|^{p/2})|^2 \, \eta^2 \, v_g$$

$$- \frac{4}{p} \int_M g(\eta \, \nabla(|\tau(\varphi)|^{p/2}), |\tau(\varphi)|^{p/2} \, \nabla\eta) \, v_g$$

$$+ C \int_M |d\varphi|^2 \, |\tau(\varphi)|^p \, \eta^2 \, v_g. \tag{6.10}$$

By the same argument as in (4.3), (4.4) and (4.5), (4.5) is changed into the following:

$$\int_M \eta^2 \, |\nabla(|\tau(\varphi)|^{p/2})|^2 \, v_g \leq \frac{p^2}{(p-1)^2} \int_M |\tau(\varphi)|^p \, |\nabla\eta|^2 \, v_g$$

$$+ C \int_M |d\varphi|^2 \, |\tau(\varphi)|^p \, \eta^2 \, v_g. \tag{6.11}$$

And then, by the same as in (4.6), we have,

$$\int_M |\nabla(|\tau(\varphi)|^{p/2} \, \eta)\,|^2 \, v_g$$

$$\leq 4 \frac{p^2}{(p-1)^2} \int_M |\tau(\varphi)|^p \, |\nabla\eta|^2 \, v_g + C \int_M |d\varphi|^2 \, |\tau(\varphi)|^p \, \eta^2 \, v_g$$

$$\leq 4 \frac{p^2}{(p-1)^2} \int_M |\tau(\varphi)|^p \, |\nabla\eta|^2 \, v_g$$

$$+ CC'' \left\{ \int_{B_r(x_0)} |d\varphi|^m \, v_g \right\}^{2/m} \int_M |\nabla(\, |\tau(\varphi)|^{p/2} \, \eta)\,|^2 \, v_g, \tag{6.12}$$

instead of (4.6). In the last inequality, we used (6.9) in Lemma 6.4.

Assume that

$$\int_{B_r(x_0)} |d\varphi|^m \, v_g \leq \epsilon_0. \tag{6.13}$$

Then, due to (6.12), we have

$$\int_M |\nabla(|\tau(\varphi)|^{p/2} \eta)|^2 \, v_g \leq 4 \frac{p^2}{(p-1)^2} \int_M |\tau(\varphi)|^p |\nabla\eta|^2 \, v_g$$
$$+ CC'' \epsilon_0{}^{2/m} \int_M |\nabla(|\tau(\varphi)|^{p/2} \eta)|^2 \, v_g. \tag{6.14}$$

If we take $\epsilon_0 > 0$ enough small such as $1 - CC'' \epsilon_0{}^{2/m} > \frac{1}{2}$, i.e., $\frac{1}{(2CC'')^{2/m}} > \epsilon_0$, then, we have

$$\int_M |\nabla(|\tau(\varphi)|^{p/2} \eta)|^2 \, v_g \leq 8 \frac{p^2}{(p-1)^2} \int_M |\tau(\varphi)|^p |\nabla\eta|^2 \, v_g. \tag{6.15}$$

Now, the proof of Theorem 3.3 works in the same way, and then we obtain Proposition 6.3.   □

(*Proof of Lemma 6.4*)   We may assume that the support of the cutoff function η is contained in $B_r(x_0)$, and then we use the Hölder inequality of this type on $B_r(x_0)$:

$$\int_{B_r(x_0)} F \, G \, g_g \leq \left(\int_{B_r(x_0)} F^{m/2} \, v_g\right)^{2/m} \left(\int_{B_r(x_0)} G^{m/(m-2)} \, v_g\right)^{(m-2)/m}.$$

Then, we have

$$\int_M |d\varphi|^2 |\tau(\varphi)|^p \eta^2 \, v_g = \int_{B_r(x_0)} |d\varphi|^2 \left(|\tau(\varphi)|^{p/2} \eta\right)^2 v_g$$

$$\leq \left(\int_{B_r(x_0)} (|d\varphi|^2)^{m/2} \, v_g\right)^{2/m} \times$$

$$\times \left(\int_{B_r(x_0)} \left((|\tau(\varphi)|^{p/2} \eta)^2\right)^{m/(m-2)} v_g\right)^{(m-2)/m}$$

$$= \left(\int_{B_r(x_0)} (|d\varphi|^2)^{m/2} \, v_g\right)^{2/m} \times$$

$$\times \left(\int_M (|\tau(\varphi)|^{p/2} \eta)^{2m/(m-2)} \, v_g\right)^{(m-2)/m}$$

$$\leq C_0 \left(\int_{B_r(x_0)} |d\varphi|^m \, v_g\right)^{2/m} \int_M |\nabla(|\tau(\varphi)|^{p/2} \eta)|^2 \, v_g, \tag{6.16}$$

where in the last inequality of (6.16), we used the Sobolev inequality for $F = |\tau(\varphi)|^{p/2} \eta$:

$$\left(\int_M F^{2m/(m-2)} \, v_g \right)^{(m-2)/m} \leq C_0 \int_M |\nabla F|^2.$$

We obtain (6.9).   □

7. Proof of Theorem 5.1

Now we are in position giving a proof of Theorem 5.1.

Take any sequence $\{\varphi_i\}$ in $\mathcal{F}$. For the $\epsilon_0 > 0$ in Proposition 6.3, and let us consider

$$\mathcal{S} := \left\{ x \in M \middle| \liminf_{i \to \infty} \int_{B_r(x)} |d\varphi_i|^m \, v_g \geq \epsilon_0 \quad (\text{for all } r > 0) \right\}. \tag{7.1}$$

Then, the set $\mathcal{S}$ is finite. Because, for every finite subset $\{x_i\}_{i=1}^k$ in $\mathcal{S}$, let us take a sufficiently small positive number $r_0 > 0$ in such a way that $B_{r_0}(x_i) \cap B_{r_0}(x_j) = \emptyset$ $(i \neq j)$. Then, we have for a sufficiently large i,

$$k \, \epsilon_0 \leq \sum_{j-1}^{k} \int_{B_{r_0}(x_j)} |d\varphi_i|^m \, v_g$$

$$= \int_{\cup_{j=1}^k B_{r_0}(x_j)} |d\varphi_i|^m \, v_g$$

$$\leq \int_M |d\varphi_i|^m \, v_g$$

$$\leq C < \infty \tag{7.2}$$

by definition of $\mathcal{F}$. Thus, we have

$$k \leq \frac{C}{\epsilon_0},$$

which implies that $\#\mathcal{S} \leq \frac{C}{\epsilon_0} < \infty$.

Then, if necessary by taking a subsequence of $\{\varphi_i\}$, we may assume that

$$\mathcal{S} = \left\{ x \in M \middle| \limsup_{i \to \infty} \int_{B_r(x)} |d\varphi_i|^m \, v_g \geq \epsilon_0 \right\}. \tag{7.3}$$

Because, if not so, let us denote the right hand side of (7.3) by $\overline{\mathcal{S}}$. Then, by definition, $\mathcal{S}$ is a proper subset of $\overline{\mathcal{S}}$. Take a point $\overline{x} \in \overline{\mathcal{S}} \backslash \mathcal{S}$. By taking a subsequence of $\{\varphi_i\}$, by the same letter, in such a way that

$$\liminf_{i \to \infty} \int_{B_r(\overline{x})} |d\varphi_i|^m \, v_g \geq \epsilon_0,$$

For this $\{\varphi_i\}$, $\bar{x}$ belongs to $\mathcal{S}$. Since $\mathcal{S}$ is a finite set, this process stops at finite times, then at last we have $\overline{\mathcal{S}} = \mathcal{S}$.

Now, let $x \in M \backslash \mathcal{S}$. Then,

$$\limsup_{i \to \infty} \int_{B_r(x)} |d\varphi_i|^m \, v_g < \epsilon_0. \tag{7.4}$$

Due to Proposition 6.3 and the definition of $\mathcal{F}$, we have

$$\sup_{B_{r/2}(x)} |\tau(\varphi_i)|^2 \leq \frac{C}{r^{m/2}} \int_{B_r(x)} |\tau(\varphi_i)|^2 \, v_g$$

$$\leq \frac{C^2}{r^{m/2}}, \tag{7.5}$$

so that we have that

(C^0):   the C^0-estimate on $B_r(x)$ of $\tau(\varphi_i)$ uniformly on i.

On the other hand, since $\varphi_i \in \mathcal{F}$, all φ_i are biharmonic, i.e., φ_i satisfy that the equations

$$\tau_2(\varphi_i) = \overline{\Delta}(\tau(\varphi_i)) - \mathcal{R}(\tau(\varphi_i)) = 0 \tag{7.6}$$

$$\Longleftrightarrow \begin{cases} (1) & \overline{\Delta}\sigma_i = \mathcal{R}(\sigma_i), \\ (2) & \tau(\varphi_i) = \sigma_i. \end{cases} \tag{7.7}$$

Notice that both the (1) and (2) of (7.7) are the (non-linear) elliptic partial differential equations. Due to (1), the C^0-estimate for σ_i means that the C^∞-estimate of σ_i, and due to (2), the C^∞-estimate of σ_i means that the C^∞-estimate of φ_i. Thus, due to the estimate (C^0) above, we obtain the C^∞-estimates on $B_r(x)$ of φ_i uniformly on i. Therefore, there exists a subsequence $\{\varphi_{i_j}\}$ of $\{\varphi_i\}$ and a smooth map $\varphi_\infty : M \backslash \mathcal{S} \to N$ such that φ_{i_j} converges to φ_∞ on $B_r(x)$ in the C^∞-topology as $j \to \infty$. Thus, $\varphi_\infty : (M \backslash \mathcal{S}, g) \to (N, h)$ is also biharmonic.

For (2) in Theorem 5.1, let us consider the Radon measures $|d\varphi_{i_j}|^m \, v_g$. Then, these have a weak limit which is also a Randon measure, say μ. Recall that μ is by definition a *Radon measure* if (1) μ is locally finite, i.e., $\mu(K) < \infty$ for every compact subset K on M, and (2) μ is Borel regular, i.e., it holds that, for all Borel subset A of M,

$\mu(A) = \sup\{\mu(K)|$ for all compact subset K of $A\}$, and

$\mu(A) = \inf\{\mu(O)|$ for all open subset O of M including $A\}$.

On the other hand, since φ_{i_j} converges to φ_∞ on $M \backslash \mathcal{S}$ in the C^∞-topology as $j \to \infty$, it holds that

$$\mu = |d\varphi_\infty|^m \, v_g \quad \text{on } M \backslash \mathcal{S}. \tag{7.8}$$

Here, $\mathcal{S}$ is a finite subset of M, say $\mathcal{S} = \{x_1, \cdots, x_k\}$. Then, the Radon measure $\mu - |d\varphi_\infty|^m v_g$ satisfies that its support contains in $\mathcal{S}$. Therefore, it holds that

$$\mu - |d\varphi_\infty|^m v_g = \sum_{i=1}^{k} a_j \, \delta_{x_j} \tag{7.9}$$

for some non-negative real numbers a_j $(j = 1, \cdots, k)$ and δ_{x_j} is the Dirac measure which satisfies by definition

$$\delta_{x_j}(A) = \begin{cases} 1 & (x_j \in A), \\ 0 & (x_j \notin A), \end{cases}$$

for every Borel subset A of M. Remark here that $a_j < \infty$ for every $j = 1, \cdots, k$. Because μ is a Radon measure, so that μ is locally finite. Therefore, the Radon measure $|d\varphi_{i_j}|^m v_g$ converges weakly to μ, and

$$\mu = |d\varphi_\infty|^m v_g + \sum_{j=1}^{k} a_j \, \delta_{x_j} \tag{7.10}$$

due to (7.9). We have (2) of Theorem 5.1. $\square$

CHAPTER 10

Conformal Change of Riemannian Metrics and Biharmonic Maps

1

ABSTRACT. For biharmonic maps between two Riemannian manifolds (M, g) and (N, h), the conformal change of g by a function f, reduces the biharmonic map equation to the third order non-linear ordinary differential equation $f^2 f''' - 2\frac{m+1}{m-2} f f' f'' + \frac{m^2}{(m-2)^2} f'^3 = 0$. We show this ODE admits no global positive solution for every $m \geq 5$. In the cases $m = 3, 4$, we also show that there exist both global positive solutions and also periodic positive solutions. For Riemannian manifolds (M, g), (N, h) both of which are the product Riemannian manifolds with a circle, every harmonic with non-constant circle component is biharmonic but not harmonic of $(M, f^2 g)$ into (N, h).

1. Introduction

Harmonic maps play a central roll in variational problems and geometry. They are critical points of the energy functional $E(\varphi) = \frac{1}{2} \int_M \|d\varphi\|^2 v_g$ for smooth maps φ of (M, g) into (N, h), and the Euler-Lagrange equation is that the tension field $\tau(\varphi)$ vanishes. By extending notion of harmonic maps, in 1983, J. Eells and L. Lemaire [40] proposed the problem to consider the bienergy

$$E_2(\varphi) = \frac{1}{2} \int_M |\tau(\varphi)|^2 v_g.$$

After G.Y. Jiang [74] studied the first and second variation formulas of E_2, whose critical maps are called biharmonic maps, there have been extensive studies in this area (for instance, see [16], [89], [90], [120], [102], [65], [73], etc.). Harmonic maps are always biharmonic maps by definition.

---

[1]This chapter is due to [107]: H. Naito and H. Urakawa, *Conformal change of Riemannian metrics and biharmonic maps*, Indiana Univ. Math. J., **63** (2014), 1631–1657.

P. Baird and D. Kamissoko [9] raised an interesting idea to produce a biharmonic but not harmonic map of $(M, \tilde{g})$ into (N, h) with conformal change of g into $\tilde{g}$, by a factor of C^∞ function f, and they reduced the ordinary differential equation on f,

$$f^2 f''' - 2\frac{m+1}{m-2} f f' f'' + \frac{m^2}{(m-2)^2} f'^3 = 0.$$

and gave some interesting examples biharmonic maps. In this paper, we give a final answer to this problem: If dim $M \geq 5$, there is no global solution of this ODE (cf. Theorem 7.1), and if dim $M = 3, 4$, there is a solution f of this ODE (cf. Theorem 7.1), and then we show

THEOREM 1.1. (cf. Theorem 9.1) For a given harmonic map $\varphi : (\Sigma^{m-1}, g) \to (P, h)$, let us define $\tilde{\varphi} : \mathbb{R} \times \Sigma^{m-1} \ni (x, y) \mapsto (ax + b, \varphi(y)) \in \mathbb{R} \times P$ $(m = 3, 4)$, where a and b are constants, and define also $\tilde{f}(x, y) := f(x)$ $((x, y) \in \mathbb{R} \times \Sigma^{m-1})$, where f is the above solution of the ODE. Then,

(1) In the case $m = 3$, the mapping $\tilde{\varphi} : (\mathbb{R} \times \Sigma^2, \tilde{f}^2 g) \to (\mathbb{R} \times P, h)$ is biharmonic, but not harmonic if $a \neq 0$.

(2) In the case $m = 4$, the mapping $\tilde{\varphi} (\mathbb{R} \times \Sigma^3, \frac{1}{\cosh x} g) \to (\mathbb{R} \times P, h)$ is biharmonic, but not harmonic if $a \neq 0$.

2. Preliminaries

In this section, we prepare materials for the first and second variation formulas for the bi-energy functional and bi-harmonic maps. Let us recall the definition of a harmonic map $\varphi : (M, g) \to (N, h)$, of a compact Riemannian manifold (M, g) into another Riemannian manifold (N, h), which is an extremal of the *energy functional* defined by

$$E(\varphi) = \int_M e(\varphi) \, v_g,$$

where $e(\varphi) := \frac{1}{2}|d\varphi|^2$ is called the energy density of φ. That is, for any variation $\{\varphi_t\}$ of φ with $\varphi_0 = \varphi$,

$$\frac{d}{dt}\bigg|_{t=0} E(\varphi_t) = -\int_M h(\tau(\varphi), V) v_g = 0, \qquad (2.1)$$

where $V \in \Gamma(\varphi^{-1}TN)$ is a variation vector field along φ which is given by $V(x) = \frac{d}{dt}|_{t=0}\varphi_t(x) \in T_{\varphi(x)}N$, $(x \in M)$, and the *tension field* is given by $\tau(\varphi) = \sum_{i=1}^m B(\varphi)(e_i, e_i) \in \Gamma(\varphi^{-1}TN)$, where $\{e_i\}_{i=1}^m$ is a locally defined frame field on (M, g), and $B(\varphi)$ is the second fundamental form of φ defined by

$$
\begin{aligned}
B(\varphi)(X, Y) &= (\widetilde{\nabla} d\varphi)(X, Y) \\
&= (\widetilde{\nabla}_X d\varphi)(Y) \\
&= \overline{\nabla}_X (d\varphi(Y)) - d\varphi(\nabla_X Y) \\
&= {}^N\nabla_{\varphi_*(X)} d\varphi(Y) - \varphi_*(\nabla_X Y),
\end{aligned} \tag{2.2}
$$

for all vector fields $X, Y \in \mathfrak{X}(M)$. Furthermore, ∇, and ${}^N\nabla$, are connections on TM, TN of (M, g), (N, h), respectively, and $\overline{\nabla}$, and $\widetilde{\nabla}$ are the induced ones on $\varphi^{-1}TN$, and $T^*M \otimes \varphi^{-1}TN$, respectively. By (1), φ is harmonic if and only if $\tau(\varphi) = 0$.

The second variation formula is given as follows. Assume that φ is harmonic. Then,

$$
\left.\frac{d^2}{dt^2}\right|_{t=0} E(\varphi_t) = \int_M h(J(V), V) v_g, \tag{2.3}
$$

where J is an elliptic differential operator, called *Jacobi operator* acting on $\Gamma(\varphi^{-1}TN)$ given by

$$
J(V) = \overline{\Delta} V - \mathcal{R}(V), \tag{2.4}
$$

where $\overline{\Delta} V = \overline{\nabla}^* \overline{\nabla} V$ is the *rough Laplacian* and $\mathcal{R}$ is a linear operator on $\Gamma(\varphi^{-1}TN)$ given by $\mathcal{R}V = \sum_{i=1}^m R^N(V, d\varphi(e_i)) d\varphi(e_i)$, and R^N is the curvature tensor of (N, h) given by $R^N(U, V)W = {}^N\nabla_U {}^N\nabla_V W - {}^N\nabla_V {}^N\nabla_U W - {}^N\nabla_{[U,V]}W$ for U, V, $W \in \mathfrak{X}(N)$.

J. Eells and L. Lemaire proposed ([**40**]) polyharmonic (k-harmonic) maps and Jiang studied ([**74**]) the first and second variation formulas of bi-harmonic maps. Let us consider the *bi-energy functional* defined by

$$
E_2(\varphi) = \frac{1}{2} \int_M |\tau(\varphi)|^2 v_g, \tag{2.5}
$$

where $|V|^2 = h(V, V)$, $V \in \Gamma(\varphi^{-1}TN)$. Then, the first and second variation formulas are given as follows.

THEOREM 2.1. (*the first variation formula*)

$$
\left.\frac{d}{dt}\right|_{t=0} E_2(\varphi_t) = -\int_M h(\tau_2(\varphi), V) v_g, \tag{2.6}
$$

where

$$
\tau_2(\varphi) = J(\tau(\varphi)) = \overline{\Delta}\tau(\varphi) - \mathcal{R}(\tau(\varphi)), \tag{2.7}
$$

J is given in (2.4).

DEFINITION 2.1. *A smooth map φ of M into N is said to be* bihar-monic *if $\tau_2(\varphi) = 0$.*

3. Formulas under conformal change of Riemannian metrics

In this section, we show several formulas under conformal changes of the domain Riemannian manifold (M, g). Let (M, g) and (N, h) be two Riemannian manifolds, and $\varphi : M \to N$, a C^∞ map from M into N. We denote the *energy functional* by

$$E_1(\varphi : g, h) := \frac{1}{2} \int_M |d\varphi|_{g,h}^2 \, v_g, \tag{3.1}$$

where $|d\varphi|_{g,h}^2$ is twice of the energy density of φ, i.e.,

$$|d\varphi|_{g,h}^2 := \sum_{i=1}^m h(\varphi_* e_i, \varphi_* e_i),$$

and $\{e_i\}_{i=1}^m$ is a local orthonormal frame field on (M, g). For a positive C^∞ function on M, we consider the conformal change of the Riemann-ian metric on M, $\tilde{g} := f^{2/(m-2)} g$, where $m = \dim M > 2$. Then, we have (cf. [**39**]) that

$$|d\varphi|_{\tilde{g},h}^2 = f^{-2/(m-2)} |d\varphi|_{g,h}^2, \tag{3.2}$$

$$v_{\tilde{g}} = f^{m/(m-2)} v_g. \tag{3.3}$$

Thus, we have ([**39**], p.161) that

$$E_1(\varphi : \tilde{g}, h) = \frac{1}{2} \int_M f \, |d\varphi|_{g,h}^2 \, v_g. \tag{3.4}$$

Let us consider the *bienergy functional* defined by

$$E_2(\varphi : g, h) := \frac{1}{2} \int_M |\tau_g(\varphi)|_{g,h}^2 \, v_g, \tag{3.5}$$

where

$$\tau_g(\varphi) := \sum_{i=1}^m \left\{ {}^N\nabla_{\varphi_* e_i} \varphi_* e_i - \varphi_*(\nabla_{e_i}^g e_i) \right\} \in \Gamma(\varphi^{-1}TN)), \tag{3.6}$$

${}^N\nabla, \nabla^g$ are the Levi-Chivita connections of $(N, h), (M, g)$, respectively.

We first see that

$$\nabla_X^{\tilde{g}} Y = \nabla_X^g Y + \frac{1}{m-2} \left\{ f^{-1}(Xf)Y + f^{-1}(Yf)X \right.$$

$$\left. - g(X, Y) f^{-1} \sum_{i=1}^m (e_i f) e_i, \right\} \tag{3.7}$$

for all $X, Y \in \mathfrak{X}(M)$. Then, we have

$$\tau_{\widetilde{g}}(\varphi) = f^{-2/(m-2)}\, \tau_g(\varphi) + f^{-m/(m-2)}\, \varphi_*(\nabla^g f) \tag{3.8}$$

$$= f^{2/(2-m)}\left\{\tau_g(\varphi) + f^{-1}\varphi_*(\nabla^g f)\right\} \tag{3.9}$$

$$= f^{m/(2-m)}\, \mathrm{div}_g(f\, d\varphi), \tag{3.10}$$

where $\nabla^g f := \sum_{j=1}^{m}(e_j f)\, e_j \in \mathfrak{X}(M)$ for $f \in C^\infty(M)$, and

$$\mathrm{div}_g(d\varphi) := \sum_{i=1}^{m}(\widetilde{\nabla}_{e_i} d\varphi)(e_i) = \sum_{i=1}^{m}\left\{\widetilde{\nabla}_{e_i}(d\varphi(e_i)) - d\varphi(\nabla_{e_i} e_i)\right\}$$

$$= \sum_{i=1}^{m}\left\{{}^{N}\nabla_{\varphi_*(e_i)} d\varphi(e_i) - \varphi_*(\nabla_{e_i} e_i)\right\}.$$

Here, recall that $\widetilde{\nabla}$ is the induced connection on $\varphi^{-1}TN \times T^*M$ from ${}^{N}\nabla$ and $\widetilde{g}$, and we have that

$$f^{m/(2-m)}\, \mathrm{div}_g(f\, d\varphi) = f^{m/(2-m)}\, d\varphi(\nabla^g f) + f^{2/(2-m)}\tau_g(\varphi). \tag{3.11}$$

Therefore, it holds (cf. [39]) that $\varphi : (M, \widetilde{g}) \to (N, h)$ is harmonic if and only if

$$f\, \tau_g(\varphi) + \varphi_*(\nabla^g f) = 0. \tag{3.12}$$

Summing up the above, we have

LEMMA 3.1. *(cf. [39], p.161) The Euler-Lagrange equation of the energy functional $E_1(\varphi : \widetilde{g}, h)$ is given by*

$$\tau(\varphi : \widetilde{g}, h) = f^{2/(2-m)}\left\{\tau(\varphi : g, h) + \varphi_*(\nabla^g \log f)\right\} \tag{3.13}$$

$$= f^{m/(2-m)}\, \mathrm{div}_g(f\, d\varphi). \tag{3.14}$$

Thus, $\varphi : (M, \widetilde{g}) \to (N, h)$ is harmonic if and only if $\mathrm{div}_g(f\, d\varphi) = 0$.

Next, we compute the Euler-Lagrange equation of the *bienergy functional*:

$$E_2(\varphi : \widetilde{g}, h) = \frac{1}{2}\int_M |\tau_{\widetilde{g}}(\varphi)|_{\widetilde{g},h}^{\,2}\, v_{\widetilde{g}}.$$

It is known (cf. [**74**]) that

$$
\begin{aligned}
\tau_2(\varphi : \widetilde{g}, h) &= J_{\widetilde{g}}(\tau_{\widetilde{g}}(\varphi)) \\
&= \overline{\Delta}_{\widetilde{g}}(\tau_{\widetilde{g}}(\varphi)) - \mathcal{R}_{\widetilde{g}}(\tau_{\widetilde{g}}(\varphi)) \\
&= -\sum_{i=1}^{m} \left\{ \overline{\nabla}_{\widetilde{e}_i}(\overline{\nabla}_{\widetilde{e}_i} \tau_{\widetilde{g}}(\varphi)) - \overline{\nabla}_{\nabla^{\widetilde{g}}_{\widetilde{e}_i} \widetilde{e}_i} \tau_{\widetilde{g}}(\varphi) \right\} \\
&\quad - \sum_{i=1}^{m} R^{N}(\tau_{\widetilde{g}}(\varphi), \varphi_* \widetilde{e}_i)\varphi_* \widetilde{e}_i,
\end{aligned}
\tag{3.15}
$$

where $\overline{\nabla}$ is the induced connection on $\varphi^{-1}TN$ from the Levi-Civita connection $^{N}\nabla$ on TN of (N, h), and $\{\widetilde{e}_i\}_{i=1}^{m}$ is the local orthonomal frame field on $(M, \widetilde{g})$ given by $\widetilde{e}_i := f^{-1/(m-2)} e_i$ $(i = 1, \cdots, m)$.

We first calculate $J_{\widetilde{g}}(V)$, $(V \in \Gamma(\varphi^{-1}TN))$ given by definition as

$$
\begin{aligned}
J_{\widetilde{g}}(V) &:= \overline{\Delta}_{\widetilde{g}}(V) - \mathcal{R}_{\widetilde{g}}(V) \\
&= -\sum_{i=1}^{m} \left\{ \overline{\nabla}_{\widetilde{e}_i}(\overline{\nabla}_{\widetilde{e}_i} V) - \overline{\nabla}_{\nabla^{\widetilde{g}}_{\widetilde{e}_i} \widetilde{e}_i} V \right\} - \sum_{i=1}^{m} R^{N}(V, \varphi_* \widetilde{e}_i)\varphi_* \widetilde{e}_i.
\end{aligned}
\tag{3.16}
$$

We have

LEMMA 3.2.

$$
J_{\widetilde{g}}(V) = f^{2/(2-m)} J_g(V) - f^{m/(2-m)} \overline{\nabla}_{\nabla^{g} f} V, \quad (V \in \Gamma(\varphi^{-1}TN)).
\tag{3.17}
$$

PROOF. Indeed, we have

$$
\begin{aligned}
\overline{\nabla}_{\widetilde{e}_i}(\overline{\nabla}_{\widetilde{e}_i} V) &= f^{1/(2-m)} \overline{\nabla}_{e_i}(f^{1/(2-m)} \overline{\nabla}_{e_i} V) \\
&= f^{1/(2-m)} e_i(f^{1/(2-m)})\overline{\nabla}_{e_i} V + f^{2/(2-m)} \overline{\nabla}_{e_i}(\overline{\nabla}_{e_i} V),
\end{aligned}
\tag{3.18}
$$

$$
\begin{aligned}
\overline{\nabla}_{\nabla^{\widetilde{g}}_{\widetilde{e}_i} \widetilde{e}_i} V &= f^{1/(2-m)} \overline{\nabla}_{\nabla^{\widetilde{g}}_{e_i}(f^{1/(2-m)} e_i)} V \\
&= f^{1/(2-m)} e_i(f^{1/(2-m)}) \overline{\nabla}_{e_i} V + f^{2/(2-m)} \overline{\nabla}_{\nabla^{g}_{e_i} e_i} V,
\end{aligned}
\tag{3.19}
$$

which implies that

$$
\overline{\Delta}_{\widetilde{g}} V = -f^{2/(2-m)} \overline{\nabla}_{e_i}(\overline{\nabla}_{e_i} V) + f^{2/(2-m)} \overline{\nabla}_{\nabla^{g}_{e_i} e_i} V.
\tag{3.20}
$$

By using (3.7) in (3.20), (3.20) is equal to

$$
f^{2/(2-m)} \overline{\Delta}_g V - f^{m/(2-m)} \sum_{i=1}^{m} (e_i f) \overline{\Delta}_{e_i} V,
\tag{3.21}
$$

and by curvature property,

$$\sum_{i=1}^{m} R^N(V, \varphi_* \tilde{e}_i)\varphi_* \tilde{e}_i = f^{2/(2-m)} \sum_{i=1}^{m} R^N(V, \varphi_* e_i)\varphi_* e_i. \tag{3.22}$$

Substituting (3.20) and (3.21) into (3.16), we have (3.17). $\square$

LEMMA 3.3. *For all* $f \in C^\infty(M)$, $V \in \Gamma(\varphi^{-1}TN)$, *real numbers* p *and* q,

$$J_g(fV) = (\Delta_g f) V - 2\overline{\nabla}_{\nabla^g f} V + f J_g V, \tag{3.23}$$

$$\nabla^g f^p = p \, f^{p-1} \, \nabla^g f, \tag{3.24}$$

$$(\nabla^g f) f^q = q \, f^{q-1} \, |\nabla^g f|_g^2, \tag{3.25}$$

$$\Delta_g f^p = p \, f^{p-1} \, \Delta_g f - p(p-1) \, f^{p-2} \, |\nabla^g f|_g^2. \tag{3.26}$$

PROOF. By a direct computation, Lemma 3.3 follows. The proof is omitted. $\square$

By Lemmas 3.1, 3.2 and 3.3, we have

LEMMA 3.4. *The bienergy tension field* $\tau_2(\varphi : \tilde{g}, h)$ *is given by*

$$\begin{aligned}
\tau_2(\varphi : \tilde{g}, h) &:= J_{\tilde{g}}(\tau_{\tilde{g}}(\varphi)) \\
&= \left\{ -\frac{4}{(2-m)^2} \, f^{2m/(2-m)} \, |\nabla^g f|_g^2 + \frac{2}{2-m} \, f^{(2+m)/(2-m)} \, \Delta_g f \right\} \tau_g(\varphi) \\
&\quad - \frac{6-m}{2-m} \, f^{(2+m)/(2-m)} \, \overline{\nabla}_{\nabla^g f} \tau_g(\varphi) + f^{4/(2-m)} \, J_g(\tau_g(\varphi)) \\
&\quad + \left\{ -\frac{m^2}{(2-m)^2} \, f^{(-2+3m)/(2-m)} \, |\nabla^g f|_g^2 + \frac{m}{2-m} \, f^{2m/(2-m)} \, \Delta_g f \right\} \\
&\quad\quad \times \varphi_*(\nabla^g f) \\
&\quad - \frac{2+m}{2-m} \, f^{2m/(2-m)} \, \overline{\nabla}_{\nabla^g f} \varphi_*(\nabla^g f) + f^{(2+m)/(2-m)} \, J_g(\varphi_*(\nabla^g f)).
\end{aligned} \tag{3.27}$$

PROOF. Indeed, we compute

$$
\begin{aligned}
&\tau_2(\varphi : \tilde{g}, h) \\
&= J_{\tilde{g}}(\tau_{\varphi g}(\varphi)) \\
&= f^{2/(2-m)} J_g(\tau_{\tilde{g}}(\varphi)) - f^{m/(2-m)} \overline{\nabla}_{\nabla^g f} \tau_{\tilde{g}}(\varphi) \\
&= f^{2/(2-m)} J_g \left(f^{2/(2-m)} \tau_g(\varphi) + f^{m/(2-m)} \varphi_*(\nabla^g f) \right) \\
&\quad - f^{m/(2-m)} \overline{\nabla}_{\nabla^g f} \left(f^{2/(2-m)} \tau_g(\varphi) + f^{m/(2-m)} \varphi_*(\nabla^g f) \right) \\
&= f^{2/(2-m)} \left\{ \Delta_g(f^{2/(2-m)}) \tau_g(\varphi) - 2 \overline{\nabla}_{\nabla^g f^{2/(2-m)}} \tau_g(\varphi) \right. \\
&\quad \left. + f^{2/(2-m)} J_g(\tau_g(\varphi)) \right\} + f^{2/(2-m)} \left\{ \Delta_g(f^{m/(2-m)}) \varphi_*(\nabla^g f) \right. \\
&\quad \left. - 2 \overline{\nabla}_{\nabla^g f^{m/(2-m)}} \varphi_*(\nabla^g f) + f^{m/(2-m)} J_g(\varphi_*(\nabla^g f)) \right\} \\
&\quad - f^{m/(2-m)} \left\{ (\nabla^g f) f^{2/(2-m)} \tau_g(\varphi) + f^{2/(2-m)} \overline{\nabla}_{\nabla^g f} \tau_g(\varphi) \right\} \\
&\quad - f^{m/(2-m)} \left\{ (\nabla^g f) f^{m/(2-m)} \varphi_*(\nabla^g f) + f^{m/(2-m)} \overline{\nabla}_{\nabla^g f} \varphi_*(\nabla^g f) \right\}.
\end{aligned}
$$
(3.28)

By using Lemma 3.3 in (3.28), and a direct computation, we have Lemma 3.4. □

Thus, we have immediately

COROLLARY 3.1.

$$
\begin{aligned}
f^{2m/(m-2)} \tau_2(\varphi : \tilde{g}, h) &= \left\{ -\frac{4}{(m-2)^2} |\nabla^g f|_g^2 - \frac{2}{m-2} f \Delta_g f \right\} \tau_g(\varphi) \\
&\quad - \frac{m-6}{m-2} f \overline{\nabla}_{\nabla^g f} \tau_g(\varphi) + f^2 J_g(\tau_g(\varphi)) \\
&\quad + f^{-1} \left\{ -\frac{m^2}{(m-2)^2} |\nabla^g f|_g^2 - \frac{m}{m-2} f \Delta_g f \right\} \varphi_*(\nabla^g f) \\
&\quad + \frac{m+2}{m-2} \overline{\nabla}_{\nabla^g f} \varphi_*(\nabla^g f) + f J_g(\varphi_*(\nabla^g f)).
\end{aligned}
$$
(3.29)

Therefore, we have also

COROLLARY 3.2. $\varphi : (M, \tilde{g}) \to (N, h)$ *is biharmonic if and only if*

$$\tau_2(\varphi : \tilde{g}, h) = 0$$

$$\Longleftrightarrow$$

$$\left\{ -\frac{4}{(m-2)^2} |\nabla^g f|_g^{\,2} - \frac{2}{m-2} f \, \Delta_g f \right\} f \, \tau_g(\varphi)$$

$$-\frac{m-6}{m-2} f^2 \, \overline{\nabla}_{\nabla^g f} \tau_g(\varphi) + f^3 \, J_g(\tau_g(\varphi))$$

$$+ \left\{ -\frac{m^2}{(m-2)^2} |\nabla^g f|_g^{\,2} - \frac{m}{m-2} f \, \Delta_g f \right\} \varphi_*(\nabla^g f)$$

$$+ \frac{m+2}{m-2} f \, \overline{\nabla}_{\nabla^g f} \varphi_*(\nabla^g f) + f^2 \, J_g(\varphi_*(\nabla^g f)) = 0.$$

$$(3.30)$$

4. Reduction of constructing proper biharmonic maps

In this section, we formulate our problem to construct proper bi-harmonic maps. A biharmonic map is said to be *proper* if it is not harmonic. Let (M, g), (N, h) be two compact Riemannian manifolds. In the following we always assume that $m = \dim(M) \le 3$. Eells and Ferreira [**39**] showed that,

> *for each homotopy class $\mathcal{H}$ in $C^\infty(M, N)$, there exist a Riemannian metric $\tilde{g}$ which is conformal to g, and a C^∞ map $\varphi \in \mathcal{H}$ such that φ is a harmonic map from $(M, \tilde{g})$ into (N, h).*

We do not assume, in general, that M and N are compact. Let us consider the following problem.

> **Problem**   *For each homotopy class $\mathcal{H}$ in $C^\infty(M, N)$, do there exist a Riemannian metric $\tilde{g}$ which is conformal to g, and a C^∞ map in $\mathcal{H}$ such that $\varphi : (M, \tilde{g}) \to (N, h)$ is a proper biharmonic map, that is, $\tau_2(\varphi, \tilde{g}, h) = 0$, but not $\tau(\varphi, \tilde{g}, h) = 0$?*

By regarding the above Eells and Ferreira's result, we fix a harmonic map $\varphi : (M, g) \to (N, h)$, that is, $\tau(\varphi) = 0$. Then, let us consider the following problem:

> **Problem'**   *does there exist a positive C^∞ function f on M such that, for $\tilde{g} = f^{2/(m-2)} g$, $\varphi : (M, \tilde{g}) \to (N, h)$ is proper biharmonic, that is, $\tau_2(\varphi, \tilde{g}, h) = 0$ and $\tau(\varphi, \tilde{g}, h) \ne 0$.*

Then, we have

THEOREM 4.1. *Assume that* $\varphi : (M, g) \to (N, h)$ *is harmonic.* *For a positive* C^∞ *function* f *on* M, *let us define* $\tilde{g} = f^{2/(m-2)} g$, *a Riemannian metric conformal to* g. *Then,*

(1) $\varphi : (M, \tilde{g}) \to (N, h)$ *is harmonic if and only if* $\varphi_*(\nabla^g f) = 0$.
(2) $\varphi : (M, \tilde{g}) \to (N, h)$ *is biharmonic if and only if the following holds:*

$$-\left\{ \frac{m^2}{(m-2)^2} |\nabla^g f|_g^2 + \frac{m}{m-2} f (\Delta_g f) \right\} \varphi_*(\nabla^g f)$$
$$+ \frac{m+2}{m-2} f \overline{\nabla}_{\nabla^g f} \varphi_*(\nabla^g f) + f^2 J_g(\varphi_*(\nabla^g f)) = 0. \qquad (4.1)$$

PROOF. For (1), due to (3.13) or (3.14) in Lemma 3.1, we have

$$\tau(\varphi : \tilde{g}, h) = f^{2/(2-m)} \tau(\varphi : g, h) + f^{m/(2-m)} \varphi_*(\nabla^g f), \qquad (4.2)$$

which implies that (1). For (2), in Corollary 3.6, sibstituting $\tau_g(\varphi) = 0$ into (3.30), we have immediately (4.1).   □

We have immediately

COROLLARY 4.1. *Let* $\varphi = \text{id} : (M, g) \to (M, g)$ *be the identity map. For a positive* C^∞ *function* f *on* M, *let us define* $\tilde{g} = f^{2/(m-2)} g$. *Then,*

(1) $\varphi = \text{id} : (M, \tilde{g}) \to (M, g)$ *is harmonic if and only if* f *is a constant.*
(2) $\varphi = \text{id} : (M, \tilde{g}) \to (M, g)$ *is biharmonic if and only if*

$$-\left\{ \frac{m^2}{(m-2)^2} |\nabla^g f|_g^2 + \frac{m}{m-2} f (\Delta_g f) \right\} \nabla^g f$$
$$+ \frac{m+2}{m-2} f \overline{\nabla}_{\nabla^g f} \nabla^g f + f^2 J_g(\nabla^g f) = 0, \qquad (4.3)$$

which is equivalent to

$$-\left\{ \frac{m^2}{(m-2)^2} |X|_g^2 + \frac{m}{m-2} f (\Delta_g f) \right\} X$$
$$+ \frac{m+2}{m-2} f \nabla_X X + f^2 (\overline{\Delta}^g(X) - \rho(X)) = 0, \qquad (4.4)$$

where $X = \nabla^g f \in \mathfrak{X}(M)$, $\rho(X) := \sum_{i=1}^{m} R^g(X, e_i)e_i$, *is the Ricci transform of* (M, g), *and* $\overline{\Delta}^g(X) := -\sum_{i=1}^{m} (\nabla_{e_i}^g \nabla_{e_i}^g X - \nabla_{\nabla_{e_i}^g e_i}^g X)$ *is the rough Laplacian on* $\mathfrak{X}(M)$, *respectively.*

PROOF. (4.3) and (4.4) follow from (4.1), and the formula $J_g(V) = \overline{\Delta}^g(V) - \rho(V)$ $(V \in \mathfrak{X}(M))$, for the identity map. $\qquad\square$

5. The identity map of the Euclidean space

Let us consider the m dimensional Euclidean space $(M, g) = (\mathbb{R}^m, g_0)$ with the standard coordinate $(x_1, \cdots, x_m)$ $(m \geq 3)$. In this case, let us take a positive C^∞ function $f = f(x_1, \cdots, x_m) \in C^\infty(\mathbb{R}^m)$. Let $X = \nabla^g f = \sum_{i=1}^m f_{x_i} \frac{\partial}{\partial x_i}$, where we denote $f_{x_i} = \frac{\partial f}{\partial x_i}$. Then, since

$$
\begin{cases}
\rho = 0, \\
\Delta_g f = -\sum_{i=1}^m f_{x_i x_i}, \\
|X|_g^2 = \sum_{i=1}^m f_{x_i}^2, \\
\nabla_X^g X = \sum_{i=1}^m \left\{ \sum_{j=1}^m f_{x_j} f_{x_i x_j} \right\} \frac{\partial}{\partial x_i}, \\
\overline{\Delta}^g(X) = \sum_{j=1}^m \Delta_g(f_{x_j}) \frac{\partial}{\partial x_j},
\end{cases}
\tag{5.1}
$$

the equation (4.3) is reduced to the following:

$$
-\left\{ \frac{m^2}{(m-2)^2} \sum_{i=1}^m f_{x_i}^2 + \frac{m}{m-2} f(\Delta_g f) f_{x_j} \right\}
$$
$$
+ f^2 \Delta_g(f_{x_j}) + \frac{m+2}{m-2} \sum_{i=1}^m f_{x_i} f_{x_i x_j} = 0 \quad (\forall j = 1, \cdots, m).
\tag{5.2}
$$

If we consider $f = f(x_1, \cdots, x_m) = f(x)$, $x = x_1$, then, the equation (5.2) is equivalent to the following ODE:

$$
f^2 f''' - 2 \frac{m+1}{m-2} f f' f'' + \frac{m^2}{(m-2)^2} (f')^3 = 0.
\tag{5.3}
$$

In the cases $m = 3$, $m = 4$, (5.3) becomes

$$
f^2 f''' - 8 f f' f'' + 9 (f')^3 = 0 \qquad (m = 3), \tag{5.4}
$$
$$
f^2 f''' - 5 f f' f'' + 4 (f')^3 = 0 \qquad (m = 4). \tag{5.5}
$$

Our problem is reduced to find a positive C^∞ solution of (5.3). In order to analyze (5.3), we put $u = \frac{f'}{f}$, then the equation (5.3) is reduced to the equation:

$$u'' + \frac{m-8}{m-2} u\, u' - \frac{2(m-4)}{(m-2)^2} u^3 = 0. \tag{5.6}$$

Then, we obtain immediately

PROPOSITION 5.1. *If a positive C^∞ solution f of (5.3) on $\mathbb{R}$, then $u = \frac{f'}{f}$ satisfies (5.6). Conversely, for every C^∞ solution u of (5.6) on $\mathbb{R}$, then $f(t) = C \exp(u(t))$ is a positive solution of (5.3) for every positive constant C.*

6. Behavior of solutions of the ODE

Due to Proposition 5.1, our problem is reduced to analyse (5.6). To do it, we need the following two lemmas.

LEMMA 6.1. *(Comparison Theorem)   Assume that a real valued function F on $\mathbb{R}$ satisfies the Lipshitz condition, i.e., there exists a positive number $L > 0$ such that*

$$|F(p) - F(q)| \le L|p - q|,$$

(1) *Two real valued functions u and v defined on the interval $[0, \epsilon)$ for some positive number $\epsilon > 0$ satisfy that*

$$\begin{cases} u'(t) \ge F(u(t)), \\ v'(t) = F(v(t)), \\ u(0) = v(0), \end{cases} \tag{6.1}$$

then it holds that

$$u(t) \ge v(t). \tag{6.2}$$

(2) *Conversely, if u and v satisfy that*

$$\begin{cases} u'(t) \le F(u(t)), \\ v'(t) = F(v(t)), \\ u(0) = v(0), \end{cases} \tag{6.3}$$

then it holds that

$$u(t) \le v(t). \tag{6.4}$$

PROOF. We first show (2). Assume that $u'(t) \le F(u(t))$. Then, we have

$$u(t) - u(0) \le \int_0^t F(u(s))ds, \text{ and } v(t) - v(0) = \int_0^t F(v(s))ds,$$

which imply that

$$u(t) - v(t) \le \int_0^t \{F(u(s)) - F(v(s))\}ds \le L \int_0^t \{u(s) - v(s)\}ds.$$

By putting $y(t) = u(t) - v(t)$, we have

$$\frac{d}{dt}\left(e^{-Lt}\int_0^t y(s)ds\right) = e^{-Lt}y(t) - Le^{-Lt}\int_0^t y(s)ds \le 0.$$

We obtain

$$e^{-Lt}\int_0^t y(s)ds \le 0, \tag{6.5}$$

in particular,

$$u(t) - v(t) = y(t) \le L \int_0^t y(s)ds \le 0, \tag{6.6}$$

that is,

$$u(t) \le v(t). \tag{6.7}$$

(1) If we have $u'(t) \ge F(u(t))$, since

$$v(t) - u(t) \le \int_0^t \{F(v(s)) - F(u(s))\}ds \le L \int_0^t \{v(s) - u(s)\}ds,$$

which implies that $v(t) \le u(t)$, we have (1).   □

Next, we have to prepare Jacobi's **sn**-function.

PROPOSITION 6.1. (1) *The solution of the initial value problem of the ordinary differential equation*

$$(y')^2 = 1 - y^4, \; y(0) = 0, \; y'(0) > 0 \tag{6.8}$$

is given by the Jacobi's elliptic function $y(t) = \mathbf{sn}(\mathbf{i}, t)$.

(2) *The function* $y(t) = \mathbf{sn}(\mathbf{i}, t)$ *is real valued in* $t \in \mathbb{R}$, *and pure imaginary valued in* $t \in \mathbf{i}\mathbb{R}$.

(3) *The function* $y(t)$ *is a double periodic function in the whole complex plane* $\mathbb{C}$ *with the two periods* $4K$ *and* $2\mathbf{i}K'$, *where* $K > 0$ *and* $K' > 0$ *are given by*

$$K := K(k) = \int_0^{\pi/2} \frac{dx}{\sqrt{1 - k^2 \sin^2 x}}, \tag{6.9}$$

$$K' := K(k'), \quad k' := \sqrt{1 - k^2}, \tag{6.10}$$

and it has the only one zeros at $2nK + 2m\mathbf{i}K'$, *and has the only poles at* $2nK + (2m+1)\mathbf{i}K'$, *where* m *and* n *run over the set of all integers.*

(4) *In particular, it has no pole in the real axis, and has no zero on the imaginary axis except* 0. *Furthermore, it has poles on the two lines through the origin with angles* $\pm\pi/4$ *in the complex plane* $\mathbb{C}$.

PROOF. For (1), we have to see the function $y(t) = \mathbf{sn}(\mathbf{i}, t)$ solves (6.8). Let us recall (cf. 281 Elliptic Functions, [**101**], pp. 873–876) the elliptic integral of the first kind $u(k, \varphi)$ with modulus k given by

$$u(k, \varphi) = \int_0^\varphi \frac{d\psi}{\sqrt{1 - k^2 \sin^2 \psi}}, \qquad (6.11)$$

and its inverse function is the amplitude function $\varphi = \mathbf{am}(k, u)$. By differentiating (6.11), we have

$$\frac{d}{d\varphi} u(k, \varphi) = \frac{1}{\sqrt{1 - k^2 \sin^2 \varphi}}. \qquad (6.12)$$

Then, we have

$$\mathbf{sn}(k, t) = \sin(\mathbf{am}(k, t)), \qquad (6.13)$$

which implies immediately that

$$\begin{aligned}
\frac{d}{dt} \mathbf{sn}(k, t) &= \left(\frac{d}{dt} \mathbf{am}(k, t) \right) \cos(\mathbf{am}(k, t)) \\
&= \left(\frac{d}{dt} \mathbf{am}(k, t) \right) \sqrt{1 - \sin^2(\mathbf{am}(k, t))} \\
&= \sqrt{1 - k^2 \sin^2(\mathbf{am}(k, t))} \sqrt{1 - \mathbf{sn}^2(k, t)} \\
&= \sqrt{(1 - k^2 \mathbf{sn}^2(k, t))(1 - \mathbf{sn}^2(k, t))}. \qquad (6.14)
\end{aligned}$$

Here, we put $k = \mathbf{i}$ in (6.14), we have

$$\frac{d}{dt} \mathbf{sn}(\mathbf{i}, t) = \sqrt{1 - \mathbf{sn}^4(\mathbf{i}, t)}, \qquad (6.15)$$

that is, the function $y(t) = \mathbf{sn}(\mathbf{i}, t)$ is a solution of the differential equation of (6.8).

Since $\mathbf{sn}(\mathbf{i}, 0) = \sin(\mathbf{am}(\mathbf{i}, 0))$ and $\mathbf{am}(\mathbf{i}, 0) = 0$, we have $\mathbf{sn}(\mathbf{i}, 0) = 0$.

To get $y'(0) > 0$, we only notice that, if we denote as the usual manner

$$\mathbf{cn}(k, t) := \cos(\mathbf{am}(k, t))),$$
$$\mathbf{dn}(k, t) := \sqrt{1 - k^2 \mathbf{sn}^2(k, t)},$$

it holds that

$$\frac{d}{dt}\mathbf{sn}(k,t) = \mathbf{cn}(k,t)\,\mathbf{dn}(k,t),\tag{6.16}$$

$$\left.\frac{d}{dt}\right|_{t=0}\mathbf{sn}(k,t) = \mathbf{cn}(k,0)\,\mathbf{dn}(k,0)$$

$$= \cos(\mathbf{am}(k,0))\sqrt{1-k^2\,\mathbf{sn}^2(k,0)} = 1,\tag{6.17}$$

that is, $y'(0) > 0$. We have (1).

For (2), $\mathbf{am}(k,t)$ is real valued if $t \in \mathbb{R}$ by definition of $\mathbf{am}(k,t)$, and then $\mathbf{sn}(k,t)$ and $\mathbf{cn}(k,t)$ are also real valued if $t \in \mathbb{R}$. On the other hand, since

$$\mathbf{sn}(k,\mathbf{i}x) = \mathbf{i}\,\frac{\mathbf{sn}(k',x)}{\mathbf{cn}(k',x)}\quad(k'=\sqrt{1-k^2}),$$

the function $\mathbf{sn}(k,t)$ is pure imaginary valued if $t \in \mathbf{i}\mathbb{R}$.

For (3) and (4), write $K' = -\tau K$ with $\tau \in \mathbb{C}$. Then $q := e^{\mathbf{i}\pi\tau} = e^{-\mathbf{i}\pi(K'/K)}$ can be written by using some series of real numbers, $\{a_\ell\}_{\ell=0}^{\infty}$, as

$$q^{1/4} = \left(\frac{k}{4}\right)^{1/2}\left(\sum_{\ell=0}^{\infty}a_\ell k^{2\ell}\right).$$

Thus, $q^{1/4} \in \mathbf{i}^{1/2}\mathbb{R}$ when $k = \mathbf{i}$, which implies that q is a negative real number. Thus, it holds that $K'/K = 1$. It is known that all the poles of $\mathbf{sn}(k,x)$ are $2nK + \mathbf{i}(2m+1)K'$, and by $K = K'$, $\mathbf{sn}(k,x)$ has poles on the lines through the origin with angles $\pm\pi/4$. The other properties have well known. □

By Proposition 6.2, we have

PROPOSITION 6.2. *For every positive integers A and C, and a real number a, all the solutions of both the ordinary differential equations*

$$v'(t) = \sqrt{A\,v(t)^4 + C},\quad v(0) = a,\tag{6.18}$$

$$v'(t) = \sqrt{A\,v(t)^4 - C},\quad v(0) = a,\,(with\,A\,a^4 > C),\tag{6.19}$$

are explosive within finite time. That is, there exist positive real numbers $T_0 > 0$ and $T_1 > 0$ depending on A, C and a such that the existence intervals of solutions of (6.18) or (6.19) are $(-T_0, T_1)$.

PROOF. Let $y(t) := \mathbf{sn}(\mathbf{i}, t)$, and $w(t) := -\mathbf{i}^{3/2} y(\mathbf{i}^{1/2} t)$. Then, we have $w' = -\mathbf{i}^{3/2+1/2} y' = y'(\mathbf{i}^{1/2} t)$ and also

$$w'(t)^2 = y'(\mathbf{i}^{1/2}t)^2 = 1 - y(\mathbf{i}t)^4 = 1 + w(t)^4$$

since $w(t)^4 = (-\mathbf{i}^{3/2})^4 y(\mathbf{i}^{1/2}t)^4 = \mathbf{i}^6 y(\mathbf{i}^{1/2}t)^4 = -y(\mathbf{i}^{1/2}t)^4$. Thus,

$$w(t) := -\mathbf{i}^{3/2} \mathbf{sn}(\mathbf{i}, \mathbf{i}^{1/2} t)$$

is a solution of

$$(w')^2 = 1 + w^4.$$

By the same way, if we put $z(t) := \mathbf{i}\, y(\mathbf{i}\, t)$, then

$$(z')^2 = \mathbf{i}^2 (y')^2 = -(1 - y^4) = y^4 - 1 = z^4 - 1.$$

Thus,

$$z(t) := \mathbf{i}\,\mathbf{sn}(\mathbf{i}, \mathbf{i}\, t)$$

is a solution of

$$(z')^2 = z^4 - 1.$$

Therefore, any solution of (6.18) or (6.19) can be obtained by scaling and/or time-shift,

$$w(t) = -\mathbf{i}^{3/2}\mathbf{sn}(\mathbf{i}, \mathbf{i}^{1/2} t), \quad z(t) = \mathbf{i}\,\mathbf{sn}(\mathbf{i}, \mathbf{i}\, t).$$

By Lemma 6.2 (4), both the obtained solutions have poles, so that solutions of (6.18) and (6.19) are explosive at finite times. ☐

REMARK 6.1. *Every solution of*

$$(v')^2 = 1 - v^4, \quad |v(0)|^4 < 1$$

exists on the whole line $t \in \mathbb{R}$. This fact follows from the fact that the poles of $\mathbf{sn}(k, t)$ do not exist on the whole real line $\mathbb{R}$ (cf. Proposition 6.2).

7. Non-existence and existence of global solutions of the ODE

7.1. Main result. In this section, we will show

THEOREM 7.1. *Let $m \geq 3$. Then, we have*

(1) In the case $m \geq 5$, there exists no C^∞ global solution u of (5.6) on the whole real line $\mathbb{R}$.

(2) In the case of $m = 4$, every solution u of (5.6) is of the form $u(t) = -b \tanh(bt + c)$ for constants b and c.

(3) In the case of $m = 3$, every solution u of (5.6) with $u(0) = 0$ and $u'(0) < 0$ is a global bounded solution on the whole line $t \in \mathbb{R}$.

There exists a solution u of (5.6) with $u(0) =$ and $u'(0) < 0$, and there exists $T > 0$ such that $u(T) = 0$ and $u'(T) > 0$. This u is a periodic solution of (5.6) with the period $2T$.

First, we write the ODE (5.6) as

$$u'' = A\,u\,u' + B\,u^2, \tag{7.1}$$

where the relations of values or signs of $A = -\frac{m-8}{m-2}$ and $B = \frac{2(m-4)}{(m-2)^2}$ are given in the following table:

	$m = 3$	$m = 4$	$m = 5, 6, 7$	$m = 8$	$m \geq 9$
A	$+$	$+$	$+$	0	$-$
B	$-$	0	$+$	$+$	$+$

Then, we have immediately

LEMMA 7.1. *Let u be a solution of (7.1). Then,*

$$\frac{d}{dt}\left(u'(t)^2 - \frac{B}{2}u(t)^4\right) = 2\,A\,u(t)\,u'(t)^2, \tag{7.2}$$

Thus, if we put $E(t) := u'(t)^2 - \frac{B}{2}u(t)^4$, we have

$$E'(t) = 2\,A\,u(t)\,u'(t)^2. \tag{7.3}$$

PROOF. By (7.1), we have

$$u'\,u'' - A\,u\,u'^2 - B\,u^3\,u' = 0, \tag{7.4}$$

so we have (7.2) or (7.3), immediately.   □

7.2. The case of $A = 0$ and $B > 0$ ($m = 8$). In this case, due to Lemma 7.2, we have immediately

LEMMA 7.2. *Assume that $A = 0$ and $B > 0$ ($m = 8$). If u is a solution of (7.1), then $E(t) := u'(t)^2 - \frac{B}{2}u(t)^4$ is constant along the solution u, that is,*

$$E(t) := u'(t)^2 - \frac{B}{2}u(t)^4 = u'(0)^2 - \frac{B}{2}u(0)^4 = E(0). \tag{7.5}$$

Then, we obtain

PROPOSITION 7.1. *In the case that $A = 0$ and $B > 0$, the equation (2.1) has no global solution defined on the whole line $\mathbb{R}$ except only the trivial solution $u(t) \equiv 0$.*

PROOF. The proof is divided into several cases on the initial conditions.

Case I: $u'(0) > 0$ and $u'(0)^2 - \frac{B}{2}u(0)^4 > 0$. In this case, due to Lemma 7.3, it holds that

$$u'(t)^2 = \frac{B}{2}u(t)^4 + \left(u'(0)^2 - \frac{B}{2}u(0)^4\right). \qquad (7.6)$$

By the assumption, we can write as $C^2 = u'(0)^2 - \frac{B}{2}u(0)^4$ for some positive constant $C > 0$. Due to the assumption $u'(0) > 0$, there exists a positive real number $\epsilon > 0$ such that $u'(t) \geq 0$ for all $t \in [-\epsilon, \epsilon]$. Then, by (7.6),

$$u'(t) = \sqrt{\frac{B}{2}u(t)^4 + C^2}, \quad (\forall t \in [-\epsilon, \epsilon]). \qquad (7.7)$$

This holds for every t which the solution u exists. Thus, due to the case (6.18) in Proposition 6.3, the solution is explosive within finite time.

Case II: $u'(0) < 0$ and $u'(0)^2 - \frac{B}{2}u(0)^4 > 0$. By the same manner as Case I, it holds that

$$u'(t) = -\sqrt{\frac{B}{2}u(t)^4 + C^2}, \quad (\forall t \in [-\epsilon, \epsilon]). \qquad (7.8)$$

In this case, the solution except the trivial solution, have to be explosive within finite time as one can show by the similar manner as Proposition 6.3.

Case III: $u(0) > 0$ and $u'(0)^2 - \frac{B}{2}u(0)^4 < 0$. In this case, by the assumption, we can find a positive constant $C > 0$ such that $-C^2 = u'(0)^2 - \frac{B}{2}u(0)^4$. Then, it holds that

$$u'(t)^2 = \frac{B}{2}u(t)^4 - C^2. \qquad (7.9)$$

Then, it turns out that our case is divided into two cases: **Case III-a** $u'(t) \geq 0$ for $t \in [-\epsilon, \epsilon]$, or **Case III-b** $u'(t) < 0$ for $t \in [-\epsilon, \epsilon]$.

Case III-a. In this case, on some interval where $u'(t) \geq 0$ holds. It holds by (7.9), that

$$u'(t) = \sqrt{\frac{B}{2}u(t)^4 - C^2}. \qquad (7.10)$$

Then, due to Proposition 6.3, the solution is explosive within finite time.

Case III-b In this case, On the interval where $u'(t) < 0$ holds, by (7.9), it holds that

$$u'(t) = -\sqrt{\frac{B}{2} u(t)^4 - C^2} \qquad (7.11)$$

on this interval. Then, again by Proposition 6.3, the solution $u(t)$ is explosive within finite time. Therefore, there is no global smooth solution defined on the whole line $\mathbb{R}$ except the trivial solution in the Case III.

Case IV $u(0) < 0$ and $u'(0)^2 - \frac{B}{2} u(0)^4 < 0$. In this case, if we consider $v(t) := -u(t)$, $v(t)$ satisfies also (7.1). And, we have also

$$v(0) = -u(0) > 0,$$

$$v'(0)^2 - \frac{B}{2} v(0)^4 = u'(0)^2 - \frac{B}{2} u(0)^4 < 0,$$

which is the Case III for $v(t)$. The solution $v(t) = -u(t)$ is again explosive within finite time.

Case V $u'(0)^2 - \frac{B}{2} u(0)^4 = 0$ and $u'(0)^2 + u(0)^2 \neq 0$. In this case, it holds that

$$u'(t)^2 = \frac{B}{2} u(t)^4. \qquad (7.12)$$

Thus, it holds that, either

$$u'(t) = \sqrt{\frac{B}{2}} u(t)^2, \quad u'(t) \geq 0,$$

or

$$u'(t) = -\sqrt{\frac{B}{2}} u(t)^2, \quad u'(t) < 0.$$

In each case, we have the initial problem:

$$u'(t) = A u(t)^2, \quad u(0) = a,$$

and its solution can be described as

$$u(t) = \frac{a}{a A t - 1},$$

thus the solution is explosive within finite time.

We have Proposition 7.4.   □

7.3. The case $A > 0$ and $B < 0$ ($m = 3$). In this case, we show the following two lemmas.

LEMMA 7.3. *In the case $A > 0$ and $B < 0$ ($m = 3$), the solution $u(t)$ of (7.1) with the initial condition $u'(0) < 0$ and $u(0) = 0$ satisfies the following:*
(1) *If $t > 0$ and $u'(t) < 0$, then $u(t) < 0$.*
(2) *If $u'(t) < 0$, the solution $u(t)$ satisfies $|u(t)| < \infty$.*
(3) *There exists a positive constant $T > 0$ such that $u'(T) = 0$ and $u(T) < 0$.*

LEMMA 7.4. *In the case $A > 0$ and $B < 0$ ($m = 3$), the solution $u(t)$ of (7.1) with the initial condition $u'(0) = 0$ and $u(0) < 0$ satisfies either one of the following:*
(1) *There exists a positive constant $S > 0$ such that $u(S) = 0$ and $u'(S) > 0$, or*
(2) *there exists $u(t)$ for all $t \in (0, \infty)$ where $u(t) < 0$.*

Due to these two lemmas, we can show

PROPOSITION 7.2. *In the case $A > 0$ and $B < 0$ ($m = 3$), the solution $u(t)$ of (7.1) with the initial condition $u(0) = 0$ and $u'(0) < 0$ is a bounded global solution on the whole line $\mathbb{R}$.*

PROOF. We first give a proof of Proposition 7.7. Let us take a solution of (7.1) with the initial condition $u(0) = 0$ and $u'(0) < 0$. Then, due to Lemma 7.5, there exists a positive constant $T_0 > 0$ such that the solution u satisfies

$$u(T_0) < 0, \quad u'(T_0) = 0. \tag{7.13}$$

So, let us consider the initial value problem of (7.1) with the initial time $t = T_0$. Then, due to Lemma 7.6, this solution $u(t)$ of (7.1) must satisfy either (1) $u(t) < 0$ for all time $t > 0$, or (2) there exists $T_1 > 0$ such that

$$u(T_0 + T_1) = 0, \quad u'(T_0 + T_1) > 0. \tag{7.14}$$

Now let us consider the reverse time solution of (7.1) with the same initial condition. The former case (1) gives a bounded global solution defined on $\mathbb{R}$, and the latter case show that there exists a periodic solution with the period $2(T_0 + T_1)$. In the both cases, there exist global solutions defined on the whole $\mathbb{R}$. □

Proof of Lemma 7.5.   (1) If $u'(t) < 0$, then u is monotone decreasing at t and $u(0) = 0$, then we have $u(t) < u(0) = 0$, we have (1). For (2), since $B < 0$, we have

$$E(t) = \frac{1}{2}u'(t)^2 - \frac{B}{4}u(t)^4 = \frac{1}{2}u'(t)^2 + \frac{|B|}{4}u(t)^4.$$
(7.15)

If we assume $u'(t) < 0$, $u(t) < 0$ by (1), and because of $A < 0$, and (7.15), we have

$$E'(t) = \frac{d}{dt}\left(\frac{1}{2}u'(t)^2 + \frac{|B|}{4}u(t)^4\right) = A\,u(t)\,u'(t)^2 < 0.$$
(7.16)

Therefore, we have

$$E(t) = \frac{1}{2}u'(t)^2 + \frac{|B|}{4}u(t)^4 \leq E(0) = \frac{1}{2}u'(0)^2 + \frac{|B|}{4}u(0)^4.$$
(7.17)

In particular, we have

$$|u'|^2 \leq 2E(0) + \frac{B}{2}|u|^4,$$
(7.18)

$$0 > u'(t) > -\sqrt{E(0) - \frac{|B|}{2}|u|^4}.$$
(7.19)

Now let us consider the initial value problem

$$v'(t) = -\sqrt{E(0) - \frac{|B|}{2}|v|^4}, \quad v(0) = 0.$$
(7.20)

The solution of (7.20) is Jacobi's **sn** function, real valued, and has no pole on the real line $\mathbb{R}$ (cf. Remark 6.4). Applying Lemma 6.1 (Comparison Theorem) to these solutions u and v, the solution u always exists during $u'(t) < 0$, and it holds that for all $t \in \mathbb{R}$,

$$0 > u(t) > v(t).$$
(7.21)

We get (2). For (3), notice that there exists a positive constant $\delta > 0$ sich that $u(\delta) < 0$ since $u'(0) < 0$ and $u(0) = 0$. Here, since $A\,u(t)\,u'(t) > 0$ and $u(t)$ is monotone decreasing, we have for all $t > \delta$,

$$u''(t) = A\,u(t)\,u'(t) + B\,u(t)^3$$
$$\geq B\,u(t)^3$$
$$\geq B\,u(\delta)^3 > 0.$$

Therefore, we have

$$u'(t) \geq u'(0) + B\,u(\delta)^3\,t \quad (\forall\, t > \delta),$$
(7.22)

so there exists a constant $T > 0$ suchthat $u'(T) = 0$. Noticing that $u'(t) < 0$ for all $t \in (0, T)$, it holds that $u(T) < 0$. We have (3).   □

Proof of Lemma 7.6. By (7.1) and our assumptions, we have

$$u''(0) = A\, u(0)\, u'(0) + B\, u(0)^3 = B\, u(0)^3 > 0,$$

so that there exists a positive constant $\delta > 0$ such that

$$u''(t) > 0. \quad (\forall t \in (0, \delta)).$$

In particular, u' is monotone increasing on $(0, \delta)$, and

$$u'(t) > 0, \quad (\forall t \in (0, \delta)).$$

Thus, u is also monotone increasing on $(0, \delta)$, and it holds that

$$u(0) < u(t), \quad (\forall t \in (0, \delta)).$$

Therefore, one of the following two cases should occur:
   (1) There exits a constant $T > 0$ $u(T) = 0$, or
   (2) for all t in the existence interval of the solution, $u(0) < u(t) < 0$.
   If the case (2) occurs, we have to show the existence of u for all time $t > 0$. Indeed, since $B < 0$, it holds that

$$|u'|^2 \leq 2\, E(0) + \frac{B}{2}\, |u|^4 \leq E(0).$$

Therefore, for all t in the existence interval of the solution u, it holds that

$$(u(t), u'(t)) \in [u(0), 0] \times [-E(0), E(0)],$$

and then the solution exists on $(0, \infty)$, and u is bounded on $(0, \infty)$.
   If the case (1) occurs, we have to show $u'(S) < 0$. Our constant S is given by

$$S = \inf\{t : t > 0, u(t) \geq 0\}.$$

Thus, there exists $\delta > 0$ such that $u(t) < 0$ for all $t \in (S - \delta, S)$, we have $u'(S) < 0$.   □

REMARK 7.1. *The case* (2) *that for some* $T > 0$, $u(T) = 0$ *and* $u'(T) > 0$ *occurs. So, we have a periodic solution* $u(t)$ *of* (7.1) *in the case* $A > 0$ *and* $B < 0$ $(m = 3)$.

8. Non-existence of global solutions of the ODE

In this section, we will show non-existence of global solutions of the ODE:

$$u'' = A u' u + B u^3 \qquad (8.1)$$

for every constant A and positive constant $B > 0$.

THEOREM 8.1. *If $B > 0$, there exists no nontrivial global solution of the initial value problem of (8.1) defined on the whole line $\mathbb{R}$.*

Since we have already shown this theorem in case of $A = 0$ in the previous section, we will treat with $A > 0$ or $A < 0$.

8.1. The simple ODE's which have only blowup solutions.
In this subsection, we prepare several propositions for the later use in the sequel subsections, which are related to blowup solutions of the initial value problem of the ODE's.

PROPOSITION 8.1. *For a positive constant $a > 0$ and every real number u_0, there exists $T(u_0) > 0$ such that every solution $u(t)$ of the initial value problem*

$$u'(t) = \sqrt{u(t)^4 + a}, \quad u(0) = u_0 \qquad (8.2)$$

is blowup at $t = T(u_0)$.

This was already proved in Proposition 6.3.   □

PROPOSITION 8.2. *For every positive constant $a > 0$ and every negative constant $u_0 < 0$, there exists $T(u_0) > 0$ such that every solution of the initial value problem*

$$u'(t) = -a u(t)^2, \quad u(0) = u_0 \qquad (8.3)$$

is blowup at $t = T(u_0)$.

PROOF. The solution of the initial value problem (8.3) is given by $u(t) = \frac{u_0}{1 + a u_0 t}$, and the interval of its existence of $u(t)$ is $(-\infty, T(u_0))$ where $T(u_0) = -\frac{1}{a u_0}$. We have Proposition 8.3.   □

PROPOSITION 8.3. *For every positive constant $a > 0$ and every real number b, let us consider the initial value problem*

$$u'(t) = u(t)^2 - \sqrt{a u(t)^4 + b}, \quad u(0) = u_0. \qquad (8.4)$$

For every u_0 satisfying $a u_0^4 + b \geq 0$ and $u_0^2 - \sqrt{a u_0^4 + b} < 0$, there exists $T(u_0) > 0$ such that every solution $u(t)$ of (8.4) is blowup at

$t = T(u_0)$. Here, if $u_0 < 0$, then $T(u_0) > 0$, and if $u_0 > 0$, then $T(u_0) < 0$.

PROOF. If $b \geq 0$, it always holds that $a\,u(t)^4 + b \geq 0$ and $\sqrt{a\,u(t)^4 + b} \geq \sqrt{a}\,u(t)^2$. Then, it holds that

$$u(t)^2 - \sqrt{a\,u(t)^4 + b} \leq (1 - \sqrt{a})\,u(t)^2.$$

Thus, $u(t)$ satisfies that

$$u'(t) \leq (1 - \sqrt{a})\,u(t)^2, \quad u(0) < 0.$$

We can apply Proposition 8.3 and Lemma 6.1, the solution $u(t)$ is blowup at $t = T(u_0)$.

if $b < 0$, $u'(0) < 0$ since $u_0^2 - \sqrt{a\,u_0^4 + b} < 0$. Thne, there exists a positive constant $\delta > 0$ such that $u'(t) < 0$ for all $t \in (0, \delta)$. On the other hand, $u(t)$ satisfies also that

$$u''(t) = 2\,u(t)\,u'(t)\left(1 - \frac{a\,u(t)^2}{\sqrt{a\,u(t)^4 + b}}\right)$$

and

$$1 - \frac{a\,u(t)^2}{\sqrt{a\,u(t)^4 + b}} < 0$$

as long as it holds that $a\,u(t)^4 + b > 0$. Thus, we get $u''(t) < 0$ for all $t \in (0, \delta)$. Therefore, $u(t)$ and $u'(t)$ are monotone decreasing on the interval $(0, \delta)$.

Thus, u and u' are monotone decreasing on the interval where the solution exists, in particular it holds that $u(t) < u_0 < 0$ and $a\,u(t)^4 - |b| \geq a\,u_0^4 - |b| \geq 0$. Let us denote $a\,(>1)$ in such a way that $a = 1 + \epsilon$. Then, we have

$$(1 + \epsilon)\,u(t)^4 - |b| \geq \left(1 + \frac{\epsilon}{2}\right)u(t)^4 \tag{8.5}$$

since $|b| < \frac{\epsilon}{2}\,u(t)^4 < a\,u(t)^2$. We have by (8.5),

$$u(t)^2 - \sqrt{a\,u(t)^4 - |b|} \leq u(t)^2 - \sqrt{\left(1 + \frac{\epsilon}{2}\right)u(t)^4} \leq -\frac{\epsilon}{2}\,u(t)^2. \tag{8.6}$$

Therefore, we have

$$u'(t) \leq \frac{1 - a}{2}\,u(t)^2, \quad u(0) < 0. \tag{8.7}$$

Then by Proposition 8.3 and Lemma 6.1 (2), there exists $T(u_0) > 0$ such that $u(t)$ is blowup at $T(u_0)$. □

8.2. The Case of $A > 0$ and $B > 0$ ($m = 5, 6, 7$). In this subsection, we will show

THEOREM 8.2. *In the case that $A > 0$ and $B > 0$, there exists no non-trivial global solution defined on the whole line $\mathbb{R}$ of the initial value problem of* (8.1)

The proof is divided into the following three cases:

1. If $u'(0) > 0$, then $u(t)$ is blowup within finite time (cf. Proposition 8.8).
    2. In the case of $u'(0) < 0$,
        (a) if $g(u(0)) < 0$, then it holds that
        either the solution $u(t)$ is blowup within finite time, or
        there exists $T > 0$ such that $u(T) < 0$ and $g(u(T)) \geq 0$
(cf. Proposition 8.10),
        (b) if $g(u(0) \geq 0$, the solution $u(t)$ is blowup within finite time
(cf. Proposition 8.11).

Here, we define the following three functions e, h and g along the solution $u(t)$ by

$$e(u(t)) := \frac{1}{2} u'(t)^2 - \frac{B}{4} u(t)^4,$$

$$h(u(t)) := \frac{1}{2} u'(t)^2 - \frac{A}{2} u'(t) u(t)^2 - \frac{B}{4} u(t)^4,$$

$$g(u(t)) := A u'(t) + B u(t)^2,$$

respectively.

By the direct computations, we have immediately

LEMMA 8.1. *Let $u(t)$ be a solution of* (8.1). *Then we have*

$$\frac{d}{dt} e(u(t)) = A u(t) u'(t)^2,$$

$$\frac{d}{dt} h(u(t)) = -\frac{A}{2} u(t)^2 u''(t) = -\frac{A}{2} u(t)^3 g(u(t)),$$

$$\frac{d}{dt} g(u(t)) = A u(t) g(u(t)) + 2 B u'(t) u(t).$$

LEMMA 8.2. *Let $u(t)$ be the solution of the initial value problem* (8.1).
    (1) *On the interval in which $u(t)$ satisfies $u(t) > 0$, it holds that*

$$u'(t)^2 \geq \frac{B}{2} u(t)^4 + E \tag{8.8}$$

for some constant E.

(2) *On the interval in which $u(t)$ satisfies $u(t) < 0$ and $g(u(t)) > 0$,*

$$u'(t)^2 \geq H - \frac{B}{4} u(t)^4 \qquad (8.9)$$

for some constant H.

(3) *On the interval in which $u(t)$ satisfies $u(t) < 0$ and $g(u(t)) \leq 0$, it holds that*

$$u'(t) \leq -\frac{B}{A} u(t)^2 < 0. \qquad (8.10)$$

PROOF. For (3), it is clear by definition of $g(u(t))$. For (1), since $\frac{d}{dt} e(u(t)) = A\, u(t)\, u'(t)^2 > 0$, $e(u(t)) = \frac{1}{2} u'(t)^2 - \frac{B}{4} u(t)^4$ is increasing, there exists some constant E such that (8.8) holds. For (2), let $u(t) < 0$ and $g(u(t)) = A\, u'(t) + B\, u(t)^2 > 0$. Then $\frac{d}{dt} h(u(t)) > 0$, there exists some constant H such that $h(u(t)) \geq H$. I.e., it holds that

$$u'(t)^2 \geq \frac{B}{4} u(t)^4 + \frac{A}{2} u'(t)\, u(t)^2 + H.$$

Here, since $A\, u'(t) > -B\, u(t)^2$ by the assumption, we have

$$u'(t)^2 \geq \frac{B}{4} u(t)^4 + \frac{1}{2} \left(Au'(t) \right) u(t)^2 + H$$

$$\geq \frac{B}{4} u(t)^4 - \frac{1}{2} \left(B\, u(t)^2 \right) u(t)^2 + H$$

$$= H - \frac{B}{4} u(t)^4,$$

which is (8.9). $\qquad \square$

PROPOSITION 8.4. *If $u'(0) \geq 0$, the solution $u(t)$ of the initial value problem of (8.1) is blowup within finite time for the initial value $u(0) \neq 0$ or $u'(0) \neq 0$.*

PROOF. The proof is divided into five cases:

(1) $u(0) > 0$ and $u'(0) > 0$; (2) $u(0) > 0$ and $u'(0) = 0$;
(3) $u(0) < 0$ and $u'(0) > 0$; (4) $u(0) < 0$ and $u'(0) = 0$;
(5) $u(0) = 0$ and $u'(0) > 0$.

For the case (1), there exists a positive number $T_0 > 0$ such that $u(t) > 0$ and $u'(t) > 0$ for all $t \in (0, T_0)$. Furthermore, since

$$u''(t) = A\, u'(t)\, u(t) + B\, u(t)^3 > 0,$$

we have $u(t) > 0$ and $u'(t) > 0$ for all t such that the solution $u(t)$ exists. Then, by Lemma 8.6, for all $t > 0$, $\frac{d}{dt} e(u(t)) = A\, u(t)\, u'(t)^2 > 0$, so that $e(u(t))$ is monotone increasing. We have $e(u(t)) \geq E_0$, where $E_0 := e(u(0))$. Thus,

$$u'(t) \geq \sqrt{\frac{B}{2} u(t)^4 + 2E_0} \,. \tag{8.11}$$

This inequality (8.11) holds for all the t which the solution $u(t)$ exists. Thus, the solution $u(t)$ is blowup within finite time due to Proposition 8.2.

For the case (3), there exists $T_0 < 0$ such that $u(t) < 0$ and $u'(t) > 0$ for all $t \in (T_0, 0)$. But, since $u''(t) = A u'(t) u(t) + B u(t)^3 < 0$, it holds that $u(t) < 0$ and $u'(t) > 0$ for all $t < 0$ in which the solution $u(t)$ exists. By Lemma 8.6, $\frac{d}{dt} e(u(t)) = A u(t) u'(t)^2 < 0$, $e(u(t))$ is monotone decreasing, so $e(u(t)) \geq E_0$ where $E_0 := e(u(0))$ for all $t < 0$. Therefore, for the backward solution $u(t)$ when t tends to $-\infty$, $u'(t) < 0$ and

$$u'(t) \leq -\sqrt{\frac{B}{2} u(t)^4 + 2E_0}. \tag{8.12}$$

(8.12) holds for all t which the solution $u(t)$ exists. Due to Proposition 8.2, the backward solution $u(t)$ is blowup within finite time when t tends to $-\infty$.

For the cases (5), there exists $\epsilon > 0$ such that $u(\epsilon) > 0$ and $u'(\epsilon) > 0$. This is the case (1) at the initial time ϵ. By the same argument of the case (1), $u(t)$ is blowup within finite time.

The remaining cases are the cases (2) and (4). For case (2), $u''(0) = B u(0)^3 > 0$, and then there exists $\epsilon >$ such that $u(\epsilon) > 0$ and $u'(\epsilon) > 0$, which is the case (1). For (4), $u''(0) = B u(0)^3 < 0$, and then there exists also $\epsilon' > 0$ such that $u(\epsilon') < 0$ and $u'(\epsilon') < 0$. But, in this case, we can go to the proof by the same way as the case (3). □

REMARK 8.1. Note that the initial value problem (8.1) with $u(0) = u'(0) = 0$ has the trivial solution, $u(t) \equiv 0$.

LEMMA 8.3. Let $u(t)$ be a solution of (8.1). Then,
(1) If $u(0) < 0$, $u'(0) < 0$ and $g(u(0)) \geq 0$, then $g(u(t)) > 0$ for all $t > 0$ satisfying that $u(t) < 0$ and $u'(t) < 0$.
(2) If $u(0) > 0$, $u'(0) < 0$ and $g(u(0)) \geq 0$, then $g(u(t)) > 0$ for all $t < 0$ satisfying that $u(t) > 0$ and $u'(t) < 0$.

PROOF. For the case (1), assume that the conclusion does not hold. Then, there exists $T > 0$ such that $u(t) < 0$ and $u'(t) < 0$ for all

$t \in [0, T]$, and $g(u(t)) > 0$ for all $t \in [0, T)$ and $g(u(T)) = 0$. Since due to Lemma 8.6,

$$\frac{d}{dt}\bigg|_{t=T} g(u(t)) = A\, u(T)\, g(u(T)) + 2\, B\, u'(T)\, u(T) = 2B\, u'(T)\, u(T) > 0.$$

Thus, $g(u(t))$ is increasing at a neighborhood around $t = T$. This contradicts to that $g(u(t)) > 0$ for all $t \in (0, T)$.

For the case 82), the proof goes in the similar way as the case (1). Assume that the conclusion does not hold. Then, there exists $T < 0$ such that $u(t) > 0$ and $u'(t) < 0$ for all $t \in [T, 0]$, $g(u(t)) > 0$ for all $t \in (T, 0]$, and $g(u(T)) = 0$. Since by Lemma 8.6,

$$\frac{d}{dt}\bigg|_{t=T} g(u(t)) = A\, u(T)\, g(u(T)) + 2\, B\, u'(T)\, u(T) = 2B\, u'(T)\, u(T) < 0.$$

Thus, $g(u(t))$ is decreasing on a neighborhood around $t = T$. But, this contradicts to that $g(u(t) > 0$ for all $t \in (T, 0)$. $\square$

Then, we have

PROPOSITION 8.5. *Let $u(t)$ be a solution of (8.1). Assume that If $g(u(0)) < 0$, then one of the following two cases occurs:*
(1) $u(t)$ is blowup within finite time, or
(2) there exists $T > 0$ such that $u(T) < 0$ and $g(u(T)) \geq 0$.

PROPOSITION 8.6. *Let $u(t)$ be a solution of (8.1). If $u'(0) < 0$ and $g(u(0)) \geq 0$, then $u(t)$ is blowup within finite time.*

(*Proof of Proposition 8.11*)   We first notice that $u'(0) < 0$. Because, since we assume that $g(u(0)) = A\, u'(0) + B\, u(0)^2 < 0$, it holds that $u'(0) < -\frac{B}{A}\, u(0)^2 < 0$ unless $u(0) = 0$. If $u(0) = 0$, our assumption $0 > g(u(0)) = A\, u'(0)$ implies also that $u'(0) < 0$.

The proof is divided into three cases: (1) $u(0) < 0$, (2) $u(0) > 0$ or (3) $u(0) = 0$.

Case (1) $u(0) < 0$. In this case, There exists $T_0 > 0$ such that $u(t) < 0$, $u'(t) < 0$, and $g(u(t)) < 0$ for all $t \in (0, T_0)$. By Lemma 8.7, for all $t \in (0, T_0)$,

$$u'(t) \leq -\frac{B}{A}\, u(t)^2 < 0, \tag{8.13}$$

and $u(0) < 0$ by our assumption. In particular, $u(t)$ is monotone decreasing on $(0, T_0)$, so that $u(t) < 0$ for all $t \in (0, T_0)$. Therefore, there exists $T > 0$ such that, either $u(T) < 0$ and $g(u(T)) \geq 0$, or $u(t) < 0$ and $g(u(t)) < 0$ for all t in the interval where the solution

$u(t)$ exists. For the latter case, since $g(u(t)) = A u'(t) + B u(t)^2 < 0$, (8.13)still holds. But, since $\frac{B}{A} > 0$ and $u(0) < 0$, due to Proposition 8.3 and Lemma 6.1, the solution $u(t)$ is blowup within finite time.

Case (2) $u(0) > 0$.There exists $T_0 < 0$ such that $u(t) > 0$, $u'(t) < 0$ and $g(u(t)) < 0$ for all $t \in (T_0, 0)$. Then, by Lemma 8.7, it holds that

$$u'(t) \leq -\frac{B}{A} u(t)^2 < 0, \tag{8.14}$$

for all $t \in (T_0, 0)$, and also $u(0) > 0$. Thus, $u(t) > 0$ for all $t \in (T_0, 0)$. Therefore, either there exists $T < 0$ such that $u(T) > 0$ and $g(u(T)) \geq 0$, or $u(t) > 0$ and $g(u(t)) < 0$ for all t in the interval where the solution exists. For the latter case, it holds that (8.14) holds since $g(u(t)) = A u'(t) + B u(t)^2 < 0$. $\frac{B}{A} > 0$ and $u(0) > 0$, by Proposition 8.3 and Lemma 6.1, the backward solution $u(t)$ is blowup within finite time when t goes to $-\infty$.

Case (3), $u(0) = 0$. In this case, since the assumption $g(u(0)) < 0$, $u'(0) < 0$, so there exists $\epsilon > 0$ such that $u(\epsilon) > 0$, $u'(\epsilon) < 0$ and $g(u(\epsilon)) < 0$. Then, we can apply Case (1) at $t = \epsilon$, and reach the same conclusion as Case (1). □

(*Proof of Proposition 8.12*)   The proof is also divided into three cases, (1) $u(0) < 0$, (2) $u(0) > 0$, or (3) $u(0) = 0$.

Case (1), $u(0) < 0$. We have $u(0) < 0$, $u'(0) < 0$, and $g(u(0)) \geq 0$. Then, by (8.1), $u''(0) = u(0) g(u(0)) \geq 0$, so that $u(t) < 0$ and $u'(t) < 0$ for all t satisfying $g(u(t)) \geq 0$. By Lemma 8.10, (2), it holds that $g(u(t)) > 0$ for all t such that $u(t)$ exists. Therefore, it holds that $u(t) < 0$, $u'(t) < 0$ and $g(u(t)) \geq 0$ for all $t > 0$ such that the solution $u(t)$ exists. Due to Lemma 8.6, we have

$$\frac{d}{dt} h(u(t)) = -\frac{A}{2} u(t)^3 g(u(t)) > 0, \tag{8.15}$$

which implies that $h(u(t))$ is monotone increasing. Therefore, it holds that $h(u(t)) \geq h(u(0))$ for all $t > 0$. Say, $h_0 := h(u(0))$. Then, by definition of $h(u(t))$, we have

$$u'(t)^2 - A u'(t) u(t)^2 - \frac{B}{2} u(t)^4 \geq h_0. \tag{8.16}$$

By (8.16), we have since $u'(t) < 0$,

$$u'(t) < \frac{1}{2} \left(A u(t)^2 - \sqrt{(a^2 + 2B) u(t)^4 + 8h_0} \right). \tag{8.17}$$

Thus, we have (8.17), and $u(0) < 0$, $u'(0) < 0$ and $2h_0 = u'(0)^2 - A\,u'(0)\,u(0)^2 - \frac{B}{2}\,u(0)^4$, which imply that all the conditions in Proposition 8.4 are satisfied. Due to Lemma 6.1, the solution $u(t)$ is blowup within finite time.

Case (2), $u(0) > 0$. In this case, we consider the backward solution. We first have $u(0) > 0$, $u'(0) < 0$ and $g(u(0)) \geq 0$. Then, since $u''(0) = u(0)\,g(u(0)) \geq 0$, it holds that $u(t) > 0$ and $u'(t) < 0$ for all the t satisfying that $g(u(t)) \geq 0$. Thus, by Lemma 8.9, $g(u(t)) > 0$ for all the t which the solution $u(t)$ exists. Therefore, it holds that $u(t) > 0$, $u'(t) < 0$ and $g(u(t)) \geq 0$ for all the $t < 0$ which the solution $u(t)$ exists. Then, due to Lemma 8.6, it holds that

$$\frac{d}{dt}h(u(t)) = -\frac{A}{2}\,u(t)^3\,g(u(t)) \leq 0, \tag{8.18}$$

so $h(u(t))$ is monotone decreasing. We have $h(u(t)) \geq h(u(0))$ for all $t < 0$. Say $h_0 := h(u(0))$. Then, by definition of $h(u(t))$, we have

$$u'(t)^2 - A\,u'(t)\,u(t)^2 - \frac{B}{2}\,u(t)^4 \geq 2\,h_0. \tag{8.19}$$

We also have that $u(0) > 0$, $u'(0) < 0$, and $2\,h_0 = u'(0)^2 - A\,u'(0)\,u(0)^2 - \frac{B}{2}\,u(0)^4$, which implies that all the conditions in Proposition 8.4 are satisfied. Thus, the backward solution $u(t)$ is blowup within finite time when t dous to $-\infty$.

Case (3), $u(0) = 0$. In this case, our assumption $u'(0) < 0$ implies that there exists $\epsilon > 0$ such that $u(\epsilon) < 0$ and $u'(\epsilon) < 0$. Thus, we can apply the case (1) at the initial value $t = \epsilon$. Then, the solution $u(t)$ is blowup within finite time. We have Proposition 8.12. $\square$

8.3. The case of $A < 0$ and $B > 0$ $(m \geq 9)$. In the previous subsection, we showed Theorem 8.1 in the case that $A > 0$ and $B > 0$. In this subsection, we have

THEOREM 8.3. *If $A < 0$ and $B > 0$, then there exists no nontrivial global solution of the initial value problem of* (8.1) *defined on the whole line* $\mathbb{R}$.

Theorem 8.13 follows directly from Theorem 8.5 and the following Proposition.

PROPOSITION 8.7. *Let $u(t)$ be a solution of the initial value problem of* (8.1) *in the case that $A < 0$ and $B > 0$. Then, $v(t) := u(-t)$ is a solution of the initial value problem of* (8.1) *in the case that $A > 0$ and $B > 0$.*

We have Theorem 8.1. □

9. Biharmonic maps between product Riemannian manifolds

Finally, we give nice applications. Let us consider the product Riemannian manifolds, $M := \mathbb{R} \times \Sigma^{m-1}$, and $N := \mathbb{R} \times P$, respectively, where $\mathbb{R}$ is a line with the standard Riemannian metric g_1, Σ^{m-1} is an $(m-1)$-dimensional manifold with a Riemannian metric g_2 ($m = 3, 4$), and P is a manifold with Riemannian metric h_2, respectively. Let us take the product Riemannian metrics $g = g_1 + g_2$ on M, and $h = g_1 + h_2$ on N, respectively.

Then, for every smooth map $\varphi = (\varphi_1, \varphi_2) : M \to N$, with $\varphi_1 : \mathbb{R} \to \mathbb{R}$, and $\varphi_2 : \Sigma^2 \to P$, the tension field $\tau(\varphi)$ is given as

$$\tau(\varphi) = (\tau(\varphi_1), \tau(\varphi_2)) \in \Gamma(\varphi^{-1}TN) = \Gamma(\varphi_1^{-1}T\mathbb{R} \times \varphi_2^{-1}TN).$$

Thus, φ is harmonic if and only if both (1) $\varphi_1 : (\mathbb{R}, g_1) \to (\mathbb{R}, g_1)$ is harmonic, and (2) $\varphi_2 : (\Sigma^{m-1}, g_2) \to (P, h_2)$ is harmonic. Notice that all the harmonic maps $\varphi_1 : (\mathbb{R}, g_1) \to (\mathbb{R}, g_1)$ are linear functions $\mathbb{R} \ni x \mapsto ax + b \in \mathbb{R}$ for some constants a and b.

Now define a conformal Riemannian metric $\tilde{g} = \tilde{f}^{2/(m-2)} g$ with $\tilde{f}(x, y) = f(t)$ ($t = x \in \mathbb{R}$, $y \in \Sigma^{m-1}$).

Then, we can easily calculate that

$$\nabla^g f = f' \frac{\partial}{\partial t},$$

$$\varphi_*(\nabla^g f) = \varphi_{1*}(f' \frac{\partial}{\partial t}) = a f' \frac{\partial}{\partial t},$$

$$\Delta^g f = -f'',$$

$$\overline{\nabla}_{\nabla^g f} \varphi_*(\nabla^g f) = a f'' f' \frac{\partial}{\partial t},$$

$$J_g(\varphi_*(\nabla^g f)) = -a f''' \frac{\partial}{\partial t}.$$

For a harmonic map $\varphi = (\varphi_1, \varphi_2) : (M, g) = (\mathbb{R} \times \Sigma^{m-1}, g) \to (N, h) = (\mathbb{R} \times P, h)$, it holds that

$\varphi : (M, \tilde{g}) : \to (N, h)$ is harmonic if and only if $\varphi_*(\nabla^g f) = a f' \frac{\partial}{\partial t} = 0$ if and only if $f(t)$ is constant in $t = x$ or φ_1 is a constant.

On the other hand, $\varphi : (M, \tilde{g}) \to (N, h)$ is biharmonic map if and only if φ_1 is a constant or the ODE (5.3) holds.

Thus, we finally obtain the following theorem which answers our Problem in the Section Four.

THEOREM 9.1. (*Final Theorem*) *For every harmonic map* $\varphi =:$ $(\Sigma^{m-1}, g) \rightarrow (P, h)$, *let us define* $\widetilde{\varphi} : \mathbb{R} \times \Sigma^{m-1} \ni (x, y) \mapsto (ax + b, \varphi(y)) \in \mathbb{R} \times P$ $(m = 3, 4)$, *where* a *and* b *are constants. Then,*

(1) *In the case* $m = 3$, *the mapping* $\widetilde{\varphi} : (\mathbb{R} \times \Sigma^2, \widetilde{f}^2 g) \rightarrow (\mathbb{R} \times P, h)$ *is biharmonic, but not harmonic if* $a \neq 0$.

(2) *In the case* $m = 4$, *the mapping* $\widetilde{\varphi} (\mathbb{R} \times \Sigma^3, \frac{1}{\cosh x} g) \rightarrow (\mathbb{R} \times P, h)$ *is biharmonic, but not harmonic if* $a \neq 0$.

Part 3

Biharmonic Submanifolds

CHAPTER 11

Biharmonic Submanifolds in a Riemannian Manifold

1

ABSTRACT. In this chapter, we study biharmonic hypersurfaces in compact symmetric spaces. Then, we classify all the homogeneous biharmonic hypersurfaces in compact symmetric spaces.

1. Introduction

Harmonic maps play a central role in geometry; they are critical points of the energy functional $E(\varphi) = \frac{1}{2} \int_M |d\varphi|^2 \, v_g$ for smooth maps φ of (M, g) into (N, h). The Euler-Lagrange equations are given by the vanishing of the tension filed $\tau(\varphi)$. In 1983, J. Eells and L. Lemaire [40] extended the notion of harmonic map to biharmonic map, which are, by definition, critical points of the bienergy functional

$$E_2(\varphi) = \frac{1}{2} \int_M |\tau(\varphi)|^2 \, v_g. \tag{1.1}$$

After G.Y. Jiang [74] studied the first and second variation formulas of E_2, extensive studies in this area have been done (for instance, see [16], [90], [102], [120], [123], [131], [63], [64], [73], etc.). Notice that harmonic maps are always biharmonic by definition. We say, for a smooth map $\varphi : (M, g) \to (N, h)$ to be *proper biharmonic* if it is biharmonic, but not harmonic. B.Y. Chen raised ([21]) so called B.Y. Chen's conjecture and later, R. Caddeo, S. Montaldo, P. Piu and C. Oniciuc raised ([16]) the generalized B.Y. Chen's conjecture.

B.Y. Chen's conjecture:
Every biharmonic submanifold of the Euclidean space $\mathbb{R}^n$ must be harmonic (minimal).

The generalized B.Y. Chen's conjecture:
Every biharmonic submanifold of a Riemannian manifold of non-positive curvature must be harmonic (minimal).

---

[1]This chapter is due to [116]: S. Ohno, T. Sakai and H. Urakawa, *Biharmoic homogeneous hypersurfaces in compact symmetric spaces*, Differential Geom. Appl. **43** (2015), 155–179.

For the generalized Chen's conjecture, Ou and Tang gave ([**123**]) a counter example in a Riemannian manifold of negative curvature. For the Chen's conjecture, affirmative answers were known for the case of surfaces in the three dimensional Euclidean space ([**21**]), and the case of hypersurfaces of the four dimensional Euclidean space ([**57**], [**34**]). Furthermore, Akutagawa and Maeta gave ([**1**]) recently a final supporting evidence to the Chen's conjecture:

THEOREM 1.1. *Any complete regular biharmonic submanifold of the Euclidean space $\mathbb{R}^n$ is harmonic (minimal).*

To the generalized Chen's conjecture, we showed ([**111**]) that

THEOREM 1.2. *Let (M,g) be a complete Riemannian manifold, and the curvature of (N,h), non-positive. Then,*
(1) every biharmonic map $\varphi : (M,g) \to (N,h)$ with finite energy and finite bienergy must be harmonic.
(2) In the case $\mathrm{Vol}(M,g) = \infty$, under the same assumtion, every biharmonic map $\varphi : (M,g) \to (N,h)$ with finite bienergy is harmonic.

We also obtained (cf. [**108**], [**109**], [**111**])

THEOREM 1.3. *Assume that (M,g) is a complete Riemannian manifold, $\varphi : (M,g) \to (N,h)$ is an isometric immersion, and the sectional curvature of (N,h) is non-positive. If $\varphi : (M,g) \to (N,h)$ is biharmonic and $\int_M |\mathbf{H}|^2 v_g < \infty$, then it is minimal. Here, $\mathbf{H}$ is the mean curvature normal vector field of the isometric immersion φ.*

Theorem 1.3 gives an affirmative answer to the generalized B.Y. Chen's conjecture under the L^2-condition and completeness of (M,g).

2. Preliminaries

We first prepare the materials for the first and second variational formulas for the bienergy functional and biharmonic maps. Let us recall the definition of a harmonic map $\varphi : (M,g) \to (N,h)$, of a compact Riemannian manifold (M,g) into another Riemannian manifold (N,h), which is an extremal of the *energy functional* defined by

$$E(\varphi) = \int_M e(\varphi) \, v_g,$$

where $e(\varphi) := \frac{1}{2}|d\varphi|^2$ is called the energy density of φ. That is, for any variation $\{\varphi_t\}$ of φ with $\varphi_0 = \varphi$,

$$\frac{d}{dt}\bigg|_{t=0} E(\varphi_t) = -\int_M h(\tau(\varphi), V)v_g = 0, \qquad (2.1)$$

where $V \in \Gamma(\varphi^{-1}TN)$ is a variation vector field along φ which is given by $V(x) = \frac{d}{dt}\big|_{t=0}\varphi_t(x) \in T_{\varphi(x)}N$, $(x \in M)$, and the *tension field* is given by $\tau(\varphi) = \sum_{i=1}^m B(\varphi)(e_i, e_i) \in \Gamma(\varphi^{-1}TN)$, where $\{e_i\}_{i=1}^m$ is a locally defined orthonormal frame field on (M, g), and $B(\varphi)$ is the second fundamental form of φ defined by

$$\begin{aligned}
B(\varphi)(X, Y) &= (\widetilde{\nabla}d\varphi)(X, Y) \\
&= (\widetilde{\nabla}_X d\varphi)(Y) \\
&= \overline{\nabla}_X(d\varphi(Y)) - d\varphi(\nabla_X Y),
\end{aligned} \qquad (2.2)$$

for all vector fields $X, Y \in \mathfrak{X}(M)$. Here, ∇, and ∇^N, are Levi-Civita connections on TM, TN of (M, g), (N, h), respectively, and $\overline{\nabla}$, and $\widetilde{\nabla}$ are the induced ones on $\varphi^{-1}TN$, and $T^*M \otimes \varphi^{-1}TN$, respectively. By (2.1), φ is *harmonic* if and only if $\tau(\varphi) = 0$.

The second variation formula is given as follows. Assume that φ is harmonic. Then,

$$\frac{d^2}{dt^2}\bigg|_{t=0} E(\varphi_t) = \int_M h(J(V), V)v_g, \qquad (2.3)$$

where J is an elliptic differential operator, called the *Jacobi operator* acting on $\Gamma(\varphi^{-1}TN)$ given by

$$J(V) = \overline{\Delta}V - \mathcal{R}(V), \qquad (2.4)$$

where $\overline{\Delta}V = \overline{\nabla}^*\overline{\nabla}V = -\sum_{i=1}^m \{\overline{\nabla}_{e_i}\overline{\nabla}_{e_i}V - \overline{\nabla}_{\nabla_{e_i}e_i}V\}$ is the *rough Laplacian* and $\mathcal{R}$ is a linear operator on $\Gamma(\varphi^{-1}TN)$ given by $\mathcal{R}(V) = \sum_{i=1}^m R^N(V, d\varphi(e_i))d\varphi(e_i)$, and R^N is the curvature tensor of (N, h) given by $R^N(U, V) = \nabla^N_U \nabla^N_V - \nabla^N_V \nabla^N_U - \nabla^N_{[U,V]}$ for $U, V \in \mathfrak{X}(N)$.

J. Eells and L. Lemaire [40] proposed polyharmonic (k-harmonic) maps and Jiang [74] studied the first and second variation formulas of biharmonic maps. Let us consider the *bienergy functional* defined by

$$E_2(\varphi) = \frac{1}{2}\int_M |\tau(\varphi)|^2 v_g, \qquad (2.5)$$

where $|V|^2 = h(V, V)$, $V \in \Gamma(\varphi^{-1}TN)$.

The first variation formula of the bienergy functional is given by

$$\frac{d}{dt}\bigg|_{t=0} E_2(\varphi_t) = -\int_M h(\tau_2(\varphi), V)v_g. \tag{2.6}$$

Here,

$$\tau_2(\varphi) := J(\tau(\varphi)) = \overline{\Delta}(\tau(\varphi)) - \mathcal{R}(\tau(\varphi)), \tag{2.7}$$

which is called the *bitension field* of φ, and J is given in (2.4).

A smooth map φ of (M, g) into (N, h) is said to be *biharmonic* if $\tau_2(\varphi) = 0$. By definition, every harmonic map is biharmonic. We say, for an immersion $\varphi : (M, g) \to (N, h)$ to be *proper biharmonic* if it is biharmonic but not harmonic (minimal).

3. Biharmonic isometric immersions

3.1. In the first part of this section, we first show a characterization theorem for an isometric immersion φ of an m dimensional Riemannian manifold (M, g) into a Riemannian manifold (N, h) whose tension field $\tau(\varphi)$ satisfies that $\overline{\nabla}_X^\perp \tau(\varphi) = 0$ ($X \in \mathfrak{X}(M)$) to be biharmonic. Let us recall the following theorem due to [**74**]:

THEOREM 3.1. *Let* $\varphi : (M^m, g) \to (N^n, h)$ *be an isometric immersion. Assume that* $\overline{\nabla}_X^\perp \tau(\varphi) = 0$ *for all* $X \in \mathfrak{X}(M)$. *Then,* φ *is biharmonic if and only if the following holds:*

$$-\sum_{j,k=1}^m h\Big(\tau(\varphi), R^h(d\varphi(e_j), d\varphi(e_k))d\varphi(e_k)\Big) d\varphi(e_j)$$

$$+\sum_{j,k=1}^m h(\tau(\varphi), B_\varphi(e_j, e_k)) B_\varphi(e_j, e_k)$$

$$-\sum_{j=1}^m R^h(\tau(\varphi), d\varphi(e_j)) d\varphi(e_j) = 0, \tag{3.1}$$

where R^h *is the curvature tensor field of* (N, h) *given by* $R^h(U, V)W = \nabla_U^h(\nabla_V^h W) - \nabla_V^h(\nabla^h U W) - \nabla_{[U,V]}^h W$, $(U, V, W \in \mathfrak{X}(N))$, *and* $B_\varphi(X, Y)$ $(X, Y \in \mathfrak{X}(M))$ *is the second fundamental form of the immersion* φ *given by* $B_\varphi(X, Y) = \nabla_{d\varphi(X)}^h d\varphi(Y) - d\varphi(\nabla_X^g Y)$, *and* $\{e_j\}$ *is a locally defined orthonormal frame field on* (M, g).

Here, let us apply the following general curvature tensorial properties ([**82**], Vol. I, Pages 198, and 201) to the first term of the left hand side of (3.1):

$$h\big(W_1, R^h(W_3, W_4)W_2\big) = h\big(W_3, R^h(W_1, W_2)W_4\big), \tag{3.2}$$

$$(W_i \in \mathfrak{X}(N), i = 1, 2, 3, 4).$$

Then, we have

$$h\big(\tau(\varphi), R^h(d\varphi(e_j), d\varphi(e_k))d\varphi(e_k)\big)$$
$$= h\big(d\varphi(e_j), R^h(\tau(\varphi), d\varphi(e_k))d\varphi(e_k)\big). \tag{3.3}$$

Therefore, for the first term of (3.1), we have that

$$\sum_{j=1}^m h\Big(d\varphi(e_j), \sum_{k=1}^m R^h(\tau(\varphi), d\varphi(e_k))d\varphi(e_k)\Big)\, d\varphi(e_j) \tag{3.4}$$

is equal to the tangential part of $\sum_{k=1}^m R^h(\tau(\varphi), d\varphi(e_k))\, d\varphi(e_k)$. Thus, we obtain

$$-\left(\sum_{k=1}^m R^h(\tau(\varphi), d\varphi(e_k))d\varphi(e_k)\right)^{\top}$$
$$+ \sum_{j,k=1}^m h(\tau(\varphi), B_\varphi(e_j, e_k))\, B_\varphi(e_j, e_k)$$
$$- \sum_{k=1}^m R^h(\tau(\varphi), d\varphi(e_k))\, d\varphi(e_k) = 0, \tag{3.5}$$

where $W^{\top}$ and $W^{\perp}$ mean the tangential part and the normal part of $W \in \mathfrak{X}(N)$, respectively. We have, by comparing the tangential part and the normal part of the equation (3.5), it is equivalent to that

$$\left(\sum_{k=1}^m R^h(\tau(\varphi), d\varphi(e_k))d\varphi(e_k)\right)^{\top} = 0, \quad \text{and} \tag{3.6}$$

$$\sum_{j,k=1}^m h(\tau(\varphi), B_\varphi(e_j, e_k))\, B_\varphi(e_j, e_k)$$
$$= \left(\sum_{k=1}^m R^h(\tau(\varphi), d\varphi(e_k))\, d\varphi(e_k)\right)^{\perp}. \tag{3.7}$$

Summarizing the above, we obtain:

THEOREM 3.2. *Let* $\varphi : (M^m, g) \to (N^n, h)$ *be an isometric immersion. Assume that* $\overline{\nabla}_X^{\perp}\tau(\varphi) = 0$ *for all* $X \in \mathfrak{X}(M)$. *Then,* φ *is biharmonic if and only if the following equations hold:*

(1) *(the tangential part)*

$$\left(\sum_{k=1}^{m} R^h(\tau(\varphi), d\varphi(e_k))d\varphi(e_k) \right)^{\top} = 0, \tag{3.8}$$

and (2) *(the normal part)*

$$\left(\sum_{k=1}^{m} R^h(\tau(\varphi), d\varphi(e_k)) d\varphi(e_k) \right)^{\perp}$$
$$= \sum_{j,k=1}^{m} h(\tau(\varphi), B_\varphi(e_j, e_k)) B_\varphi(e_j, e_k). \tag{3.9}$$

As a corollary of Theorem 3.2, we obtain:

COROLLARY 3.1. *Assume that the sectional curvature of the target space (N^n, h) is non-positive. Let $\varphi : (M^m, g) \to (N^n, h)$ be an isometric immersion whose tension field satisfies $\overline{\nabla}_X^{\perp} \tau(\varphi) = 0$ for all $X \in \mathfrak{X}(M)$. Then, if φ is biharmonic, then it is harmonic.*

Proof. Due to (3.9) in (2) of Theorem 3.2, taking the inner product to the both hand side of (3.9) with $\tau(\varphi)$ which is normal,

$$\sum_{j,k=1}^{m} h(\tau(\varphi), B_\varphi(e_j, e_k))^2 = h\left(\tau(\varphi), \sum_{k=1}^{m} R^h(\tau(\varphi), d\varphi(e_k)), d\varphi(e_k)\right)$$
$$= \sum_{k=1}^{m} h\left(\tau(\varphi), R^h(\tau(\varphi), d\varphi(e_k)), d\varphi(e_k)\right)$$
$$\leq 0. \tag{3.10}$$

Because the quantity $h\left(\tau(\varphi), R^h(\tau(\varphi), d\varphi(e_k)), d\varphi(e_k)\right)$ is a multiple of the non-negative number $h(\tau(\varphi), \tau(\varphi))$ times the sectional curvature of (N, h) along the plane $\{\tau(\varphi), d\varphi(e_k)\}$ $(k = 1, \cdots, m)$ which are also non-positive by our assumption. Therefore, the both hand sides of (3.10) must be zero, i.e.,

$$\begin{cases} h\left(\tau(\varphi), R^h(\tau(\varphi), d\varphi(e_k)) d\varphi(e_k)\right) = 0 & (\forall \ k = 1, \cdots, m), \quad \text{and} \\ h(\tau(\varphi), B_\varphi(e_j, e_k)) = 0 & (\forall \ j, \ k = 1, \cdots, m). \end{cases}$$

Therefore, we obtain

$$h(\tau(\varphi), \tau(\varphi)) = h\left(\tau(\varphi), \sum_{j=1}^{m} B_\varphi(e_j, e_j)\right) = 0 \tag{3.11}$$

which implies that $\tau(\varphi) = 0$. □

REMARK 3.1. *Corollary 3.3 give a partial evidence to the generalized B.-Y. Chen's conjecture: every biharmonic isometric immersion into a non-positive curvature manifold must be harmonic. On the other hand, notice that the generalized B.-Y. Chen's conjecture was given by a counter example due to Y. Ou and L. Tang, in 2012, [123] .*

3.2. In the second part of this section, we apply Theorem 3.2, to an isometric immersion into an Einstein manifold (N^n, h) whose Ricci transform $\rho^h(u) := \sum_{i=1}^n R^h(u, e_i')e_i'$ $(u \in T_y N, \ y \in N)$, where $\{e_i'\}_{i=1}^n$ is a locally defined orthonormal frame field on (N^n, h).

Let $\{(\xi_i)_x\}_{i=1}^p$ $(x \in M)$ be an orthonormal basis of the orthogonal complement $T_{\varphi(x)}^{\perp} M$ of the tangent space $d\varphi_x(T_{\varphi(x)} M)$ of M in the one $T_{\varphi(x)} N$ of N with respect to $h_{\varphi(x)}$ $(x \in M)$:

$$T_{\varphi(x)} N = d\varphi(T_x M) \oplus \sum_{i=1}^p \mathbb{R}\,(\xi_i)_x \qquad (x \in M),$$

where $(p := \dim N - \dim M = n - m)$. Then, we have

$$\rho^h(u) = \sum_{k=1}^m R^h(u, d\varphi_k(e_k))d\varphi(e_k) + \sum_{i=1}^p R^h(u, \xi_i)\xi_i \tag{3.12}$$

for every $u \in T_y N$ $(y \in N)$ since φ is an isometric immersion of (M^m, g) into (N^n, h). Therefore, for (3.8) and (3.9) in Theorem 3.2, we have that (3.8) is equivalent to the following:

$$\left(\rho^h(\tau(\varphi)) - \sum_{i=1}^p R^h(\tau(\varphi), \xi_i)\xi_i\right)^{\top} = 0, \tag{3.13}$$

and (3.9) is equivalent to the following:

$$\left(\rho^h(\tau(\varphi)) - \sum_{i=1}^p R^h(\tau(\varphi), \xi_i)\xi_i\right)^{\perp}$$

$$= \sum_{j,k=1}^m h\big(\tau(\varphi), B_\varphi(e_j, e_k)\big) B_\varphi(e_j, e_k). \tag{3.14}$$

Now assume that (N, h) is Einstein, namely, the Ricci transform ρ^h satisfies that $\rho^h(u) = c\,\mathbb{I}$ for some constant c, where $\mathbb{I}$ is the identity transform. Then, since $\rho^h(\tau(\varphi)) = c\,\tau(\varphi)$, the left hand sides of (3.13) and (3.14) are

$$\left(\sum_{i=1}^p R^h(\tau(\varphi), \xi_i)\xi_i\right)^{\top}, \qquad c\,\tau(\varphi) - \left(\sum_{i=1}^p R^h(\tau(\varphi), \xi_i)\xi_i\right)^{\perp},$$

respectively. Therefore, the two equations (3.13) and (3.14) hold if and only if the following single equation holds:

$$c\,\tau(\varphi) - \sum_{i=1}^{p} R^h(\tau(\varphi),\xi_i)\xi_i = \sum_{j,k=1}^{m} h\big(\tau(\varphi), B_\varphi(e_j,e_k))\big)\, B_\varphi(e_j,e_k). \tag{3.15}$$

Thus we obtain

THEOREM 3.3. *Assume that* $\varphi : (M^m,g) \to (N^n,h)$ *is an isometric immersion whose tension field* $\tau(\varphi)$ *satisfies that* $\overline{\nabla}^{\perp}_X \tau(\varphi) = 0$, *and the target space* (N,h) *is an Einstein, i.e., the Ricci transform* ρ^h *of* (N,h) *satisfies* $\rho^h = c\,\mathrm{Id}$ *for some constant* c. *Then,* φ *is biharmonic if and only if* (3.15) *holds.*

3.3. In the following, we treat with a hypersurface $\varphi : (M^m,g) \to (N^n,h)$, i.e., $p=1$, and $m = \dim M = \dim N - 1 = n-1$. In this case, let us $\xi = \xi_1$ be a unit normal vector field along φ, and denote the second fundamental form B_φ as $B_\varphi(e_j,e_k) = H_{jk}\,\xi$ $(j,k=1,\cdots,m)$. Then,

$$\tau(\varphi) = \sum_{j=1}^{m} B_\varphi(e_j,e_j) = \left(\sum_{j=1}^{m} H_{jj}\right)\xi, \tag{3.16}$$

$$R^h(\tau(\varphi),\xi)\xi = R^h\left(\sum_{j=1}^{m} H_{jj}\,\xi,\xi\right)\xi = 0, \tag{3.17}$$

$$\sum_{j,k=1}^{m} h\big(\tau(\varphi),B_\varphi(e_j,e_k))\big)\, B_\varphi(e_j,e_k)$$
$$= \left(\sum_{i=1}^{m} H_{ii}\right)\left(\sum_{j,k=1}^{m} H_{jk}{}^2\right)\xi$$
$$= \left(\sum_{i=1}^{m} H_{ii}\right)\|B_\varphi\|^2\,\xi, \tag{3.18}$$

where $\|B_\varphi\|^2 = \sum_{j,k=1}^{m}\|B_\varphi(e_j,e_k)\|^2 = \sum_{j,k=1}^{m} H_{jk}{}^2$.
Therefore, (3.15) holds if and only if

$$c\left(\sum_{j=1}^{m} H_{jj}\right)\xi = \left(\sum_{j=1}^{m} H_{jj}\right)\|B_\varphi\|^2\,\xi \tag{3.19}$$

which is equivalent to that, either φ is harmonic, i.e., $\sum_{j=1}^{m} H_{jj} = 0$, or

$$\|B_\varphi\|^2 = c. \tag{3.20}$$

Thus, we obtain the following theorem:

THEOREM 3.4. *Assume that $\varphi : (M^m, g) \to (N^n, h)$ is an isometric immersion whose tension field $\overline{\nabla}^{\perp}_X \tau(\varphi) = 0$ $(\forall\, X \in \mathfrak{X}(M))$ and φ is hypersurface, i.e., $m = n - 1$.*

(1) If φ is not harmonic, then φ is biharmonic if and only if

$$\rho^h(\xi) = \|B_\varphi\|^2\, \xi, \tag{3.21}$$

where ρ^h is the Ricci transform of (N, h), and ξ is a unit normal vector field along φ.

(2) In particular, if (N, h) is an Einstein manifold, i.e., $\rho^h = c\,\mathrm{Id}$, and φ is not harmonic, then φ is biharmonic if and only if $\|B_\varphi\|^2 = c$.

Furthermore, we have

THEOREM 3.5. *Assume that $\varphi : (M, g) \to (N, h)$ is an isometric immersion into a Riemannian manifold (N, h) whose Ricci curvature is non-positive, $\dim M = \dim N - 1$, and $\overline{\nabla}^{\perp}_X \tau(\varphi) = 0$ for all C^∞ vector field X on M. Then, if φ is biharmonic, it is harmonic.*

Proof. If we assume φ is not harmonic, then due to (1) of Theorem 3.6, we have

$$\rho^h(\xi) = \|B_\varphi\|^2\, \xi. \tag{3.22}$$

Together with the assumption of non-positivity of the Ricci curvature of (N, h), we have

$$0 \le \|B_\varphi\|^2\, h(\xi, \xi) = h(\rho^h(\xi), \xi) \le 0. \tag{3.23}$$

THerefore, we have

$$h(\rho^h(\xi), \xi) = \|B_\varphi\|^2 = 0. \tag{3.24}$$

In particular, we have $B_\varphi \equiv 0$, in particular, we have that $\tau(\varphi) = \sum_{i=1}^{m} B_\varphi(e_i, e_i) = 0$ which contradicts the assumption. $\qquad\square$

Finally, in this section, on the condition $\overline{\nabla}^{\perp}_X \tau(\varphi) = 0$ $(\forall\, X \in \mathfrak{X}(M))$, we give the following criterion:

PROPOSITION 3.1. *Assume that $\varphi : (M^m, g) \to (N^n, h)$ is an isometric immersion with $m = \dim M = \dim N - 1 = n - 1$. Then, the following equivalence holds: The condition that $\overline{\nabla}^{\perp}_X \tau(\varphi) = 0$ $(\forall\, X \in \mathfrak{X}(M)$ holds if and only if the mean curvature $\mathbf{H} = \frac{1}{m} \sum_{i=1}^{m} H_{ii}$, is constant on M. Here, $B_\varphi(e_i, e_j) = H_{ij}\, \xi$, and ξ is a unit normal vector field along φ.*

Proof. Since $\tau(\varphi) = m\,\mathbf{H}\,\xi$ where $\mathbf{H} = \frac{1}{m}\sum_{i=1}^{m} H_{ii}$, we have, for all C^{∞} vector field X on M,

$$0 = \overline{\nabla}_X^{\perp}\tau(\varphi) = m\,(X\mathbf{H})\,\xi + m\,\mathbf{H}\,\overline{\nabla}_X^{\perp}\xi. \tag{3.25}$$

Since $h(\xi,\xi) = 1$, we have

$$0 = \frac{1}{2}X_x\,h(\xi,\xi)_{\varphi(x)} = h(\nabla^h_{d\varphi(X)}\xi,\xi) = h(\overline{\nabla}_X^{\perp}\xi,\xi). \tag{3.26}$$

By taking the inner product (3.25) and ξ, we have, due to (3.26),

$$0 = m\,(X\,\mathbf{H}) + m\,\mathbf{H}\,h(\overline{\nabla}_X^{\perp}\xi,\xi) = m\,(X\,\mathbf{H}). \tag{3.27}$$

We have $X\,\mathbf{H} = 0$ for all C^{∞} vector field X on M, which implies that $\mathbf{H}$ is constant on M.

Conversely, if $\mathbf{H}$ is constant on M, then we have

$$\overline{\nabla}_X^{\perp}\tau(\varphi) = m\,(X\,\mathbf{H})\,\xi + m\,\mathbf{H}\,\overline{\nabla}_X^{\perp}\xi = m\,\mathbf{H}\,\overline{\nabla}_X^{\perp}\xi,$$

so that we have, due to (3.26),

$$h(\overline{\nabla}_X^{\perp}\tau(\varphi),\xi) = m\,\mathbf{H}\,h(\overline{\nabla}_X^{\perp}\xi,\xi) = 0,$$

which implies that $\overline{\nabla}_X^{\perp}\tau(\varphi) = 0$ since it is normal. □

Summarizing Theorems 3.6 and 3.7, and Proposition 3.8, we obtain

COROLLARY 3.2. *Let $\varphi : (M^m, g) \to (N^n, h)$ be an isometric immersion. Assume that $m = \dim M = \dim N - 1 = n - 1$, and the mean curvature of φ, $\mathbf{H} = (1/m)\sum_{i=1}^{m} H_{ii} = (1/m)h(\tau(\varphi),\xi)$, is constant. Then, the following hold:*

(1) *Assume that $\mathbf{H} \neq 0$, i.e., φ is not harmonic. Then, it holds that φ is biharmonic if and only if $\rho^h(\xi) = \|B_{\varphi}\|^2\,\xi$, where ρ^h is the Ricci transform of (N, h), ξ is a unit normal vector field along φ and B_{φ} is the second fundamental form of φ.*

(2) *Assume that $\mathbf{H} \neq 0$ and (N, h) is Einstein, i.e., $\rho^h = c\,\mathrm{Id}$ for some constant c. Then, φ is biharmonic if and only if $\|B_{\varphi}\|^2 = c$.*

(3) *Assume that $\mathbf{H} \neq 0$ and the Ricci curvature of (N, h) is nonpositive. Then, if φ is biharmonic, it is harmonic.*

4. Hermann actions and symmetric triads

From this section, we apply the results in Section 3 to the orbits of Hermann actions using symmetric triads (cf. [67], [68], [70]), and determine biharmonic regular orbits of cohomogeneity one Hermann actions. For this purpose, we express the tension field and the square norm of the second fundamental form of orbits of Hermann actions in terms of symmetric triads.

4.1. First we recall the notions of root system and symmetric triad. See [67] for details. Let $(\mathfrak{a}, \langle \cdot, \cdot \rangle)$ be a finite dimensional inner product space over $\mathbb{R}$. For each $\alpha \in \mathfrak{a}$, we define an orthogonal transformation $s_\alpha : \mathfrak{a} \to \mathfrak{a}$ by

$$s_\alpha(H) = H - \frac{2\langle \alpha, H \rangle}{\langle \alpha, \alpha \rangle} \alpha \quad (H \in \mathfrak{a}),$$

namely s_α is the reflection with respect to the hyperplane $\{H \in \mathfrak{a} \mid \langle \alpha, H \rangle = 0\}$.

DEFINITION 4.1. *A finite subset Σ of $\mathfrak{a} \setminus \{\mathfrak{o}\}$ is a root system of $\mathfrak{a}$, if it satisfies the following three conditions:*

(1) *Span$(\Sigma) = \mathfrak{a}$.*
(2) *If $\alpha, \beta \in \Sigma$, then $s_\alpha(\beta) \in \Sigma$.*
(3) *$2\langle \alpha, \beta \rangle / \langle \alpha, \alpha \rangle \in \mathbb{Z} \quad (\alpha, \beta \in \Sigma)$.*

A root system of $\mathfrak{a}$ is said to be irreducible if it cannot be decomposed into two disjoint nonempty orthogonal subsets.

Let Σ be a root system of $\mathfrak{a}$. The Weyl group $W(\Sigma)$ of Σ is the finite subgroup of the orthogonal group $O(\mathfrak{a})$ of $\mathfrak{a}$ generated by $\{s_\alpha \mid \alpha \in \Sigma\}$.

DEFINITION 4.2. ([67]) *A triple $(\tilde{\Sigma}, \Sigma, W)$ of finite subsets of $\mathfrak{a} \setminus \{\mathfrak{o}\}$ is a symmetric triad of $\mathfrak{a}$, if it satisfies the following six conditions:*

(1) *$\tilde{\Sigma}$ is an irreducible root system of $\mathfrak{a}$.*
(2) *Σ is a root system of $\mathfrak{a}$.*
(3) *$(-1)W = W$, $\tilde{\Sigma} = \Sigma \cup W$.*
(4) *$\Sigma \cap W$ is a nonempty subset. If we put $l := \max\{\|\alpha\| \mid \alpha \in \Sigma \cap W\}$, then $\Sigma \cap W = \{\alpha \in \tilde{\Sigma} \mid \|\alpha\| \leq l\}$.*
(5) *For $\alpha \in W$ and $\lambda \in \Sigma \setminus W$,*

$$2\frac{\langle \alpha, \lambda \rangle}{\langle \alpha, \alpha \rangle} \text{ is odd if and only if } s_\alpha(\lambda) \in W \setminus \Sigma.$$

(6) *For $\alpha \in W$ and $\lambda \in W \setminus \Sigma$,*

$$2\frac{\langle \alpha, \lambda \rangle}{\langle \alpha, \alpha \rangle} \text{ is odd if and only if } s_\alpha(\lambda) \in \Sigma \setminus W.$$

We define an open subset $\mathfrak{a}_r$ of $\mathfrak{a}$ by

$$\mathfrak{a}_r = \bigcap_{\lambda \in \Sigma, \alpha \in W} \left\{ H \in \mathfrak{a} \mid \langle \lambda, H \rangle \notin \pi\mathbb{Z}, \ \langle \alpha, H \rangle \notin \frac{\pi}{2} + \pi\mathbb{Z} \right\}.$$

A point in $\mathfrak{a}_r$ is called a regular point, and a point in the complement of $\mathfrak{a}_r$ in $\mathfrak{a}$ is called a singular point. A connected component of $\mathfrak{a}_r$ is called a cell. The affine Weyl group $\tilde{W}(\tilde{\Sigma}, \Sigma, W)$ of $(\tilde{\Sigma}, \Sigma, W)$ is a subgroup of the affine group of $\mathfrak{a}$, i.e. the semidirect product $O(\mathfrak{a}) \times \mathfrak{a}$, generated by

$$\left\{ \left(s_\lambda, \frac{2n\pi}{\|\lambda\|^2}\lambda \right) \ \Big| \ \lambda \in \Sigma, n \in \mathbb{Z} \right\} \cup \left\{ \left(s_\alpha, \frac{(2n+1)\pi}{\|\alpha\|^2}\alpha \right) \ \Big| \ \alpha \in W, n \in \mathbb{Z} \right\}.$$

The action of $(s_\lambda, (2n\pi/\|\lambda\|^2)\lambda)$ on $\mathfrak{a}$ is the reflection with respect to the hyperplane $\{H \in \mathfrak{a} \mid \langle \lambda, H \rangle = n\pi\}$, and the action of $(s_\alpha, ((2n+1)\pi/\|\alpha\|^2)\alpha)$ on $\mathfrak{a}$ is the reflection with respect to the hyperplane $\{H \in \mathfrak{a} \mid \langle \alpha, H \rangle = (n+1/2)\pi\}$. The affine Weyl group $\tilde{W}(\tilde{\Sigma}, \Sigma, W)$ acts transitively on the set of all cells. More precisely, for each cell P, it holds that

$$\mathfrak{a} = \bigcup_{s \in \tilde{W}(\tilde{\Sigma}, \Sigma, W)} s\overline{P}.$$

We take a fundamental system $\tilde{\Pi}$ of $\tilde{\Sigma}$. We denote by $\tilde{\Sigma}^+$ the set of positive roots in $\tilde{\Sigma}$. Set $\Sigma^+ = \tilde{\Sigma}^+ \cap \Sigma$ and $W^+ = \tilde{\Sigma}^+ \cap W$. Denote by Π the set of simple roots of Σ. We set

$$W_0 = \{\alpha \in W^+ \mid \alpha + \lambda \notin W \ (\lambda \in \Pi)\}.$$

From the classification of symmetric triads, we have that W_0 consists of only one element, denoted by $\tilde{\alpha}$. We define an open subset P_0 of $\mathfrak{a}$ by

$$P_0 = \left\{ H \in \mathfrak{a} \ \Big| \ \langle \tilde{\alpha}, H \rangle < \frac{\pi}{2}, \ \langle \lambda, H \rangle > 0 \ (\lambda \in \Pi) \right\}. \tag{4.1}$$

Then P_0 is a cell.

DEFINITION 4.3. ([**67**]) *Let $(\tilde{\Sigma}, \Sigma, W)$ be a symmetric triad of $\mathfrak{a}$. Consider two mappings m and n from $\tilde{\Sigma}$ to $\mathbb{R}_{\geq 0} := \{a \in \mathbb{R} \mid a \geq 0\}$ which satisfy the following four conditions:*

 (1) *For any $\lambda \in \tilde{\Sigma}$,*
  (1-1) $m(\lambda) = m(-\lambda)$, $n(\lambda) = n(-\lambda)$,
  (1-2) $m(\lambda) > 0$ *if and only if* $\lambda \in \Sigma$,

(1-3) $n(\lambda) > 0$ *if and only if* $\lambda \in W$.
(2) *When* $\lambda \in \Sigma, \alpha \in W, s \in W(\Sigma)$, *then* $m(\lambda) = m(s(\lambda))$, $n(\alpha) = n(s(\alpha))$.
(3) *When* $\lambda \in \tilde{\Sigma}$, $\sigma \in W(\tilde{\Sigma})$, *then* $m(\lambda) + n(\lambda) = m(\sigma(\lambda)) + n(\sigma(\lambda))$.
(4) *Let* $\lambda \in \Sigma \cap W$, $\alpha \in W$. *If* $2\langle \alpha, \lambda \rangle / \langle \alpha, \alpha \rangle$ *is even, then* $m(\lambda) = m(s_\alpha(\lambda))$. *If* $2\langle \alpha, \lambda \rangle / \langle \alpha, \alpha \rangle$ *is odd, then* $m(\lambda) = n(s_\alpha(\lambda))$.

We call $m(\lambda)$ and $n(\alpha)$ the multiplicities of λ and α, respectively.

4.2. We will review some basics of the theory of compact symmetric spaces.

Let G be a compact connected Lie group and K a closed subgroup of G. Assume that there exists an involutive automorphism θ of G which satisfies $(G_\theta)_0 \subset K \subset G_\theta$, where G_θ is the set of fixed points of θ and $(G_\theta)_0$ is the identity component of G_θ. Then the pair (G, K) is called a compact symmetric pair. We denote the Lie algebras of G and K by $\mathfrak{g}$ and $\mathfrak{k}$, respectively. The involutive automorphism θ of G induces an involutive automorphism of $\mathfrak{g}$, which is also denoted by the same symbol θ. We can see that

$$\mathfrak{k} = \{X \in \mathfrak{g} \mid \theta(X) = X\},$$

and we define

$$\mathfrak{m} = \{X \in \mathfrak{g} \mid \theta(X) = -X\}.$$

Take an inner product $\langle \cdot, \cdot \rangle$ on $\mathfrak{g}$ which is invariant under the actions of $\mathrm{Ad}(G)$ and θ. The inner product $\langle \cdot, \cdot \rangle$ induces a bi-invariant Riemannian metric on G and a G-invariant Riemannian metric on $N = G/K$, which are denoted by the same symbol $\langle \cdot, \cdot \rangle$. Then $(N, \langle \cdot, \cdot \rangle)$ is a compact symmetric space. Conversely, any compact symmetric space can be constructed in this way. Since θ is involutive, we have an orthogonal direct sum decomposition of $\mathfrak{g}$:

$$\mathfrak{g} = \mathfrak{k} \oplus \mathfrak{m}.$$

This decomposition is called the canonical decomposition of (G, K). We denote by π the natural projection from G onto N. The tangent space $T_{\pi(e)} N$ of N at the origin $\pi(e)$ is identified with $\mathfrak{m}$ in a natural way. The Ricci tensor $\mathrm{Ric}(\cdot, \cdot)$ of N is given by

$$\mathrm{Ric}(X, Y) = -\frac{1}{2}\mathrm{Killing}(X, Y) \quad (X, Y \in \mathfrak{m}),$$

where $\mathrm{Killing}(\cdot, \cdot)$ is the Killing form of $\mathfrak{g}$. If G is semisimple, then we can give an $\mathrm{Ad}(G)$-invariant inner product on $\mathfrak{g}$ by $\langle \cdot, \cdot \rangle = -\mathrm{Killing}(\cdot, \cdot)$, hence N is an Einstein manifold with Einstein constant $c = 1/2$.

Here, let us recall the notion of hyperpolar actions (cf. [**85**]). An isometric action of a compact Lie group on a Riemannian manifold is said to be hyperpolar if there exists a closed, connected submanifold that is flat in the induced metric and meets all orbits orthogonally. Such a submanifold is called a section of the Lie group action. Kollross [**85**] classified hyperpolar actions on irreducible symmetric spaces of compact type.

4.3. Our aim is to apply the theory of symmetric triad due to Ikawa [**67**] in order to express the second fundamental form of orbits of Hermann actions.

Let (G, K_1) and (G, K_2) be compact symmetric pairs with respect to involutive automorphisms θ_1 and θ_2 of a compact Lie group G, respectively. Then the triple (G, K_1, K_2) is called a compact symmetric triad. We denote the Lie algebras of G, K_1 and K_2 by $\mathfrak{g}, \mathfrak{k}_1$ and $\mathfrak{k}_2$, respectively. The involutive automorphism of $\mathfrak{g}$ induced from θ_i will be also denoted by θ_i. Take an $\mathrm{Ad}(G)$-invariant inner product $\langle \cdot, \cdot \rangle$ on $\mathfrak{g}$. Then the inner product $\langle \cdot, \cdot \rangle$ induces a bi-invariant Riemannian metric on G and G-invariant Riemannian metrics on the coset manifolds $N_1 = G/K_1$ and $N_2 = G/K_2$. We denote these Riemannian metrics on G, N_1 and N_2 by the same symbol $\langle \cdot, \cdot \rangle$. The isometric action of K_2 on N_1 and the action of K_1 on N_2 are called Hermann actions. Now we have two canonical decompositions of $\mathfrak{g}$:

$$\mathfrak{g} = \mathfrak{k}_1 \oplus \mathfrak{m}_1 = \mathfrak{k}_2 \oplus \mathfrak{m}_2,$$

where $\mathfrak{m}_i = \{X \in \mathfrak{g} \mid \theta_i(X) = -X\}$ $(i = 1, 2)$. We define a closed subgroup G_{12} of G by

$$G_{12} = \{k \in G \mid \theta_1(k) = \theta_2(k)\},$$

and we denote the identity component of G_{12} by $(G_{12})_0$. Then θ_1 induces an involutive automorphism of $(G_{12})_0$. Hence $((G_{12})_0, K_{12})$ is a compact symmetric pair, where K_{12} is a closed subgroup of $(G_{12})_0$ defined by

$$K_{12} = \{k \in (G_{12})_0 \mid \theta_1(k) = k\}.$$

The canonical decomposition of the Lie algebra $\mathfrak{g}_{12}$ of $(G_{12})_0$ is given by

$$\mathfrak{g}_{12} = (\mathfrak{k}_1 \cap \mathfrak{k}_2) \oplus (\mathfrak{m}_1 \cap \mathfrak{m}_2).$$

Fix a maximal abelian subspace $\mathfrak{a}$ in $\mathfrak{m}_1 \cap \mathfrak{m}_2$. Then $\exp \mathfrak{a}$ is a torus subgroup in $(G_{12})_0$. We denote by π_i the natural projection from G onto N_i $(i = 1, 2)$. Then, the totally geodesic flat torus $\pi_1(\exp \mathfrak{a})$ is a section of K_2-action on N_1, hence the action is hyperpolar. Similarly

$\pi_2(\exp \mathfrak{a})$ is a section of K_1-action on N_2. The cohomogeneity of K_2-action on N_1 and that of K_1-action on N_2 are equal to $\dim \mathfrak{a}$. We call an orbit of the maximal dimension a regular orbit. For $k \in G$, we denote the left transformation of G by L_k. The isometries on N_1 and N_2 induced by L_k will be also denoted by the same symbol L_k.

We should prepare several terminologies to determine the second fundamental forms of the regular orbits of Hermann actions.

DEFINITION 4.4. *A compact symmetric triad* (G, K_1, K_2) *is said to be commutative if* $\theta_1 \theta_2 = \theta_2 \theta_1$. *Then* K_2-*action on* N_1 *and* K_1-*action on* N_2 *are called commutative Hermann actions.*

Hereafter we assume that (G, K_1, K_2) is a commutative compact symmetric triad where G is semisimple. Then we have

$$\mathfrak{g} = (\mathfrak{k}_1 \cap \mathfrak{k}_2) \oplus (\mathfrak{m}_1 \cap \mathfrak{m}_2) \oplus (\mathfrak{k}_1 \cap \mathfrak{m}_2) \oplus (\mathfrak{m}_1 \cap \mathfrak{k}_2).$$

We define subspaces in $\mathfrak{g}$ as follows:

$$\mathfrak{k}_0 = \{X \in \mathfrak{k}_1 \cap \mathfrak{k}_2 \mid [\mathfrak{a}, X] = \{0\}\},$$
$$V(\mathfrak{k}_1 \cap \mathfrak{m}_2) = \{X \in \mathfrak{k}_1 \cap \mathfrak{m}_2 \mid [\mathfrak{a}, X] = \{0\}\},$$
$$V(\mathfrak{m}_1 \cap \mathfrak{k}_2) = \{X \in \mathfrak{m}_1 \cap \mathfrak{k}_2 \mid [\mathfrak{a}, X] = \{0\}\},$$

and for $\lambda \in \mathfrak{a}$

$$\mathfrak{k}_\lambda = \{X \in \mathfrak{k}_1 \cap \mathfrak{k}_2 \mid [H, [H, X]] = -\langle \lambda, H \rangle^2 X \ (H \in \mathfrak{a})\},$$
$$\mathfrak{m}_\lambda = \{X \in \mathfrak{m}_1 \cap \mathfrak{m}_2 \mid [H, [H, X]] = -\langle \lambda, H \rangle^2 X \ (H \in \mathfrak{a})\},$$
$$V_\lambda^\perp(\mathfrak{k}_1 \cap \mathfrak{m}_2) = \{X \in \mathfrak{k}_1 \cap \mathfrak{m}_2 \mid [H, [H, X]] = -\langle \lambda, H \rangle^2 X \ (H \in \mathfrak{a})\},$$
$$V_\lambda^\perp(\mathfrak{m}_1 \cap \mathfrak{k}_2) = \{X \in \mathfrak{m}_1 \cap \mathfrak{k}_2 \mid [H, [H, X]] = -\langle \lambda, H \rangle^2 X \ (H \in \mathfrak{a})\}.$$

We set

$$\Sigma = \{\lambda \in \mathfrak{a} \setminus \{0\} \mid \mathfrak{k}_\lambda \neq \{0\}\},$$
$$W = \{\alpha \in \mathfrak{a} \setminus \{0\} \mid V_\alpha^\perp(\mathfrak{k}_1 \cap \mathfrak{m}_2) \neq \{0\}\},$$
$$\tilde{\Sigma} = \Sigma \cup W.$$

It is known that $\dim \mathfrak{k}_\lambda = \dim \mathfrak{m}_\lambda$ and $\dim V_\lambda^\perp(\mathfrak{k}_1 \cap \mathfrak{m}_2) = \dim V_\lambda^\perp(\mathfrak{m}_1 \cap \mathfrak{k}_2)$ for each $\lambda \in \tilde{\Sigma}$. Thus we define $m(\lambda) := \dim \mathfrak{k}_\lambda$ and $n(\lambda) := \dim V_\lambda^\perp(\mathfrak{k}_1 \cap \mathfrak{m}_2)$. Notice that Σ is the root system of the symmetric pair $((G_{12})_0, K_{12})$ with respect to $\mathfrak{a}$.

PROPOSITION 4.1. *Let* (G, K_1, K_2) *be a commutative compact symmetric triad where* G *is semisimple. Then* $\tilde{\Sigma}$ *is a root system of* $\mathfrak{a}$. *In addition, if* G *is simple and* $\theta_1 \not\sim \theta_2$, *then* $(\tilde{\Sigma}, \Sigma, W)$ *is a symmetric triad of* $\mathfrak{a}$, *moreover* $m(\lambda)$ *and* $n(\alpha)$ *are multiplicities of* $\lambda \in \Sigma$ *and*

$\alpha \in W$. Here $\theta_1 \not\sim \theta_2$ means that θ_1 and θ_2 cannot be transformed each other by an inner automorphism of $\mathfrak{g}$.

We take a basis of $\mathfrak{a}$ and define a lexicographic ordering $>$ on $\mathfrak{a}$ with respect to the basis. We set

$$\tilde{\Sigma}^+ = \{\lambda \in \tilde{\Sigma} \mid \lambda > 0\}, \quad \Sigma^+ = \Sigma \cap \tilde{\Sigma}^+, \quad W^+ = W \cap \tilde{\Sigma}^+.$$

Then we have an orthogonal direct sum decomposition of $\mathfrak{g}$:

$$\mathfrak{g} = \mathfrak{k}_0 \oplus \sum_{\lambda \in \Sigma^+} \mathfrak{k}_\lambda \oplus \mathfrak{a} \oplus \sum_{\lambda \in \Sigma^+} \mathfrak{m}_\lambda \oplus V(\mathfrak{k}_1 \cap \mathfrak{m}_2) \oplus \sum_{\alpha \in W^+} V_\alpha^\perp(\mathfrak{k}_1 \cap \mathfrak{m}_2)$$

$$\oplus V(\mathfrak{m}_1 \cap \mathfrak{k}_2) \oplus \sum_{\alpha \in W^+} V_\alpha^\perp(\mathfrak{m}_1 \cap \mathfrak{k}_2).$$

Furthermore we have the following lemma.

LEMMA 4.1. ([**67**])

(1) For each $\lambda \in \Sigma^+$, there exist orthonormal bases $\{S_{\lambda,i}\}_{i=1}^{m(\lambda)}$ and $\{T_{\lambda,i}\}_{i=1}^{m(\lambda)}$ of $\mathfrak{k}_\lambda$ and $\mathfrak{m}_\lambda$ respectively such that for any $H \in \mathfrak{a}$

$$[H, S_{\lambda,i}] = \langle \lambda, H \rangle T_{\lambda,i}, \quad [H, T_{\lambda,i}] = -\langle \lambda, H \rangle S_{\lambda,i}, \quad [S_{\lambda,i}, T_{\lambda,i}] = \lambda,$$

$$\mathrm{Ad}(\exp H)S_{\lambda,i} = \cos\langle \lambda, H \rangle S_{\lambda,i} + \sin\langle \lambda, H \rangle T_{\lambda,i},$$
$$\mathrm{Ad}(\exp H)T_{\lambda,i} = -\sin\langle \lambda, H \rangle S_{\lambda,i} + \cos\langle \lambda, H \rangle T_{\lambda,i}.$$

(2) For each $\alpha \in W^+$, there exist orthonormal bases $\{X_{\alpha,j}\}_{j=1}^{n(\alpha)}$ and $\{Y_{\alpha,j}\}_{j=1}^{n(\alpha)}$ of $V_\alpha^\perp(\mathfrak{k}_1 \cap \mathfrak{m}_2)$ and $V_\alpha^\perp(\mathfrak{m}_1 \cap \mathfrak{k}_2)$ respectively such that for any $H \in \mathfrak{a}$

$$[H, X_{\alpha,j}] = \langle \alpha, H \rangle Y_{\alpha,j}, \quad [H, Y_{\alpha,j}] = -\langle \alpha, H \rangle X_{\alpha,j}, \quad [X_{\alpha,j}, Y_{\alpha,j}] = \alpha,$$

$$\mathrm{Ad}(\exp H)X_{\alpha,j} = \cos\langle \alpha, H \rangle X_{\alpha,j} + \sin\langle \alpha, H \rangle Y_{\alpha,j},$$
$$\mathrm{Ad}(\exp H)Y_{\alpha,j} = -\sin\langle \alpha, H \rangle X_{\alpha,j} + \cos\langle \alpha, H \rangle Y_{\alpha,j}.$$

Now we consider the second fundamental form of an orbit $K_2\pi_1(x)$ of the action of K_2 on $N_1 = G/K_1$ for $x \in G$ (cf. Lemma 4.2). Without loss of generalities we can assume that $x = \exp H$ where $H \in \mathfrak{a}$, since $\pi_1(\exp \mathfrak{a})$ is a section of the action. We identify the tangent space $T_{\pi_1(e)}N_1$ with $\mathfrak{m}_1$ via $(d\pi_1)_e$. For $x = \exp H$ $(H \in \mathfrak{a})$, the tangent space and the normal space of $K_2\pi_1(x)$ at $\pi_1(x)$ are given as

$$dL_x^{-1}(T_{\pi_1(x)}(K_2\pi_1(x))) \cong (\mathrm{Ad}(\mathrm{x}^{-1})\mathfrak{k}_2)_{\mathfrak{m}_1}$$

$$= \sum_{\substack{\lambda\in\Sigma^+ \\ \langle\lambda,H\rangle\notin\pi\mathbb{Z}}} \mathfrak{m}_\lambda \oplus V(\mathfrak{m}_1\cap\mathfrak{k}_2) \oplus \sum_{\substack{\alpha\in W^+ \\ \langle\alpha,H\rangle\notin(\pi/2)+\pi\mathbb{Z}}} V_\alpha^\perp(\mathfrak{m}_1\cap\mathfrak{k}_2),$$

$$dL_x^{-1}(T_{\pi_1(x)}^\perp(K_2\pi_1(x))) \cong (\mathrm{Ad}(\mathrm{x}^{-1})\mathfrak{m}_2)\cap\mathfrak{m}_1$$

$$= \mathfrak{a} \oplus \sum_{\substack{\lambda\in\Sigma^+ \\ \langle\lambda,H\rangle\in\pi\mathbb{Z}}} \mathfrak{m}_\lambda \oplus \sum_{\substack{\alpha\in W^+ \\ \langle\alpha,H\rangle\in(\pi/2)+\pi\mathbb{Z}}} V_\alpha^\perp(\mathfrak{m}_1\cap\mathfrak{k}_2),$$

where $X_{\mathfrak{m}_1}$ denotes $\mathfrak{m}_1$-component of $X \in \mathfrak{g}$ with respect to the canonical decomposition $\mathfrak{g} = \mathfrak{k}_1 \oplus \mathfrak{m}_1$. Using the above decompositions of the tangent space and the normal space of the orbit $K_2\pi_1(x)$ and the orthonormal basis given in Lemma 4.1, we can apply Ikawa's results (cf. Lemma 4.22 in [67]) to our cases. Let us denote the second fundamental form and the tension field of the orbit $K_2\pi_1(x)$ in N_1 by B_H and τ_H, respectively. Then we have the following lemma.

LEMMA 4.2. Let $x = \exp H$ for $H \in \mathfrak{a}$. Then we have:

(1) $dL_x^{-1}B_H(dL_x(T_{\lambda,i}), dL_x(T_{\mu,j})) = \cot(\langle\mu,H\rangle)[T_{\lambda,i}, S_{\mu,j}]^\perp$,
(2) $dL_x^{-1}B_H(dL_x(Y_{\alpha,i}), dL_x(Y_{\beta,j})) = -\tan(\langle\beta,H\rangle)[Y_{\alpha,i}, X_{\beta,j}]^\perp$,
(3) $B_H(dL_x(Y_1), dL_x(Y_2)) = 0$,
(4) $B_H(dL_x(T_{\lambda,i}), dL_x(Y_2)) = 0$,
(5) $B_H(dL_x(Y_{\alpha,i}), dL_x(Y_2)) = 0$,
(6) $dL_x^{-1}B_H(dL_x(T_{\lambda,i}), dL_x(Y_{\beta,j})) = -\tan(\langle\beta,H\rangle)[T_{\lambda,i}, X_{\beta,j}]^\perp$,

for $\lambda,\mu \in \Sigma^+$ with $\langle\lambda,H\rangle \notin \pi\mathbb{Z}$, $\langle\mu,H\rangle \notin \pi\mathbb{Z}$, and $1 \le i \le m(\lambda)$, $1 \le j \le m(\mu)$; and for $\alpha,\beta \in W^+$ with $\langle\alpha,H\rangle$, $\langle\beta,H\rangle \notin \frac{\pi}{2} + \pi\mathbb{Z}$, and $1 \le i \le n(\alpha)$, $1 \le j \le n(\beta)$; and for $Y_1,Y_2 \in V(\mathfrak{m}_1\cap\mathfrak{k}_2)$.

Here $X^\perp$ is the normal component, i.e. $(\mathrm{Ad}(x^{-1})\mathfrak{m}_2)\cap\mathfrak{m}_1$-component, of a tangent vector $X \in \mathfrak{m}_1$.

Due to Lemma 4.2, we have the following.

THEOREM 4.1. If $K_2\pi_1(x)$ is a regular orbit, then

$$\|B_H\|^2 = \sum_{\lambda\in\Sigma^+} m(\lambda)(\cot\langle\lambda,H\rangle)^2\langle\lambda,\lambda\rangle + \sum_{\alpha\in W^+} n(\alpha)(\tan\langle\alpha,H\rangle)^2\langle\alpha,\alpha\rangle, \tag{4.2}$$

$$dL_x^{-1}(\tau_H) = -\sum_{\lambda\in\Sigma^+} m(\lambda)\cot\langle\lambda,H\rangle\lambda + \sum_{\alpha\in W^+} n(\alpha)\tan\langle\alpha,H\rangle\alpha. \tag{4.3}$$

Proof. For (4.2), since the orbit $K_2\pi_1(x)$ is regular, its tangent space and normal space are given as:

$$dL_x^{-1}(T_{\pi_1(x)}(K_2\pi_1(x))) = \sum_{\lambda \in \Sigma^+} \mathfrak{m}_\lambda \oplus V(\mathfrak{m}_1 \cap \mathfrak{k}_2) \oplus \sum_{\alpha \in W^+} V_\alpha^\perp(\mathfrak{m}_1 \cap \mathfrak{k}_2),$$

$$dL_x^{-1}(T_{\pi_1(x)}^\perp(K_2\pi_1(x))) = \mathfrak{a}.$$

For each $\lambda, \mu \in \Sigma^+, 1 \le i \le m(\lambda), 1 \le j \le m(\mu)$, we have

$$\langle [T_{\lambda,i}, S_{\mu,j}], H' \rangle = \langle T_{\lambda,i}, [S_{\mu,j}, H'] \rangle = \langle T_{\lambda,i}, -\langle \mu, H' \rangle T_{\mu,j} \rangle$$

$$= -\delta_{\lambda,\mu} \delta_{i,j} \langle \mu, H' \rangle$$

for all $H' \in \mathfrak{a}$. Thus, we have

$$[T_{\lambda,i}, S_{\mu,j}]^\perp = -\delta_{\lambda,\mu} \delta_{i,j} \mu.$$

Similarly, we have

$$[Y_{\alpha,i}, X_{\beta,j}]^\perp = -\delta_{\alpha,\beta} \delta_{i,j} \beta \quad (\alpha, \beta \in W^+, 1 \le i \le n(\alpha), 1 \le j \le n(\beta)),$$

$$[T_{\lambda,i}, X_{\beta,j}]^\perp = 0 \quad (\lambda \in \Sigma^+, \beta \in W^+, 1 \le i \le m(\lambda), 1 \le j \le n(\beta)).$$

From Lemma 4.2, we obtain

$$\|B_H\|^2 = \sum_{\lambda,\mu,i,j} \| \cot\langle \mu, H \rangle [T_{\lambda,i}, S_{\mu,j}]^\perp \|^2 + \sum_{\alpha,\beta,i,j} \|$$

$$- \tan\langle \beta, H \rangle [Y_{\alpha,i}, X_{\beta,j}]^\perp \|^2$$

$$= \sum_{\lambda,\mu,i,j} \| - \cot\langle \mu, H \rangle \delta_{\lambda,\mu} \delta_{i,j} \mu \|^2 + \sum_{\alpha,\beta,i,j} \| \tan\langle \beta, H \rangle \delta_{\alpha,\beta} \delta_{i,j} \beta \|^2$$

$$= \sum_{\lambda \in \Sigma^+} \sum_{i=1}^{m(\lambda)} (\cot\langle \lambda, H \rangle)^2 \langle \lambda, \lambda \rangle + \sum_{\alpha \in W^+} \sum_{i=1}^{n(\alpha)} (\tan\langle \alpha, H \rangle)^2 \langle \alpha, \alpha \rangle$$

$$= \sum_{\lambda \in \Sigma^+} m(\lambda)(\cot\langle \lambda, H \rangle)^2 \langle \lambda, \lambda \rangle + \sum_{\alpha \in W^+} n(\alpha)(\tan\langle \alpha, H \rangle)^2 \langle \alpha, \alpha \rangle.$$

The formula (4.3) was proved in Corollary 4.23, [**67**].   □

5. Biharmonic orbits of cohomogeneity one Hermann actions

In this section, applying Theorem 3.6, we will study biharmonic regular orbits of cohomogeneity one Hermann actions.

Let (G, K_1, K_2) be a commutative compact symmetric triad where G is semisimple, and let us define an inner product $\langle \cdot, \cdot \rangle$ on $\mathfrak{g}$ by $\langle \cdot, \cdot \rangle = -\text{Killing}(\cdot, \cdot)$. Then, $(N_1, \langle \cdot, \cdot \rangle)$ and $(N_2, \langle \cdot, \cdot \rangle)$ are Einstein manifolds with Einstein constant $c = 1/2$. It is known that the tension field of an orbit of a Hermann action is parallel in the normal bundle (see [**70**]), i.e. $\overline{\nabla}_X^\perp \tau_H = 0$ for every vector field X on the orbit $K_2\pi_1(x)$.

Hereafter we assume that $\dim \mathfrak{a} = 1$. Since the cohomogeneity of K_2-action on N_1 and that of K_1-action on N_2 are equal to $\dim \mathfrak{a}$, regular orbits of K_2-actions (resp. K_1-action) are homogeneous hypersurfaces in N_1 (resp. N_2). Hence we can apply (2) of Theorem 3.4 for regular orbits of these actions. Clearly, $K_2\pi_1(x)$ is a regular orbit if and only if $K_1\pi_2(x)$ is also a regular orbit. Therefore, we have the following proposition.

PROPOSITION 5.1. *Let* $x = \exp H$ *for* $H \in \mathfrak{a}$. *Suppose that* $K_2\pi_1(x)$ *is a regular orbit of* K_2-*action on* N_1, *so* $K_1\pi_2(x)$ *is also a regular orbit of* K_1-*action on* N_2. *Then,*

   (1) *An orbit* $K_2\pi_1(x)$ *is harmonic if and only if* $K_1\pi_2(x)$ *is harmonic.*

   (2) *An orbit* $K_2\pi_1(x)$ *is proper biharmonic if and only if* $K_1\pi_2(x)$ *is proper biharmonic.*

Proof. Analogous to Lemma 4.2, we can express the second fundamental form B'_H of $K_1\pi_2(x)$ in N_2 using the orthonormal basis given in Lemma 4.1. Then easily we can verify

$$\|B'_H\|^2 = \|B_H\|^2, \qquad dL_x^{-1}(\tau'_H) = dL_x^{-1}(\tau_H),$$

where τ'_H denotes the tension field of $K_1\pi_2(x)$ in N_2. Therefore, from Theorem 3.4, we have the consequence. □

If G is simple and $\theta_1 \not\sim \theta_2$, then for a commutative compact symmetric triad (G, K_1, K_2) the triple $(\tilde{\Sigma}, \Sigma, W)$ is a symmetric triad with multiplicities $m(\lambda)$ and $n(\alpha)$ (cf. Proposition 4.1). In this case, for $x = \exp H$ ($H \in \mathfrak{a}$), the orbit $K_2\pi_1(x)$ is regular if and only if H is a regular point with respect to $(\tilde{\Sigma}, \Sigma, W)$.

All the symmetric triads with $\dim \mathfrak{a} = 1$ are classified into the following four types ([**67**]):

	Σ^+	W^+	$\tilde{\alpha}$
III-B$_1$	$\{\alpha\}$	$\{\alpha\}$	α
I-BC$_1$	$\{\alpha, 2\alpha\}$	$\{\alpha\}$	α
II-BC$_1$	$\{\alpha\}$	$\{\alpha, 2\alpha\}$	2α
III-BC$_1$	$\{\alpha, 2\alpha\}$	$\{\alpha, 2\alpha\}$	2α

Let $\vartheta := \langle \tilde{\alpha}, H \rangle$ for $H \in \mathfrak{a}$. Then, by (4.1), $P_0 = \{H \in \mathfrak{a} \mid 0 < \vartheta < \pi/2\}$ is a cell in these types. If N_1 is simply connected, then the orbit space of K_2-action on N_1 is identified with $\overline{P_0} = \{H \in \mathfrak{a} \mid 0 \le \vartheta \le \pi/2\}$, more precisely, each orbit meets $\pi_1(\exp \overline{P_0})$ at one point. A point in the interior of the orbit space corresponds to a regular orbit, and there exists a unique minimal (harmonic) orbit among regular orbits. On the other hand, two endpoints of the orbit space correspond to singular orbits. These singular orbits are minimal (harmonic), moreover these are weakly reflective ([**71**]).

In the following, we express the two equations $\|B_H\|^2 = 1/2$ and $\tau_H = 0$ in terms of ϑ for each type. For this purpose, here we should calculate $\langle \alpha, \alpha \rangle$.

$$\langle \alpha, \alpha \rangle = -\mathrm{Killing}(\alpha, \alpha) = -\mathrm{tr}(\mathrm{ad}(\alpha)^2)$$

$$= - \left\{ \sum_{\lambda \in \Sigma^+} \sum_{i=1}^{m(\lambda)} \langle \mathrm{ad}(\alpha)^2 S_{\lambda,i}, S_{\lambda,i} \rangle + \sum_{\lambda \in \Sigma^+} \sum_{i=1}^{m(\lambda)} \langle \mathrm{ad}(\alpha)^2 T_{\lambda,i}, T_{\lambda,i} \rangle \right.$$

$$\left. + \sum_{\beta \in W^+} \sum_{j=1}^{n(\beta)} \langle \mathrm{ad}(\alpha)^2 X_{\beta,j}, X_{\beta,j} \rangle + \sum_{\beta \in W^+} \sum_{j=1}^{n(\beta)} \langle \mathrm{ad}(\alpha)^2 Y_{\beta,j}, Y_{\beta,j} \rangle \right\}$$

$$= \sum_{\lambda \in \Sigma^+} 2m(\lambda) \langle \alpha, \lambda \rangle^2 + \sum_{\beta \in W^+} 2n(\beta) \langle \alpha, \beta \rangle^2.$$

In the case of type III-BC$_1$, i.e. $\Sigma^+ = \{\alpha, 2\alpha\}$ and $W^+ = \{\alpha, 2\alpha\}$, we can see that

$$\langle \alpha, \alpha \rangle = 2m(\alpha)\langle \alpha, \alpha \rangle^2 + 2m(2\alpha)\langle \alpha, 2\alpha \rangle^2 + 2n(\alpha)\langle \alpha, \alpha \rangle^2 + 2n(2\alpha)\langle \alpha, 2\alpha \rangle^2$$

$$= 2\langle \alpha, \alpha \rangle^2 (m(\alpha) + 4m(2\alpha) + n(\alpha) + 4n(2\alpha)).$$

Therefore, we obtain

$$\langle \alpha, \alpha \rangle = \frac{1}{2(m(\alpha) + 4m(2\alpha) + n(\alpha) + 4n(2\alpha))}. \tag{5.1}$$

In the cases of other types, we have $\langle \alpha, \alpha \rangle$ by letting $m(2\alpha) = 0$ (resp. $n(2\alpha) = 0$) if $2\alpha \notin \Sigma^+$ (resp. $2\alpha \notin W^+$).

5.1. Type III-B$_1$. By (4.2), the biharmonic condition $\|B_H\|^2 = 1/2$ is equivalent to

$$m(\alpha) + n(\alpha) = m(\alpha)(\cot \vartheta)^2 + n(\alpha)(\tan \vartheta)^2$$

for $H \in P_0$. Thus we have

$$\tan \vartheta = 1, \text{ or } \sqrt{\frac{m(\alpha)}{n(\alpha)}}.$$

On the other hand, by (4.3), the harmonic condition $\tau_H = 0$ is equivalent to

$$-m(\alpha)\cot \vartheta + n(\alpha)\tan \vartheta = 0.$$

Thus we have

$$\tan \vartheta = \sqrt{\frac{m(\alpha)}{n(\alpha)}}.$$

By (2) of Theorem 3.4, the situation is divided into the following two cases:

(1) When $m(\alpha) = n(\alpha)$, if an orbit $K_{2\pi_1}(x)$ is biharmonic, then it is harmonic.

(2) When $m(\alpha) \neq n(\alpha)$, an orbit $K_{2\pi_1}(x)$ is proper biharmonic if and only if $(\tan \vartheta)^2 = 1$ for $H \in P_0$. In this case, a unique proper biharmonic orbit exists at the center of P_0, namely $\vartheta = \pi/4$.

5.2. Type I-BC$_1$. We denote $m_1 := m(\alpha)$, $m_2 := m(2\alpha)$ and $n_1 := n(\alpha)$ for short. Then, by (4.2), the biharmonic condition $\|B_H\|^2 = 1/2$ is equivalent to

$$m_1 + n_1 + 4m_2 = m_1(\cot \vartheta)^2 + n_1(\tan \vartheta)^2 + 4m_2(\cot 2\vartheta)^2.$$

Thus, we have

$$(\tan \vartheta)^2 = \frac{m_1 + n_1 + 6m_2 \pm \sqrt{(m_1 + n_1 + 6m_2)^2 - 4(n_1 + m_2)(m_1 + m_2)}}{2(n_1 + m_2)}.$$

By (4.3), the harmonic condition $\tau_H = 0$ is equivalent to

$$-m_1 \cot \vartheta + n_1 \tan \vartheta - 4m_2 \cot 2\vartheta = 0.$$

Thus, we have

$$(\tan \vartheta)^2 = \frac{m_1 + m_2}{n_1 + m_2}.$$

Since

$$0 < \frac{m_1 + n_1 + 6m_2 - \sqrt{(m_1 + n_1 + 6m_2)^2 - 4(n_1 + m_2)(m_1 + m_2)}}{2(n_1 + m_2)}$$

$$< \frac{m_1 + m_2}{n_1 + m_2}$$

$$< \frac{m_1 + n_1 + 6m_2 + \sqrt{(m_1 + n_1 + 6m_2)^2 - 4(n_1 + m_2)(m_1 + m_2)}}{2(n_1 + m_2)},$$

by (2) of Theorem 3.4, an orbit $K_{2\pi_1}(x)$ is proper biharmonic if and only if

$$(\tan \vartheta)^2 = \frac{m_1 + n_1 + 6m_2 \pm \sqrt{(m_1 + n_1 + 6m_2)^2 - 4(n_1 + m_2)(m_1 + m_2)}}{2(n_1 + m_2)}$$

holds for $H \in P_0$. Furthermore, a unique harmonic regular orbit exists between two proper biharmonic orbits in P_0.

5.3. Type II-BC$_1$**.** By the definition of multiplicities, if $2\alpha \in W^+$, then $m(\alpha) = n(\alpha)$. Hence we denote $m_1 := m(\alpha) = n(\alpha)$ and $n_2 := n(2\alpha)$. Then, by (4.2), the biharmonic condition $\|B_H\|^2 = 1/2$ is equivalent to

$$2m_1 + 4n_2 = m_1\left((\cot(\vartheta/2))^2 + (\tan(\vartheta/2))^2\right) + 4n_2(\tan\vartheta)^2.$$

Thus, we have

$$(\tan\vartheta)^2 = \frac{n_2 \pm \sqrt{n_2^2 - 4n_2m_1}}{2n_2} = \frac{1}{2} \pm \sqrt{\frac{n_2 - 4m_1}{4n_2}}.$$

By (4.3), the harmonic condition $\tau_H = 0$ is equivalent to

$$m_1\left(-\cot(\vartheta/2) + \tan(\vartheta/2)\right) + 2n_2\tan\vartheta = 0.$$

Thus, we have

$$(\tan\vartheta)^2 = \frac{m_1}{n_2}.$$

By (2) of Theorem 3.4, the situation is divided into the following three cases:

(1) When $n_2 < 4m_1$, if $K_2\pi_1(x)$ is biharmonic, then it is harmonic.
(2) When $n_2 = 4m_1$, an orbit $K_2\pi_1(x)$ is proper biharmonic if and only if $(\tan\vartheta)^2 = 1/2$ for $H \in P_0$.
(3) When $n_2 > 4m_1$, an orbit $K_2\pi_1(x)$ is proper biharmonic if and only if

$$(\tan\vartheta)^2 = \frac{n_2 \pm \sqrt{n_2^2 - 4n_2m_1}}{2n_2}$$

holds for $H \in P_0$, since

$$0 < \frac{m_1}{n_2} < \frac{n_2 - \sqrt{n_2^2 - 4n_2m_1}}{2n_2} < \frac{n_2 + \sqrt{n_2^2 - 4n_2m_1}}{2n_2}.$$

5.4. Type III-BC$_1$**.** By the definition of multiplicities, if $2\alpha \in W^+$, then $m(\alpha) = n(\alpha)$. Hence we denote $m_1 := m(\alpha) = n(\alpha)$, $m_2 := m(2\alpha)$ and $n_2 := n(2\alpha)$. Then, by (4.2), the biharmonic condition $\|B_H\|^2 = 1/2$ is equivalent to

$$\begin{aligned}2m_1 + 4m_2 + 4n_2 = {}& m_1\left((\cot(\vartheta/2))^2 + (\tan(\vartheta/2))^2\right) + 4m_2(\cot\vartheta)^2 \\ & + 4n_2(\tan\vartheta)^2.\end{aligned}$$

Thus, we have

$$(\tan \vartheta)^2 = \frac{m_2 + n_2 \pm \sqrt{(m_2 + n_2)^2 - 4n_2(m_1 + m_2)}}{2n_2}$$

$$= \frac{m_2 + n_2 \pm \sqrt{(m_2 - n_2)^2 - 4n_2 m_1}}{2n_2}.$$

By (4.3), the harmonic condition $\tau_H = 0$ is equivalent to

$$m_1\left(\tan(\vartheta/2) - \cot(\vartheta/2)\right) - 2m_2 \cot \vartheta + 2n_2 \tan \vartheta = 0.$$

Thus, we have

$$(\tan \vartheta)^2 = \frac{m_1 + m_2}{n_2}.$$

By (2) of Theorem 3.4, we obtain the following results:

(1) When $(m_2 - n_2)^2 - 4n_2 m_1 < 0$, if $K_2\pi_1(x)$ is biharmonic, then it is harmonic.
(2) When $(m_2 - n_2)^2 - 4n_2 m_1 = 0$, an orbit $K_2\pi_1(x)$ is proper biharmonic if and only if $(\tan \vartheta)^2 = (m_2 + n_2)/2n_2$ for $H \in P_0$.
(3) When $(m_2 - n_2)^2 - 4n_2 m_1 > 0$, an orbit $K_2\pi_1(x)$ is proper biharmonic if and only if

$$(\tan \vartheta)^2 = \frac{m_2 + n_2 \pm \sqrt{(m_2 - n_2)^2 - 4n_2 m_1}}{2n_2}$$

for $H \in P_0$.

For the proof of (2), we will show that

$$\frac{m_1 + m_2}{n_2} \neq \frac{m_2 + n_2}{2n_2}.$$

If $(m_1 + m_2)/n_2 = (m_2 + n_2)/(2n_2)$, then $2m_1 + m_2 - n_2 = 0$. Hence $(m_2 - n_2)^2 - 4n_2 m_1 = -4m_1(m_1 + m_2) < 0$, which is a contradiction.

For the proof of (3), we will show that

$$\frac{m_1 + m_2}{n_2} \neq \frac{m_2 + n_2 \pm \sqrt{(m_2 - n_2)^2 - 4n_2 m_1}}{2n_2}.$$

If the equality holds, then we have $(2m_1 + m_2 - n_2)^2 = (m_2 - n_2)^2 - 4n_2 m_1$. Hence $4m_1(m_1 + m_2) = 0$, which is a contradiction.

In fact, in the cases of type III-BC$_1$, a compact symmetric triad which is not (1) is only $(E_6, SO(10) \cdot U(1), F_4)$ in the list below. In this case,

$$\frac{m_1 + m_2}{n_2} < \frac{m_2 + n_2 - \sqrt{(m_2 - n_2)^2 - 4n_2 m_1}}{2n_2}$$

$$< \frac{m_2 + n_2 + \sqrt{(m_2 - n_2)^2 - 4n_2 m_1}}{2n_2}$$

holds.

Let $b > 0$, $c > 1$ and $q > 1$. Each commutative compact symmetric triad (G, K_1, K_2) where G is simple, $\theta_1 \not\sim \theta_2$ and $\dim \mathfrak{a} = 1$ is one of the following (see [**68**]):

Type III-B$_1$

(G, K_1, K_2)	$(m(\alpha), n(\alpha))$
$(\mathrm{SO}(1 + b + c), \mathrm{SO}(1 + b) \times \mathrm{SO}(c), \mathrm{SO}(b + c))$	$(c - 1, b)$
$(\mathrm{SU}(4), \mathrm{Sp}(2), \mathrm{SO}(4))$	$(2, 2)$
$(\mathrm{SU}(4), \mathrm{S}(\mathrm{U}(2) \times \mathrm{U}(2)), \mathrm{Sp}(2))$	$(3, 1)$
$(\mathrm{Sp}(2), \mathrm{U}(2), \mathrm{Sp}(1) \times \mathrm{Sp}(1))$	$(1, 2)$

Type I-BC$_1$

(G, K_1, K_2)	$(m(\alpha), m(2\alpha), n(\alpha))$
$(\mathrm{SO}(2 + 2q), \mathrm{SO}(2) \times \mathrm{SO}(2q), \mathrm{U}(1 + q))$	$(2(q - 1), 1, 2(q - 1))$
$(\mathrm{SU}(1 + b + c), \mathrm{S}(\mathrm{U}(1 + b) \times \mathrm{U}(c)), \mathrm{S}(\mathrm{U}(1) \times \mathrm{U}(b + c)))$	$(2(c - 1), 1, 2b)$
$(\mathrm{Sp}(1 + b + c), \mathrm{Sp}(1 + b) \times \mathrm{Sp}(c), \mathrm{Sp}(1) \times \mathrm{Sp}(b + c))$	$(4(c - 1), 3, 4b)$
$(\mathrm{SO}(8), \mathrm{U}(4), \mathrm{U}(4)')$	$(4, 1, 1)$

Type II-BC$_1$

(G, K_1, K_2)	$(m(\alpha), n(\alpha), n(2\alpha))$
$(\mathrm{SO}(6), \mathrm{U}(3), \mathrm{SO}(3) \times \mathrm{SO}(3))$	$(2, 2, 1)$
$(\mathrm{SU}(1 + q), \mathrm{SO}(1 + q), \mathrm{S}(\mathrm{U}(1) \times \mathrm{U}(q)))$	$(q - 1, q - 1, 1)$

Type III-BC$_1$

(G, K_1, K_2)	$(m(\alpha), m(2\alpha), n(\alpha), n(2\alpha))$
$(\mathrm{SU}(2 + 2q), \mathrm{S}(\mathrm{U}(2) \times \mathrm{U}(2q)), \mathrm{Sp}(1 + q))$	$(4(q - 1), 3, 4(q - 1), 1)$
$(\mathrm{Sp}(1 + q), \mathrm{U}(1 + q), \mathrm{Sp}(1) \times \mathrm{Sp}(q))$	$(2(q - 1), 1, 2(q - 1), 2)$
$(\mathrm{E}_6, \mathrm{SU}(6) \cdot \mathrm{SU}(2), \mathrm{F}_4)$	$(8, 3, 8, 5)$
$(\mathrm{E}_6, \mathrm{SO}(10) \cdot \mathrm{U}(1), \mathrm{F}_4)$	$(8, 7, 8, 1)$
$(\mathrm{F}_4, \mathrm{Sp}(3) \cdot \mathrm{Sp}(1), \mathrm{Spin}(9))$	$(4, 3, 4, 4)$

Here, we define $\mathrm{U}(4)' = \{g \in \mathrm{SO}(8) \mid JgJ^{-1} = g\}$ where

$$J = \left[\begin{array}{c|c} & I_3 \\ & \quad -1 \\ \hline -I_3 & \\ & 1 \end{array} \right]$$

and I_l denotes the identity matrix of $l \times l$.

6. Main result and examples

6.1. Summarising up all the results in the previous sections, we can classify all the biharmonic hypersurfaces in irreducible compact symmetric spaces which are orbits of commutative Hermann actions. Namely, we obtain the following theorem.

THEOREM 6.1. *Let (G, K_1, K_2) be a commutative compact symmetric triad where G is simple, and suppose that K_2-action on $N_1 = G/K_1$ is cohomogeneity one (hence K_1-action on $N_2 = G/K_2$ is also cohomogeneity one). Then all the proper biharmonic hypersurfaces which are regular orbits of K_2-action (resp. K_1-action) in the compact symmetric space N_1 (resp. N_2) are classified into the following lists:*

(1) *When (G, K_1, K_2) is one of the following cases, there exists a unique proper biharmonic hypersurface which is a regular orbit of K_2-action on N_1 (resp. K_1-action on N_2).*

  (1-1) $(\mathrm{SO}(1 + b + c), \ \mathrm{SO}(1 + b) \times \mathrm{SO}(c), \ \mathrm{SO}(b + c))$
  $$(b > 0, \ c > 1, \ c - 1 \neq b)$$

  (1-2) $(\mathrm{SU}(4), \ \mathrm{S}(\mathrm{U}(2) \times \mathrm{U}(2)), \ \mathrm{Sp}(2))$

  (1-3) $(\mathrm{Sp}(2), \ \mathrm{U}(2), \ \mathrm{Sp}(1) \times \mathrm{Sp}(1))$

(2) *When (G, K_1, K_2) is one of the following cases, there exist exactly two distinct proper biharmonic hypersurfaces which are regular orbits of of K_2-action on N_1 (resp. K_1-action on N_2).*

  (2-1) $(\mathrm{SO}(2 + 2q), \ \mathrm{SO}(2) \times \mathrm{SO}(2q), \ \mathrm{U}(1 + q))$ $\quad (q > 1)$

  (2-2) $(\mathrm{SU}(1 + b + c), \ \mathrm{S}(\mathrm{U}(1 + b) \times \mathrm{U}(c)), \ \mathrm{S}(\mathrm{U}(1) \times \mathrm{U}(b + c)))$
  $$(b \geq 0, \ c > 1)$$

  (2-3) $(\mathrm{Sp}(1 + b + c), \ \mathrm{Sp}(1 + b) \times \mathrm{Sp}(c), \ \mathrm{Sp}(1) \times \mathrm{Sp}(b + c))$
  $$(b \geq 0, \ c > 1)$$

  (2-4) $(\mathrm{SO}(8), \ \mathrm{U}(4), \ \mathrm{U}(4)')$

  (2-5) $(\mathrm{E}_6, \ \mathrm{SO}(10) \cdot \mathrm{U}(1), \ \mathrm{F}_4)$

  (2-6) $(\mathrm{SO}(1 + q), \ \mathrm{SO}(q), \ \mathrm{SO}(q))$ $\quad (q > 1)$

  (2-7) $(\mathrm{F}_4, \ \mathrm{Spin}(9), \ \mathrm{Spin}(9))$

(3) *When (G, K_1, K_2) is one of the following cases, any biharmonic regular orbit of K_2-action on N_1 (resp. K_1-action on N_2) is harmonic.*

  (3-1) $(\mathrm{SO}(2c), \ \mathrm{SO}(c) \times \mathrm{SO}(c), \ \mathrm{SO}(2c - 1))$ $\quad (c > 1)$

  (3-2) $(\mathrm{SU}(4), \ \mathrm{Sp}(2), \ \mathrm{SO}(4))$

  (3-3) $(\mathrm{SO}(6), \ \mathrm{U}(3), \ \mathrm{SO}(3) \times \mathrm{SO}(3))$

  (3-4) $(\mathrm{SU}(1 + q), \ \mathrm{SO}(1 + q), \ \mathrm{S}(\mathrm{U}(1) \times \mathrm{U}(q)))$ $\quad (q > 1)$

  (3-5) $(\mathrm{SU}(2 + 2q), \ \mathrm{S}(\mathrm{U}(2) \times \mathrm{U}(2q)), \ \mathrm{Sp}(1 + q))$ $\quad (q > 1)$

  (3-6) $(\mathrm{Sp}(1 + q), \ \mathrm{U}(1 + q), \ \mathrm{Sp}(1) \times \mathrm{Sp}(q))$ $\quad (q > 1)$

  (3-7) $(\mathrm{E}_6, \ \mathrm{SU}(6) \cdot \mathrm{SU}(2), \ \mathrm{F}_4)$

  (3-8) $(\mathrm{F}_4, \ \mathrm{Sp}(3) \cdot \mathrm{Sp}(1), \ \mathrm{Spin}(9))$

REMARK 6.1. *In Theorem 6.1, we determined all the biharmonic hypersurfaces in irreducible compact symmetric spaces which are orbits of commutative Hermann actions.*

(1) *In the previous section we assumed $\theta_1 \not\sim \theta_2$. If $\theta_1 \sim \theta_2$, then the action of K_2 on N_1 is orbit equivalent to the isotropy action of K_1 on N_1. We will discuss these cases in Section 6.3.*

(2) *The commutative condition $\theta_1\theta_2 = \theta_2\theta_1$ is essential for our discussion. Indeed, there exist some Hermann actions where $\theta_1\theta_2 \neq \theta_2\theta_1$. Moreover there exist some hyperpolar actions of cohomogeneity one on irreducible compact symmetric spaces which are not Hermann actions (cf. [85]).*

6.2. We shall explain details of the cases (1-1), (2-2) and (3-1) in Theorem 6.1, and give new examples of proper biharmonic orbits. By Proposition 5.1, a proper biharmonic orbit $K_2\pi_1(x)$ in N_1 corresponds to a proper biharmonic orbit $K_1\pi_2(x)$ in N_2. In particular, we can obtain new examples of proper biharmonic orbits corresponding to some known examples.

We consider the isotropy subgroups of orbits of Hermann actions. For $x = \exp H$ ($H \in \mathfrak{a}$), we define the isotropy subgroups

$$(K_2)_{\pi_1(x)} = \{k \in K_2 \mid k\pi_1(x) = \pi_1(x)\},$$

$$(K_1)_{\pi_2(x)} = \{k \in K_1 \mid k\pi_2(x) = \pi_2(x)\}.$$

Then we can show that $(K_2)_{\pi_1(x)} \cong (K_1)_{\pi_2(x)}$ by an inner automorphism of G. The orbit $K_2\pi_1(x)$ (resp. $K_1\pi_2(x)$) is diffeomorphic to the homogeneous space $K_2/((K_2)_{\pi_1(x)})$ (resp. $K_1/((K_1)_{\pi_2(x)})$). If $K_2\pi_1(x)$ is a regular orbit, then $K_1\pi_2(x)$ is also a regular orbit, and we have $\text{Lie}((K_2)_{\pi_1(x)}) = \text{Lie}((K_1)_{\pi_2(x)}) = \mathfrak{k}_0$.

Example 1. $(SO(1 + b + c), SO(1 + b) \times SO(c), SO(b + c))$ Let $(G, K_1, K_2) = (SO(1+b+c), SO(1+b) \times SO(c), SO(b+c))$ $(b > 0, c > 1)$. This is the case of (3-1) when $c - 1 = b$, otherwise the case of (1-1) in Theorem 6.1. In this case, the involutions θ_1 and θ_2 are given by

$$\theta_1(k) = I'_{1+b}kI'_{1+b}, \quad \theta_2(k) = I'_1kI'_1 \quad (k \in G),$$

where

$$I'_l = \begin{bmatrix} -I_l & 0 \\ 0 & I_{1+b+c-l} \end{bmatrix} \quad (1 \leq l \leq b + c).$$

Then, we have the canonical decompositions $\mathfrak{g} = \mathfrak{k}_1 \oplus \mathfrak{m}_1 = \mathfrak{k}_2 \oplus \mathfrak{m}_2$ as

$$\mathfrak{k}_1 = \left\{ \begin{bmatrix} X & 0 \\ 0 & Y \end{bmatrix} \Bigg| \begin{matrix} X \in \mathfrak{so}(1+b) \\ Y \in \mathfrak{so}(c) \end{matrix} \right\}, \quad \mathfrak{m}_1 = \left\{ \begin{bmatrix} 0 & X \\ -{}^tX & 0 \end{bmatrix} \Bigg| X \in M_{1+b,c}(\mathbb{R}) \right\},$$

$$\mathfrak{k}_2 = \left\{ \begin{bmatrix} 0 & 0 \\ 0 & X \end{bmatrix} \,\middle|\, X \in \mathfrak{so}(b+c) \right\}, \quad \mathfrak{m}_2 = \left\{ \begin{bmatrix} 0 & X \\ -{}^tX & 0 \end{bmatrix} \,\middle|\, X \in M_{1,b+c}(\mathbb{R}) \right\}.$$

Thus, we have

$$\mathfrak{k}_1 \cap \mathfrak{k}_2 = \left\{ \begin{bmatrix} 0 & 0 & 0 \\ 0 & X & 0 \\ 0 & 0 & Y \end{bmatrix} \,\middle|\, \begin{array}{l} X \in \mathfrak{so}(b) \\ Y \in \mathfrak{so}(c) \end{array} \right\},$$

$$\mathfrak{m}_1 \cap \mathfrak{m}_2 = \left\{ \begin{bmatrix} 0 & 0 & X \\ 0 & 0 & 0 \\ -{}^tX & 0 & 0 \end{bmatrix} \,\middle|\, X \in M_{1,c}(\mathbb{R}) \right\},$$

$$\mathfrak{k}_1 \cap \mathfrak{m}_2 = \left\{ \begin{bmatrix} 0 & X & 0 \\ -{}^tX & 0 & 0 \\ 0 & 0 & 0 \end{bmatrix} \,\middle|\, X \in M_{1,b}(\mathbb{R}) \right\},$$

$$\mathfrak{m}_1 \cap \mathfrak{k}_2 = \left\{ \begin{bmatrix} 0 & 0 & 0 \\ 0 & 0 & X \\ 0 & -{}^tX & 0 \end{bmatrix} \,\middle|\, X \in M_{b,c}(\mathbb{R}) \right\}.$$

We take a maximal abelian subspace $\mathfrak{a}$ in $\mathfrak{m}_1 \cap \mathfrak{m}_2$ as

$$\mathfrak{a} = \left\{ H(\vartheta) = \begin{bmatrix} 0 & 0 & X \\ 0 & 0 & 0 \\ -{}^tX & 0 & 0 \end{bmatrix} \,\middle|\, \begin{array}{l} X = [0,\dots,0,\vartheta] \\ \vartheta \in \mathbb{R} \end{array} \right\}.$$

Then we have

$$\mathfrak{k}_0 = \left\{ \begin{bmatrix} 0 & 0 & 0 & 0 \\ 0 & X & 0 & 0 \\ 0 & 0 & Y & 0 \\ 0 & 0 & 0 & 0 \end{bmatrix} \,\middle|\, \begin{array}{l} X \in \mathfrak{so}(b) \\ Y \in \mathfrak{so}(c-1) \end{array} \right\},$$

$$V(\mathfrak{k}_1 \cap \mathfrak{m}_2) = \{0\},$$

$$V(\mathfrak{m}_1 \cap \mathfrak{k}_2) = \left\{ \begin{bmatrix} 0 & 0 & 0 & 0 \\ 0 & 0 & X & 0 \\ 0 & -{}^tX & 0 & 0 \\ 0 & 0 & 0 & 0 \end{bmatrix} \,\middle|\, X \in M_{b,c-1}(\mathbb{R}) \right\}.$$

Let E_i^j be a matrix whose (i,j)-entry is one and all the other entries are zero. We define $A_i^j := E_i^j - E_j^i$. Then, we can see

$$[H(\vartheta), A_1^j] = -\vartheta A_{1+b+c}^j \qquad (2 \le j \le b+c),$$
$$[H(\vartheta), A_{1+b+c}^j] = \vartheta A_1^j \qquad (2 \le j \le b+c).$$

We define a vector $\alpha \in \mathfrak{a}$ by $\langle H(\vartheta), \alpha \rangle = \vartheta$ $(\vartheta \in \mathbb{R})$. Then

$$
\begin{aligned}
\mathfrak{k}_\alpha &= \mathrm{Span}\{A_{1+b+c}^{2+b}, \ldots, A_{1+b+c}^{b+c}\}, \\
\mathfrak{m}_\alpha &= \mathrm{Span}\{A_1^{2+b}, \ldots, A_1^{b+c}\}, \\
V_\alpha^\perp(\mathfrak{k}_1 \cap \mathfrak{m}_2) &= \mathrm{Span}\{A_1^2, \ldots, A_1^{1+b}\}, \\
V_\alpha^\perp(\mathfrak{m}_1 \cap \mathfrak{k}_2) &= \mathrm{Span}\{A_{1+b+c}^2, \ldots, A_{1+b+c}^{1+b}\}.
\end{aligned}
$$

Hence, in this case, we have

$$
\Sigma^+ = \{\alpha\}, \quad W^+ = \{\alpha\}, \quad m(\alpha) = c - 1, \quad n(\alpha) = b.
$$

Let $x_0 = \exp(H(\pi/4))$. By the computation in Section 5.1, we can see that $K_2\pi_1(x_0)$ and $K_1\pi_2(x_0)$ are biharmonic hypersurfaces in N_1 and N_2, respectively. These orbits exist at the center of the orbit space $\overline{P_0} = \{H(\vartheta) \mid 0 \leq \vartheta \leq \pi/2\}$. When $c - 1 = b$, these orbits are harmonic. When $c - 1 \neq b$, these are not harmonic, hence proper biharmonic. The orbit $K_2\pi_1(x_0)$ is the Clifford hypersurface $S^b(1/\sqrt{2}) \times S^{c-1}(1/\sqrt{2}) \cong (\mathrm{SO}(1+b) \times \mathrm{SO}(c))/(\mathrm{SO}(b) \times \mathrm{SO}(c-1))$ embedded in the sphere $S^{b+c}(1) \cong \mathrm{SO}(1+b+c)/\mathrm{SO}(b+c) = N_2$ ([64]). On the other hand, the orbit $K_2\pi_1(x_0)$ is diffeomorphic to $\mathrm{SO}(b+c)/(\mathrm{SO}(b) \times \mathrm{SO}(c-1))$, i.e. the universal covering of a real flag manifold, and embedded in the oriented real Grassmannian manifold $\widetilde{G}_{1+b}(\mathbb{R}^{1+b+c}) \cong \mathrm{SO}(1+b+c)/(\mathrm{SO}(1+b) \times \mathrm{SO}(c)) = N_1$ as the tube of radius $\pi/4$ over the totally geodesic sub-Grassmannian $\widetilde{G}_b(\mathbb{R}^{b+c})$. The orbit $K_2\pi_1(x_0)$ in N_1 gives a new example of a proper biharmonic hypersurface in the oriented real Grassmannian manifold.

Example 2. $(\mathrm{SU}(1+b+c), \mathrm{S}(\mathrm{U}(1+b) \times \mathrm{U}(c)), \mathrm{S}(\mathrm{U}(1) \times \mathrm{U}(b+c)))$. Let $(G, K_1, K_2) = (\mathrm{SU}(1+b+c), \mathrm{S}(\mathrm{U}(1+b) \times \mathrm{U}(c)), \mathrm{S}(\mathrm{U}(1) \times \mathrm{U}(b+c)))$ $(b > 0, c > 1)$. This is the case of (2-2) except for $b = 0$ in Theorem 6.1. In this case, the involutions θ_1 and θ_2 are given by

$$
\theta_1(k) = I'_{1+b} k I'_{1+b}, \quad \theta_2(k) = I'_1 k I'_1 \quad (k \in G).
$$

Analogous to the previous example, in this case, we have

$$
\Sigma^+ = \{\alpha, 2\alpha\}, \ W^+ = \{\alpha\}, \ m(\alpha) = 2(c-1), \ m(2\alpha) = 1, \ n(\alpha) = 2b.
$$

Therefore, the symmetric triad $(\tilde{\Sigma}, \Sigma, W)$ is of type I-BC$_1$. By the computation in Section 5, we have two distinct proper biharmonic hypersurfaces in N_1, and also in N_2. More precisely, let $x_\pm = \exp(H(\vartheta_\pm))$

where $\vartheta_\pm$ is a solution of the equation

$$(\tan\vartheta)^2 = \frac{m_1 + n_1 + 6m_2 \pm \sqrt{(m_1 + n_1 + 6m_2)^2 - 4(n_1 + m_2)(m_1 + m_2)}}{2(n_1 + m_2)}$$

$$= \frac{(c-1) + b + 3 \pm \sqrt{((c-1) + b + 3)^2 - (2b+1)(2(c-1)+1)}}{2b+1}.$$

Then $K_2\pi_1(x_\pm)$ and $K_1\pi_2(x_\pm)$ are proper biharmonic hypersurfaces in N_1 and N_2, respectively. The orbit $K_1\pi_2(x_\pm) \cong \mathrm{S}(\mathrm{U}(1+b) \times \mathrm{U}(c))/\mathrm{S}(\mathrm{U}(b) \times \mathrm{U}(c-1) \times \mathrm{U}(1))$ is the tube of radius $\vartheta_\pm$ over the totally geodesic $\mathbb{C}P^b$ in the complex projective space $\mathbb{C}P^{b+c} \cong \mathrm{SU}(1+b+c)/\mathrm{S}(\mathrm{U}(1) \times \mathrm{U}(b+c)) = N_2$ (see Theorem 5 in [64]). On the other hand, the orbit $K_2\pi_1(x_\pm) \cong \mathrm{S}(\mathrm{U}(1) \times \mathrm{U}(b+c))/\mathrm{S}(\mathrm{U}(b) \times \mathrm{U}(c-1) \times \mathrm{U}(1))$ is the tube of radius $\vartheta_\pm$ over the totally geodesic sub-Grassmannian $G_b(\mathbb{C}^{b+c})$ in the complex Grassmannian manifold $G_{1+b}(\mathbb{C}^{1+b+c}) \cong \mathrm{SU}(1 + b + c)/\mathrm{S}(\mathrm{U}(1+b) \times \mathrm{U}(c)) = N_1$. The orbit $K_2\pi_1(x_\pm)$ in N_1 gives a new example of a proper biharmonic hypersurface in the complex Grassmannian manifold.

6.3. In the above argument, we supposed that $\theta_1 \nsim \theta_2$ in order to use the classification of commutative compact symmetric triads. However, we can apply our method to the cases of $\theta_1 \sim \theta_2$. When $\theta_1 \sim \theta_2$, a Hermann action is orbit equivalent to the isotropy action of a compact symmetric space (see [67]). Hence, it is sufficient to discuss the cases of isotropy actions, that is, $\theta_1 = \theta_2$. When $\theta_1 = \theta_2$, we have $W = \emptyset$, since $\mathfrak{k}_1 \cap \mathfrak{m}_2 = \mathfrak{m}_1 \cap \mathfrak{k}_2 = \{0\}$. Thus we have $\tilde{\Sigma} = \Sigma$. Moreover, Σ is the root system of the compact symmetric space N_1 with respect to $\mathfrak{a}$. Since we consider the cases of $\dim \mathfrak{a} = 1$, the rank of N_1 equals to one. All the simply connected, rank one symmetric spaces of compact type are classified as follows:

$$S^q, \ \mathbb{C}P^q, \ \mathbb{H}P^q, \ \mathbb{O}P^2 \quad (q \geq 2).$$

The isotropy actions of these symmetric spaces correspond to the cases (2-6), (2-2) with $b = 0$, (2-3) with $b = 0$, and (2-9) in Theorem 6.1, respectively. Except for the case of $\mathbb{O}P^2$, homogeneous biharmonic hypersurfaces in compact, rank one symmetric spaces were classified ([63], [64]). Therefore, we consider the octonionic projective plane $\mathbb{O}P^2 \cong F_4/\mathrm{Spin}(9)$.

Let $(G, K_1, K_2) = (F_4, \mathrm{Spin}(9), \mathrm{Spin}(9))$ with $\theta_1 = \theta_2$. This is the case of (2-9) in Theorem 6.1. Since $K_1 = K_2$, we denote

$$\mathfrak{k} := \mathfrak{k}_1 = \mathfrak{k}_2, \quad \mathfrak{m} := \mathfrak{m}_1 = \mathfrak{m}_2.$$

We define an $\mathrm{Ad}(G)$-invariant inner product on $\mathfrak{g}$ by $\langle \cdot, \cdot \rangle = -\mathrm{Killing}(\cdot, \cdot)$. Fix a maximal abelian subspace $\mathfrak{a}$ in $\mathfrak{m}$. Then we have $\Sigma^+ = \{\alpha, 2\alpha\}$ and $m(\alpha) = 8$, $m(2\alpha) = 7$ ([**59**], Page 534). By letting $n(\alpha) = n(2\alpha) = 0$ in (5.1) since $W^+ = \emptyset$, we can see that $\langle \alpha, \alpha \rangle = 1/\{2(8 + 4 \cdot 7)\}$. Let $x = \exp H$ for $H \in \mathfrak{a}$. By Theorem 6.1, we have the following:

$$\|B_H\|^2 = 8(\cot\langle \alpha, H \rangle)^2 \langle \alpha, \alpha \rangle + 7(\cot\langle 2\alpha, H \rangle)^2 \langle 2\alpha, 2\alpha \rangle$$

$$= \frac{1}{18}\{2(\cot\langle \alpha, H \rangle)^2 + 7(\cot\langle 2\alpha, H \rangle)^2\},$$

$$dL_x^{-1}(\tau_H) = -(8 \cot\langle \alpha, H \rangle \alpha + 7 \cot\langle 2\alpha, H \rangle 2\alpha)$$

$$= -2(4 \cot\langle \alpha, H \rangle + 7 \cot\langle 2\alpha, H \rangle)\alpha.$$

Then, the biharmonic condition $\|B_H\|^2 = 1/2$ is equivalent to

$$9 = 2(\cot\langle \alpha, H \rangle)^2 + 7(\cot\langle 2\alpha, H \rangle)^2.$$

Thus we have

$$(\cot\langle \alpha, H \rangle)^2 = \frac{25 \pm 2\sqrt{130}}{15}.$$

The harmonic condition $\tau_H = 0$ is equivalent to

$$4 \cot\langle \alpha, H \rangle + 7 \cot\langle 2\alpha, H \rangle = 0.$$

Thus we have

$$(\cot\langle \alpha, H \rangle)^2 = \frac{7}{15}.$$

Since

$$0 < \frac{25 - 2\sqrt{130}}{15} < \frac{7}{15} < \frac{25 + 2\sqrt{130}}{15},$$

by (2) ofTheorem 3. 4, an orbit $K_2\pi_1(x)$ is proper biharmonic if and only if

$$(\cot\langle \alpha, H \rangle)^2 = \frac{25 \pm 2\sqrt{130}}{15}$$

holds for $H \in \mathfrak{a}$ with $0 < \langle \alpha, H \rangle < \pi/2$. Furthermore, a unique harmonic regular orbit exists between two proper biharmonic orbits in $\{H \in \mathfrak{a} \mid 0 < \langle \alpha, H \rangle < \pi/2\}$. These regular orbits are diffeomorphic to S^{15} embedded in $\mathbb{O}P^2$.

CHAPTER 12

Sasaki Manifolds, Kähler Cone Manifolds and Biharmonic Submanifolds

[1]

ABSTRACT. For every Legendrian submanifold M of a Sasaki manifold N, we study harmonicity and biharmonicity of the corresponding Lagrangian cone submanifold $C(M)$ of a Kähler manifold $C(N)$. We will show that, if $C(M)$ is biharmonic in $C(N)$, then it is harmonic. Furthermore, we will show that M is proper biharmonic in N if and only if $C(M)$ has a non-zero eigen-section of the Jacobi operator with the eigenvalue $m = \dim M$.

1. Introduction

Harmonic maps play a central role in geometry; they are critical points of the energy functional $E(\varphi) = \frac{1}{2} \int_M |d\varphi|^2 \, v_g$ for smooth maps φ of (M, g) into (N, h). The Euler-Lagrange equations are given by the vanishing of the tension filed $\tau(\varphi)$. In 1983, J. Eells and L. Lemaire [40] extended the notion of harmonic map to biharmonic map, which are, by definition, critical points of the bienergy functional

$$E_2(\varphi) = \frac{1}{2} \int_M |\tau(\varphi)|^2 \, v_g. \qquad (1.1)$$

After G.Y. Jiang [74] studied the first and second variation formulas of E_2, extensive studies in this area have been done (for instance, see [16], [90], [102], [123], [131], [63], [64], [73], etc.). Notice that harmonic maps are always biharmonic by definition. We say, for a smooth map $\varphi : (M, g) \to (N, h)$ to be *proper biharmonic* if it is biharmonic, but not harmonic. B.Y. Chen raised ([21]) so called B.Y. Chen's conjecture and later, R. Caddeo, S. Montaldo, P. Piu and C. Oniciuc raised ([16]) the generalized B.Y. Chen's conjecture.

B.Y. Chen's conjecture:

Every biharmonic submanifold of the Euclidean space $\mathbb{R}^n$ must be harmonic (minimal).

---

[1]This chapter is due to [155]: H. Urakawa, *Sasaki manifolds, Kähler cone manifolds and biharmonic submanifolds*, Illinois J. Math., **58** (2014), 521–535.

The generalized B.Y. Chen's conjecture:
Every biharmonic submanifold of a Riemannian manifold of non-positive curvature must be harmonic (minimal).

For the generalized Chen's conjecture, Ou and Tang gave ([**123**]) a counter example in a Riemannian manifold of negative curvature. For the Chen's conjecture, affirmative answers were known for the case of surfaces in the three dimensional Euclidean space ([**21**]), and the case of hypersurfaces of the four dimensional Euclidean space ([**57**], [**34**]). Furthermore, Akutagawa and Maeta gave ([**1**]) recently a final supporting evidence to the Chen's conjecture:

THEOREM 1.1. *Any complete regular biharmonic submanifold of the Euclidean space $\mathbb{R}^n$ is harmonic (minimal).*

To the generalized Chen's conjecture, we showed ([**111**]) that

THEOREM 1.2. *Let (M, g) be a complete Riemannian manifold, and the curvature of (N, h), non-positive. Then,*
(1) *every biharmonic map $\varphi : (M, g) \to (N, h)$ with finite energy and finite bienergy must be harmonic.*
(2) *In the case* $\mathrm{Vol}(M, g) = \infty$, *under the same assumtion, every biharmonic map $\varphi : (M, g) \to (N, h)$ with finite bienergy is harmonic.*

We also obtained (cf. [**108**], [**109**], [**111**])

THEOREM 1.3. *Assume that (M, g) is a complete Riemannian manifold, $\varphi : (M, g) \to (N, h)$ is an isometric immersion, and the sectional curvature of (N, h) is non-positive. If $\varphi : (M, g) \to (N, h)$ is biharmonic and $\int_M |\mathbf{H}|^2 v_g < \infty$, then it is minimal. Here, $\mathbf{H}$ is the mean curvature normal vector field of the isometric immersion φ.*

Theorem 1.3 gives an affirmative answer to the generalized B.Y. Chen's conjecture under the L^2-condition and completeness of (M, g).

In this paper, for every Legendrian submanifold $\varphi : (M^m, g) \to (N^{2m+1}, h)$ of a Sasaki manifold (N^{2m+1}, h), and the Lagrangian cone submanifold $\overline{\varphi} : (C(M), \overline{g}) \to (C(N), \overline{h})$ of a Kähler cone manifold $(C(N), \overline{h})$, we show (Theorems 3.3 and 4.4) that (1) $\overline{\varphi} : (C(M), \overline{g}) \to (C(N), \overline{h})$ is biharmonic if and only if it is harmonic, which is equivalent to that $\varphi : (M, g) \to (N, h)$ is harmonic. (2) $\varphi : (M, g) \to (N, h)$ is proper biharmonic if and only if $\tau(\overline{\varphi})$ is a non-zero eigen-section of the Jacobi operator $J_{\overline{\varphi}}$ with the eigenvalue $m = \dim M$. The assertion

(2) can be regarded as a biharmonic map version of T. Takahashi's theorem (cf. Theorem 4.5) which claims that each coordinate function of the isometric immersion of (M^m, g) into the unit sphere $S^n \hookrightarrow \mathbb{R}^{n+1}$ is the eigenfunction of the Laplacian of (M, g) with the eigenvalue $m = \dim M$.

Acknowledgement.   This work was finished during the stay at the University of Basilicata, Potenza, Italy, June of 2013. The author would like to express his sincere gratitude to Professors Sorin Dragomir and Elisabetta Barletta for their hospitality and helpful discussions, and also Dr. Shun Maeta for his helpful comments on Sasahara's works. The author also express his gratitude to Professor T. Sasahara who pointed several errors in the first draft.

2. Preliminaries

We first prepare the materials for the first and second variational formulas for the bienergy functional and biharmonic maps. Let us recall the definition of a harmonic map $\varphi : (M, g) \to (N, h)$, of a compact Riemannian manifold (M, g) into another Riemannian manifold (N, h), which is an extremal of the *energy functional* defined by

$$E(\varphi) = \int_M e(\varphi)\, v_g,$$

where $e(\varphi) := \frac{1}{2}|d\varphi|^2$ is called the energy density of φ. That is, for any variation $\{\varphi_t\}$ of φ with $\varphi_0 = \varphi$,

$$\frac{d}{dt}\bigg|_{t=0} E(\varphi_t) = -\int_M h(\tau(\varphi), V) v_g = 0, \tag{2.1}$$

where $V \in \Gamma(\varphi^{-1}TN)$ is a variation vector field along φ which is given by $V(x) = \frac{d}{dt}\big|_{t=0}\varphi_t(x) \in T_{\varphi(x)}N$, $(x \in M)$, and the *tension field* is given by $\tau(\varphi) = \sum_{i=1}^m B(\varphi)(e_i, e_i) \in \Gamma(\varphi^{-1}TN)$, where $\{e_i\}_{i=1}^m$ is a locally defined orthonormal frame field on (M, g), and $B(\varphi)$ is the second fundamental form of φ defined by

$$B(\varphi)(X, Y) = (\widetilde{\nabla} d\varphi)(X, Y)$$
$$= (\widetilde{\nabla}_X d\varphi)(Y)$$
$$= \overline{\nabla}_X(d\varphi(Y)) - d\varphi(\nabla_X Y), \tag{2.2}$$

for all vector fields $X, Y \in \mathfrak{X}(M)$. Here, ∇, and ∇^N, are Levi-Civita connections on TM, TN of (M, g), (N, h), respectively, and $\overline{\nabla}$, and $\widetilde{\nabla}$

are the induced ones on $\varphi^{-1}TN$, and $T^*M \otimes \varphi^{-1}TN$, respectively. By (2.1), φ is *harmonic* if and only if $\tau(\varphi) = 0$.

The second variation formula is given as follows. Assume that φ is harmonic. Then,

$$\frac{d^2}{dt^2}\bigg|_{t=0} E(\varphi_t) = \int_M h(J(V), V)v_g, \tag{2.3}$$

where J is an elliptic differential operator, called the *Jacobi operator* acting on $\Gamma(\varphi^{-1}TN)$ given by

$$J(V) = \overline{\Delta}V - \mathcal{R}(V), \tag{2.4}$$

where $\overline{\Delta}V = \overline{\nabla}^*\overline{\nabla}V = -\sum_{i=1}^{m}\{\overline{\nabla}_{e_i}\overline{\nabla}_{e_i}V - \overline{\nabla}_{\nabla_{e_i}e_i}V\}$ is the *rough Laplacian* and $\mathcal{R}$ is a linear operator on $\Gamma(\varphi^{-1}TN)$ given by $\mathcal{R}(V) = \sum_{i=1}^{m} R^N(V, d\varphi(e_i))d\varphi(e_i)$, and R^N is the curvature tensor of (N, h) given by $R^N(U, V) = \nabla^N_U\nabla^N_V - \nabla^N_V\nabla^N_U - \nabla^N_{[U,V]}$ for $U, V \in \mathfrak{X}(N)$.

J. Eells and L. Lemaire [**40**] proposed polyharmonic (k-harmonic) maps and Jiang [**74**] studied the first and second variation formulas of biharmonic maps. Let us consider the *bienergy functional* defined by

$$E_2(\varphi) = \frac{1}{2}\int_M |\tau(\varphi)|^2 v_g, \tag{2.5}$$

where $|V|^2 = h(V, V)$, $V \in \Gamma(\varphi^{-1}TN)$.

The first variation formula of the bienergy functional is given by

$$\frac{d}{dt}\bigg|_{t=0} E_2(\varphi_t) = -\int_M h(\tau_2(\varphi), V)v_g. \tag{2.6}$$

Here,

$$\tau_2(\varphi) := J(\tau(\varphi)) = \overline{\Delta}(\tau(\varphi)) - \mathcal{R}(\tau(\varphi)), \tag{2.7}$$

which is called the *bitension field* of φ, and J is given in (2.4).

A smooth map φ of (M, g) into (N, h) is said to be *biharmonic* if $\tau_2(\varphi) = 0$. By definition, every harmonic map is biharmonic. We say, for an immersion $\varphi : (M, g) \to (N, h)$ to be *proper biharmonic* if it is biharmonic but not harmonic (minimal).

3. Legendrian submanifolds and Lagrangian submanifolds

In this section, we first show a correspondence between the set of all Legendrian submanifolds of a Sasakian manifold and the one of all Lagrangian submanifolds of a Kähler cone manifold.

An $n(= 2m + 1)$ dimensional contact Riemannian manifold (N, h) with a contact form η is said to be a *contact metric manifold* if there exist a smooth $(1,1)$ tensor field J and a smooth vector field ξ on N, called a *basic vector field*, satisfying that

$$J^2 = -\mathrm{Id} + \eta \otimes \xi, \tag{3.1}$$

$$\eta(\xi) = 1, \tag{3.2}$$

$$J\xi = 0, \tag{3.3}$$

$$\eta \circ J = 0, \tag{3.4}$$

$$h(JX, JY) = h(X, Y) - \eta(X)\,\eta(Y), \tag{3.5}$$

$$\eta(X) = h(X, \xi), \tag{3.6}$$

$$d\eta(X, Y) = h(X, JY), \tag{3.7}$$

for all smooth vector fields X, Y on N. Here, Id is the identity transformation of $T_x N$ ($x \in N$). A contact metric manifold (N, h, J, ξ, η) is *Sasakian* if $(C(N), \overline{h}, I)$ is a Kähler manifold. Here, a cone manifold $C(N) := N \times \mathbb{R}^+$ where $\mathbb{R}^+ := \{r \in \mathbb{R} \mid r > 0\}$, $\overline{h}$ is a cone metric on $C(N)$, $\overline{h} := dr^2 + r^2\, h$, which is a Hermitian metric with respect to an almost complex structure I on $C(N)$ given by

$$\begin{cases} IY := JY + \eta(Y)\,\Psi, & (Y \in \mathfrak{X}(N)), \\ I\Psi := -\xi, \end{cases} \tag{3.8}$$

where $\Psi := r\,\frac{\partial}{\partial r}$ is called the *Liouville vector field* on $C(N)$. We denote by $\mathfrak{X}(N)$, the set of all smooth vector fields on N. A contact metric manifold (N, h, J, ξ, η) is Sasakian if and only if

$$(\nabla_X^N J)(Y) = h(X, Y)\,\xi - \eta(Y)\,X \quad (X, Y \in \mathfrak{X}(N)). \tag{3.9}$$

Let us recall the definition

DEFINITION 3.1. *Let M^m be an m-dimensional manifold, an immersion $\varphi : M^m \to N^{2m+1}$. M^m is called to be a* Legendrian *submanifold of an $(2m+1)$-dimensional Sasakian manifold (N, h, J, ξ, η) if $\varphi^* \eta \equiv 0$ which is equivalent to that*

$$\varphi_{*x}(X_x) \in \mathrm{Ker}(\eta_{\varphi(x)}) \tag{3.10}$$

for all $X_x \in T_x M$ ($x \in M$).

A Legendrian submanifold M^m satisfies the following two conditions:

(1) $\varphi_*(T_x M)$ is orthogonal $J(\varphi_*(T_x M))$ with respect to h for all $x \in M$. This is equivalent to that the normal bundle $T^\perp M$ of $\varphi : M \to N$ has the following splitting:

$$T_x M^\perp = \mathbb{R}\xi_{\varphi(x)} \oplus J\,\varphi_* T_x M \quad (x \in M).$$

(2) The second fundamental form B of $\varphi(M) \subset N$ has its value at $\mathrm{Ker}(\eta)$, that is,

$$B(\varphi_* X, \varphi_* Y) = \nabla^N_X \varphi_* Y - \varphi_* (\nabla_X Y) \in \varphi_* (T_x M)^\perp,$$

where $T_x M^\perp$ is $\varphi_* (T_x M)^\perp$, which is

$$\{W_{\varphi(x)} \in T_{\varphi(x)} N \,|\, h(W_{\varphi(x)}, \varphi_{**} X_x) = 0 \,(\forall\, X_x \in T_x M)\}.$$

Here, ∇, ∇^N are Levi-Civita connections of (M, g), (N, h) where g is the induced metric on M by $g := \varphi^* h$.

In the following, we identify $\varphi(M)$ with M, itself. The following theorem is well known, but essentially important for us.

THEOREM 3.1. *Let M^m be an m-dimensional submanifold of a Sasakian manifold $(N^{2m+1}, h, J, \xi, \eta)$. Then, M is a Legendrian submanifold of a Sasaki manifold N if and only if $C(M) \subset C(N)$ is a Lagrangian submanifold of a Kähler cone manifold $(C(N), \overline{h}, I)$.*

Proof We have the equivalence that $M \subset N$ is Legendrian if and only if

$$\begin{cases} \xi_x{}^\perp = T_x M \oplus J T_x M, \\ h(T_x M, J T_x M) = \{0\} \end{cases} \tag{3.11}$$

for all $x \in M$. That is, $h(\xi, X) = 0$ and $h(X, JY) = 0$ for all $X, Y \in \mathfrak{X}(M)$. Then, (3.11) is equivalent to that

$$\Omega(f_1 \Phi + X, f_2 \Phi + Y) = r^2 \left\{ f_1 \, h(\xi, Y) - f_2 \, h(\xi, X) + h(X, JY) \right\}$$
$$= 0 \tag{3.12}$$

for all smooth functions f_1, f_2 on $C(M)$ and X, $Y \in \mathfrak{X}(M)$. Here, Ω is the Kähler form of $C(N)$ which is given by $\Omega = 2\, r \, dr \wedge \eta + r^2 \, d\eta$. Finally, (3.12) is equivalent to that $C(M) \subset C(N)$ is Lagrangian. □

Now our main theorem is as follows:

THEOREM 3.2. *Let $\varphi : (M, g) \to (N, h)$ be a Legendrian submanifold of a Sasakian manifold (N^n, h, J, ξ, η) ($n = 2m + 1$) and $\overline{\varphi} : (C(M), \overline{g}) \ni (r, x) \mapsto (r, \varphi(x)) \in (C(N), \overline{h}, I)$, a Lagrangian submanifold of a Kähler cone manifold. Here $C(M) := M \times \mathbb{R}^+ \subset C(N) := N \times \mathbb{R}^+$, $\overline{g} = dr^2 + r^2 \, g$, and $\overline{h} = dr^2 + r^2 \, h$. Then,*
(1) it holds that

$$\tau(\overline{\varphi}) = \frac{1}{r^2} \, \tau(\varphi). \tag{3.13}$$

Thus, we have the equivalence that $\varphi : (M, g) \to (N, h)$ is harmonic if and only if $\overline{\varphi}(C(M), \overline{g}) \to (C(N), \overline{h})$ is also harmonic.

(2) *Secondly, it holds that*

$$\tau_2(\overline{\varphi}) = \frac{1}{r^4}\tau_2(\varphi) + \frac{m}{r^2}\tau(\varphi). \tag{3.14}$$

Then, we have the equivalence that $\varphi : (M,g) \to (N,h)$ *is proper biharmonic if and only if for* $\overline{\varphi} : (C(M),\overline{g}) \to (C(N),\overline{h})$, *the tension field* $\tau(\overline{\varphi})$ *is a non-zero eigen-section of the Jacobi operator* $J_{\overline{\varphi}}$ *with the eigenvalue* $m = \dim M$. *And we have the equivalence that* $\overline{\varphi} : (C(M),\overline{g}) \to (C(N),\overline{h})$ *is biharmonic if and only if it is harmonic, which is equivalent to that* $\varphi : (M,g) \to (N,h)$ *is harmonic.*

(3) *Thirdly, it holds that*

$$\tau_2(\overline{\varphi})^\perp = \frac{1}{r^4}\tau_2(\varphi)^\perp + \frac{m}{r^2}\tau(\varphi). \tag{3.15}$$

Then, we have the equivalence that $\varphi : (M,g) \to (N,h)$ *is minimal if and only if* $\overline{\varphi} : (C(M),\overline{g}) \to (C(N),\overline{h})$ *is bi-minimal.*

(4) *Finally, it holds that*

$$\operatorname{div}_{\overline{g}}(I\,\tau(\overline{\varphi})) = \frac{1}{r^2}\operatorname{div}_g(J\,\tau(\varphi)). \tag{3.16}$$

Then, we have also the equivalence that $\varphi : (M,g) \to (N,h,J,\xi,\eta)$ *is Legendrian minimal if and only if* $\overline{\varphi} : (C(M),\overline{g}) \to (C(N),\overline{h},I)$ *is also Lagrangian minimal.*

To prove Theorem 3.3, we need the following lemma.

LEMMA 3.1. *The Levi-Civia connection* $\nabla^{C(M)}$ *of the cone manifold* $(C(M),\overline{g})$ *of a Riemannian manifold* (M,g), *where the cone metric* $\overline{g} = dr^2 + r^2\,g$, *is given as follows:*

$$\begin{cases} \nabla_X^{C(M)}Y = \nabla_X Y - r\,g(X,Y)\dfrac{\partial}{\partial r}, \\[2mm] \nabla_X^{C(M)}\dfrac{\partial}{\partial r} = \dfrac{1}{r}X, \\[2mm] \nabla_{\frac{\partial}{\partial r}}^{C(M)}Y = \dfrac{1}{r}Y, \\[2mm] \nabla_{\frac{\partial}{\partial r}}^{C(M)}\dfrac{\partial}{\partial r} = 0. \end{cases} \tag{3.17}$$

Here, $X, Y \in \mathfrak{X}(M)$, *and* ∇ *is the Levi-Civita connection of* (M,g).

The proof of Lemma 3.4 is a direct computation which is omitted.

To proceed to give a proof of Theorem 3.3, we first take a locally defined orthonormal frame field $\{e_i\}_{i=1}^m$ on (M,g). Define $\overline{e}_i := \frac{1}{r}e_i$

$(i = 1, \ldots, m)$, and $\bar{e}_{m+1} := \frac{\partial}{\partial r}$. Then, $\{\bar{e}_i\}_{i=1}^{m+1}$ is a locally defined orthonormal frame field on the cone manifold $(C(M), \bar{g})$.

Let $\varphi : (M^m, g) \to (N^n, h)$ $(n = 2m + 1)$ be a Legendrian submanifold of a Sasakian manifold, and $\bar{\varphi} : (C(M), \bar{g}) \to (C(N), \bar{h})$, the corresponding cone submanifold of a Kähler cone $(C(N), \bar{h})$. We should see a relation between the induced bundles $\varphi^{-1}TN$ and $\bar{\varphi}^{-1}TC(N)$. We denote by $\Gamma(E)$, the space of all smooth sections of the vector bundle E. Then, every smooth section W of the induced bundle $\bar{\varphi}^{-1}TC(N)$ can be written as

$$W = V + B \frac{\partial}{\partial r} \tag{3.18}$$

where V is a smooth section of the induced bundle $\varphi^{-1}TN$ and B is a smooth function on $C(M) = M \times \mathbb{R}^+$. Because, for every point $(x, r) \in C(M) = M \times \mathbb{R}^+$, $\bar{\varphi}(x, r) = (\varphi(x), r)$, and $W_{(x,r)} \in T_{\bar{\varphi}(x,r)}C(N) = T_{(\varphi(x),r)}(N \times \mathbb{R}^+) = T_{\varphi(x)}N \oplus T_r\mathbb{R}^+$, so we can write as $W_{(x,r)} = V_x + B(x, r) \frac{\partial}{\partial r}$, where $V_x \in T_{\varphi(x)}N$ and $B(x, r) \in \mathbb{R}$.

Then, if we denote by $\bar{\nabla}$, and $\bar{\bar{\nabla}}$, the induced connections of the induced bundles $\varphi^{-1}TN$, and $\bar{\varphi}^{-1}TC(N)$ from the connections ∇^N, $\nabla^{C(N)}$ of (N, h) and $(C(N), \bar{h})$, respectively, then we have for every $W \in \Gamma(\bar{\varphi}^{-1}TC(N))$, with $W = V + B \frac{\partial}{\partial r}$ and $V \in \Gamma(\varphi^{-1}TN)$ and $B \in C^\infty(M \times \mathbb{R}^+)$,

$$\begin{cases} \bar{\bar{\nabla}}_X W = \bar{\nabla}_X V + \dfrac{B}{r} X + (XB) \dfrac{\partial}{\partial r}, & (X \in \mathfrak{X}(M)), \\ \bar{\bar{\nabla}}_{\frac{\partial}{\partial r}} W = \dfrac{\partial B}{\partial r} \dfrac{\partial}{\partial r}. \end{cases} \tag{3.19}$$

Proof of Theorem 3.3.
(1) We have, for $i = 1, \ldots, m$, $(m = \dim M)$,

$$\bar{\varphi}_* \nabla_{\bar{e}_i}^{C(M)} \bar{e}_i = \bar{\varphi}_* \left(\frac{1}{r^2} \nabla_{e_i}^{C(M)} e_i \right)$$

$$= \frac{1}{r^2} \bar{\varphi}_* \left(\nabla_{e_i} e_i - r\, g(e_i, e_i) \frac{\partial}{\partial r} \right) \qquad \text{(by Lemma 3.4 (3.17))}$$

$$= \frac{1}{r^2} \left(\nabla_{e_i} e_i - r \frac{\partial}{\partial r} \right) \tag{3.20}$$

since $\bar{\varphi}$ is the inclusion map of $C(M)$ into $C(N)$. For $i = m + 1$, we have

$$\bar{\varphi}_* \left(\nabla_{\bar{e}_{m+1}}^{C(M)} \bar{e}_{m+1} \right) = \bar{\varphi}_* \left(\nabla_{\frac{\partial}{\partial r}}^{C(M)} \frac{\partial}{\partial r} \right) = 0. \tag{3.21}$$

Furthermore, we have, for $i = 1, \ldots, m$,

$$\overline{\nabla}_{\overline{e}_*}\overline{\varphi}_*\overline{e}_i = \nabla^{C(N)}_{\frac{1}{r}e_i} \frac{1}{r} e_i$$

$$= \frac{1}{r^2} \left\{ \nabla^N_{e_i} e_i - r\, h(e_i, e_i) \frac{\partial}{\partial r} \right\}$$

$$= \frac{1}{r^2} \left\{ \nabla^N_{e_i} e_i - r \frac{\partial}{\partial r} \right\} \tag{3.22}$$

since $\overline{\varphi}^*\overline{h} = \overline{g}$ and $\varphi^*h = g$. For $i = m+1$, we have also

$$\overline{\nabla}_{\overline{e}_{m+1}}\overline{\varphi}_*\overline{e}_{m+1} = \nabla^{C(N)}_{\frac{\partial}{\partial r}} \frac{\partial}{\partial r} = 0. \tag{3.23}$$

Thus, we have

$$\tau(\overline{\varphi}) = \sum_{i=1}^{m+1} \left\{ \overline{\nabla}_{\overline{e}_i}\overline{\varphi}_*\overline{e}_i - \overline{\varphi}_* \left(\nabla^{C(M)}_{\overline{e}_i} \overline{e}_i \right) \right\}$$

$$= \frac{1}{r^2} \sum_{i=1}^{m} \left\{ \nabla^N_{e_i} e_i - \nabla_{e_i} e_i \right\} \qquad \text{(by (3.20), (3.21), (3.22), (3.23))}$$

$$= \frac{1}{r^2} \tau(\varphi), \tag{3.24}$$

which is (3.13).

For (2), we have to see relations between

$$J_\varphi(V) = \overline{\Delta}_\varphi V - \sum_{i=1}^{m} R^N(V, \varphi_* e_i)\varphi_* e_i, \qquad (V \in \Gamma(\varphi^{-1}TN)), \tag{3.25}$$

$$J_{\overline{\varphi}}(W) = \overline{\overline{\Delta}}_{\overline{\varphi}} W - \sum_{i=1}^{m+1} R^{C(N)}(W, \overline{\varphi}_*\overline{e}_i)\overline{\varphi}_*\overline{e}_i, \qquad (W \in \Gamma(\overline{\varphi}^{-1}TC(N)). \tag{3.26}$$

where

$$\overline{\Delta}_\varphi V := -\sum_{i=1}^{m} \{ \overline{\nabla}_{e_i}(\overline{\nabla}_{e_i} V) - \overline{\nabla}_{\nabla_{e_i} e_i} V \}, \tag{3.27}$$

$$\overline{\overline{\Delta}}_{\overline{\varphi}} W := -\sum_{i=1}^{m+1} \{ \overline{\overline{\nabla}}_{\overline{e}_i}(\overline{\overline{\nabla}}_{\overline{e}_i} W) - \overline{\overline{\nabla}}_{\nabla^{C(M)}_{\overline{e}_i}\overline{e}_i} W \}. \tag{3.28}$$

Here, $\overline{\nabla}$, and $\overline{\overline{\nabla}}$ are the induced connections of $\varphi^{-1}TN$ and $\overline{\varphi}^{-1}TC(N)$ from the Levi-Civita connections ∇^N and $\nabla^{C(N)}$ of (N, h) and $(C(N), \overline{h})$ with $\overline{h} = dr^2 + r^2 h$, respectively.

(*The first step*)   By (3.19), we have

$$
\begin{cases}
\overline{\overline{\nabla}}_X(\overline{\nabla}_Y W) = \overline{\nabla}_X(\overline{\nabla}_Y V) + \dfrac{B}{r}\,\nabla_X^N Y + \dfrac{XB}{r}\,Y + \dfrac{YB}{r}\,X \\[2mm]
\qquad\qquad + X(YB)\dfrac{\partial}{\partial r}, \qquad (X,Y \in \mathfrak{X}(M)), \\[3mm]
\overline{\overline{\nabla}}_{\frac{\partial}{\partial r}}\left(\overline{\nabla}_{\frac{\partial}{\partial r}} W\right) = \dfrac{\partial^2 B}{\partial r^2}\dfrac{\partial}{\partial r},
\end{cases}
\tag{3.29}
$$

where we used that $\overline{\overline{\nabla}}_X(\overline{\nabla}_Y V) = \overline{\nabla}_X(\overline{\nabla}_Y V)$, $\overline{\overline{\nabla}}_X Y = \overline{\nabla}_X Y = \nabla_X^N Y$ and $\overline{\overline{\nabla}}_X \frac{\partial}{\partial r} = \frac{1}{r}X$ for every $X,Y \in \mathfrak{X}(M)$. Thus, we obtain, for $W = V + B\frac{\partial}{\partial r} \in \Gamma(\overline{\varphi}^{-1} TC(N))$ with $V \in \Gamma(\varphi^{-1} TN)$ and $B \in C^\infty(M \times \mathbb{R}^+)$,

$$
\overline{\Delta}_{\overline{\varphi}} W = \dfrac{1}{r^2}\,\overline{\Delta}_\varphi V - \dfrac{B}{r^3}\,\tau(\varphi) - \dfrac{2}{r^3}\,\mathrm{grad}_M B
$$
$$
+ \left(\dfrac{1}{r^2}\,\Delta_M B - \dfrac{\partial^2 B}{\partial r^2} - \dfrac{m}{r}\dfrac{\partial B}{\partial r}\right)\dfrac{\partial}{\partial r},
\tag{3.30}
$$

where let us recall

$$
\overline{\Delta}_\varphi V = -\sum_{i=1}^m \{\overline{\nabla}_{e_i}(\overline{\nabla}_{e_i} V) - \overline{\nabla}_{\nabla_{e_i} e_i} V\} \qquad (V \in \Gamma(\varphi^{-1} TN)),
$$
$$
\tau(\varphi) = \sum_{i=1}^m (\nabla_{e_i}^N e_i - \nabla_{e_i} e_i), \qquad \mathrm{grad}_M B = \sum_{i=1}^m (e_i B)\, e_i,
$$
$$
\Delta_M B = -\sum_{i=1}^m \{e_i(e_i B) - \nabla_{e_i} e_i\, B\} \qquad (B \in C^\infty(M \times \mathbb{R}^+)).
$$

(*The second step*)   By a direct computation, we have the curvature tensor field $R^{C(N)}$ of $(C(N),\overline{h})$:

$$
\begin{cases}
R^{C(N)}(X,Y)Z = R^N(X,Y)Z - h(Y,Z)\,X + h(X,Z)Y, \\[2mm]
R^{C(N)}\left(X,\dfrac{\partial}{\partial r}\right)\dfrac{\partial}{\partial r} = 0, \\[3mm]
R^{C(N)}\left(\dfrac{\partial}{\partial r},Y\right)Z = 0,
\end{cases}
\tag{3.31}
$$

for every $X,Y,Z \in \mathfrak{X}(M)$. Therefore, we obtain

$$
\sum_{i=1}^m R^{C(N)}(W,\overline{\varphi}_* \overline{e}_i)\overline{\varphi}_* \overline{e}_i = \dfrac{1}{r^2}\sum_{i=1}^m R^N(V,\varphi_* e_i)\varphi_* e_i - \dfrac{m}{r^2}\,V + \dfrac{1}{r^2} V^{\mathrm{T}},
\tag{3.32}
$$

for $W = V + B\frac{\partial}{\partial r} \in \Gamma(\overline{\varphi}^{-1} TC(N))$, where V^{T} is the tangential part of V.

(*The third step*)   Therefore, we have

$$J_{\overline{\varphi}}(W) = \overline{\overline{\Delta}}_{\overline{\varphi}} W - \sum_{i=1}^{m} R^{C(N)}(W, \overline{\varphi}_* \overline{e}_i) \overline{\varphi}_* \overline{e}_i$$

$$= \frac{1}{r^2} \left(\overline{\Delta}_\varphi V - \sum_{i=1}^{m} R^N(V, \varphi_* e_i) \varphi_* e_i \right) + \frac{m}{r^2} V - \frac{1}{r^2} V^{\mathrm{T}}$$

$$- \frac{B}{r^3} \tau(\varphi) - \frac{2}{r^3} \mathrm{grad}_M B$$

$$+ \left(\frac{1}{r^2} \Delta_M B - \frac{\partial^2 B}{\partial r^2} - \frac{m}{r} \frac{\partial B}{\partial r} \right) \frac{\partial}{\partial r}. \tag{3.33}$$

Here, we have already $\tau(\overline{\varphi}) = \frac{1}{r^2} \tau(\varphi)$ in Theorem 3.3 (1) (3.13). For this $W := \tau(\overline{\varphi})$, we have $V = \frac{1}{r^2} \tau(\varphi)$, $B = 0$ and $V^{\mathrm{T}} = 0$, and we have

$$J_{\overline{\varphi}}(\tau(\overline{\varphi})) = \frac{1}{r^4} \left(\overline{\Delta}_\varphi(\tau(\varphi)) - \sum_{i=1}^{m} R^N(\tau(\varphi), \varphi_* e_i) \varphi_* e_i \right) + \frac{m}{r^2} \tau(\varphi)$$

$$= \frac{1}{r^4} J_\varphi(\tau(\varphi)) + \frac{m}{r^2} \tau(\varphi). \tag{3.34}$$

We have (3.14) in (2). By (3.34), we have the equivalence between the bi-harmonicity of φ and that $\tau(\overline{\varphi})$ is a non-zero eigen-section of the Jacobi operator $J_{\overline{\varphi}}$ with eigenvalue $m = \dim M$. Furthermore, $\tau_2(\overline{\varphi}) = 0$ if and only if $\tau_2(\varphi) + m r^2 \tau(\varphi) = 0$ for all $r > 0$, which is equivalent to that $\tau(\varphi) = 0$.

For (3) in Theorem 3.3, we only observe the following orthogonal decompositions:

$$T_x N = T_x M \oplus T_x M^\perp, \quad T_x M^\perp = J T_x M \oplus \mathbb{R} \, \xi_x, \tag{3.35}$$

$$T_{(x,r)} C(N) = T_x N \oplus T_r \mathbb{R}^+$$
$$= T_x M \oplus J T_x M \oplus \mathbb{R} \, \xi_x \oplus T_r \mathbb{R}^+$$
$$= T_{(x,r)} C(M) \oplus J T_x M \oplus \mathbb{R} \, \xi_x$$
$$= T_{(x,r)} C(M) \oplus T_x M^\perp, \tag{3.36}$$

for every $x \in M \subset N$. So let us decompose $\tau_2(\overline{\varphi}) = \frac{1}{r^4} \tau_2(\varphi)$ following (3.35) and (3.36). Then, we have

$$\tau_2(\overline{\varphi}) = \tau_2(\overline{\varphi})^{\mathrm{T}} + \tau_2(\overline{\varphi})^\perp \tag{3.37}$$

where $\tau_2(\overline{\varphi})^{\mathrm{T}} \in T_{(x,r)} C(M)$ and $\tau_2(\overline{\varphi})^\perp \in T_x M^\perp$, and also we have

$$\frac{1}{r^4} \tau_2(\varphi) + \frac{m}{r^2} \tau(\varphi) = \frac{1}{r^4} \tau_2(\varphi)^{\mathrm{T}} + \frac{1}{r^4} \tau_2(\varphi)^\perp + \frac{m}{r^2} \tau(\varphi), \tag{3.38}$$

where $\tau_2(\varphi)^{\mathrm{T}} \in T_x M$ and $\tau_2(\varphi)^{\perp} \in T_x M^{\perp}$. But, since we have $T_x M \subset T_{(x,r)} C(M)$, we have

$$
\begin{cases}
\tau_2(\overline{\varphi})^{\mathrm{T}} = \dfrac{1}{r^4}\, \tau_2(\varphi)^{\mathrm{T}}, \\[3mm]
\tau_2(\overline{\varphi})^{\perp} = \dfrac{1}{r^4}\, \tau_2(\varphi)^{\perp} + \dfrac{m}{r^2}\, \tau(\varphi).
\end{cases}
\tag{3.39}
$$

Then, we have $\tau_2(\varphi)^{\perp} = 0$ if and only if $\tau_2(\varphi)^{\perp} + mr^2\, \tau(\varphi) = 0$ for all $r > 0$, which is equivalent to that $\tau(\varphi) = 0$.

For (4), we first show that

$$
\begin{aligned}
I\,\tau(\overline{\varphi}) &= J\,\tau(\overline{\varphi}) + \eta(\tau(\overline{\varphi}))\,\Psi \\[2mm]
&= \frac{1}{r^2}\, J\,\tau(\varphi) + \frac{1}{r^2}\, \eta(\tau(\varphi))\,\Psi \\[2mm]
&= \frac{1}{r^2}\, J\,\tau(\varphi).
\end{aligned}
\tag{3.40}
$$

Because for a Legendrian submanifold of a Sasaki manifold, the second fundamental form B takes its value in $\mathrm{Ker}(\eta)$, so $\tau(\varphi) = \mathrm{Trace}(B) \subset \mathrm{Ker}(\eta)$, that is,

$$
\eta(\tau(\varphi)) = 0.
\tag{3.41}
$$

Then, we have

$$
\begin{aligned}
\mathrm{div}_{\overline{g}}(I\,\tau(\overline{\varphi})) &= \sum_{i=1}^{m+1} \overline{g}(\overline{e}_i, \nabla^{C(M)}_{\overline{e}_i}(I\,\tau(\overline{\varphi}))) \\[2mm]
&= \frac{1}{r^4} \sum_{i=1}^{m} \overline{g}(e_i, \nabla^{C(M)}_{e_i}(J\,\tau(\varphi))) \\[2mm]
&\quad + \frac{1}{r^2}\, \overline{g}\Big(\frac{\partial}{\partial r}, \nabla^{C(M)}_{\frac{\partial}{\partial r}}(J\,\tau(\varphi))\Big).
\end{aligned}
\tag{3.42}
$$

But, the first term of the right hand side of (3.42) coincides with

$$
\begin{aligned}
\frac{1}{r^4} \sum_{i=1}^{m} \overline{g}\Big(e_i, \nabla_{e_i}(J\,\tau(\varphi)) - r\,g(e_i, J\,\tau(\varphi))\frac{\partial}{\partial r}\Big) \\[2mm]
= \frac{1}{r^2} \sum_{i=1}^{m} g(e_i, \nabla_{e_i}(J\,\tau(\varphi))) \\[2mm]
= \frac{1}{r^2}\, \mathrm{div}_g(J\,\tau(\varphi)).
\end{aligned}
\tag{3.43}
$$

On the other hand, the second term of the right hand side of (3.42) coincides with

$$\frac{1}{r^2}\,\overline{g}\Big(\frac{\partial}{\partial r}, \nabla^{C(M)}_{\frac{\partial}{\partial r}}(J\,\tau(\varphi))\Big) = \frac{1}{r^3}\,\overline{g}\Big(\frac{\partial}{\partial r}, J\,\tau(\varphi)\Big) = 0 \tag{3.44}$$

because $J\,\tau(\varphi)$ is tangential to $T_x M$ for the Legendrian immersion $\varphi : (M,g) \to (N,h,J)$. Therefore, we obtain the desired formula:

$$\mathrm{div}_{\overline{g}}(I\,\tau(\overline{\varphi})) = \frac{1}{r^2}\,\mathrm{div}_g(J\,\tau(\varphi)).$$

We obtain Theorem 3.3. $\square$

REMARK 3.1. *The assertion (4) in Theorem 3.3 was given by I. Castro, H.Z. Li and F. Urbano ([19]), and H. Iriyeh ([72]), independently in a different manner from ours.*

4. Biharmonic Legendrian submanifolds of Sasakian manifolds

By Theorem 3.3, we turn to review studies of a proper biharmonic Legendrian submanifold of a Sasaki manifold (N^n, h, J, ξ, η) and give Takahashi-type theorem (cf. Theorem 4.4). First let us recall the equations of biharmonicity of an isometric immersions (cf. [95]).

LEMMA 4.1. *Let $\varphi : (M^m, g) \to (N^n, h)$ be an isometric immersion. Then φ is biharmonic if and only if*

$$\begin{cases} \displaystyle\sum_{i=1}^{m}(\nabla_{e_i} A_{\mathbf{H}})(e_i) + \sum_{i=1}^{m} A_{\nabla^{\perp}_{e_i}\mathbf{H}}(e_i) - \sum_{i=1}^{m}\Big(R^N(\mathbf{H}, e_i)e_i\Big)^{\mathrm{T}} = 0, \\[2mm] \displaystyle\Delta^{\perp}\mathbf{H} + \sum_{i=1}^{m} B(A_{\mathbf{H}}(e_i), e_i) - \sum_{i=1}^{m}\Big(R^N(\mathbf{H}, e_i)e_i\Big)^{\perp} = 0, \end{cases} \tag{4.1}$$

where $\mathbf{H} = \frac{1}{m}\sum_{i=1}^{m} B(e_i, e_i)$ is the mean curvature vector field along φ, $(\ )^{\mathrm{T}}$, $(\ )^{\perp}$ are the tangential part and normal part, respectively, B is the second fundamental form, and A is the shape operator for the isometric immersion $\varphi : (M,g) \to (N,h)$.

For an isometric immersion of a Legendrian submanifold into a Sasakian manifold, we have

THEOREM 4.1. *Let $\varphi : (M^m, g) \to (N^n, h, J, \xi, \eta)$ $(n = 2m+1)$ be an isometric immersion of a Legendrian submanifold of a Sasakian* *manifold. Then φ is biharmonic if and only if*

$$\sum_{i=1}^{m}(\nabla_{e_i}A_{\mathbf{H}})(e_i) + \sum_{i=1}^{m}A_{\nabla^{\perp}_{e_i}\mathbf{H}}(e_i)$$

$$-\sum_{i,j=1}^{m}h\big((\nabla^{\perp}_{e_j}B)(e_i,e_i)-(\nabla^{\perp}_{e_i}B)(e_j,e_i),\mathbf{H}\big)\,e_j$$

$$= 0, \tag{4.2}$$

$$\Delta^{\perp}\mathbf{H} + \sum_{i=1}^{m}B(A_{\mathbf{H}}(e_i),e_i)$$

$$+\sum_{j=1}^{m}\mathrm{Ric}^{N}(J\mathbf{H},e_j)\,Je_j - \sum_{j=1}^{m}\mathrm{Ric}^{M}(J\mathbf{H},e_j)\,Je_j$$

$$-\sum_{i=1}^{m}J\,A_{B(J\mathbf{H},e_i)}(e_i) + m\,J\,A_{\mathbf{H}}(J\mathbf{H}) + \mathbf{H}$$

$$= 0. \tag{4.3}$$

In the case that (N^{2m+1},h,J,ξ,η) is a Sasaki space form $N^{2m+1}(\epsilon)$ of constant J-sectional curvature ϵ whose curvature tensor R^{N} is given by

$$R^{N}(X,Y)Z = \frac{\epsilon+3}{4}\big\{h(Y,Z)\,X - h(Z,X)\,Y\big\}$$

$$+\frac{\epsilon-1}{4}\big\{\eta(X)\eta(Z)Y - \eta(Y)\eta(Z)X + h(X,Z)\eta(Y)\xi - h(Y,Z)\eta(X)\xi$$

$$+ h(Z,JY)\,JX - h(Z,JX)\,JY + 2h(X,JY)\,JZ\big\}, \tag{4.4}$$

for all $X,\,Y,\,Z \in \mathfrak{X}(N)$, we have ([**50**], [**67**], [**133**])

THEOREM 4.2. *Let $\varphi : (M^m,g) \to N^{2m+1}(\epsilon)$ be a Legendrian submanifold of a Sasaki space form of constant J-sectional curvature ϵ. Then φ is biharmonic if and only if*

$$\overline{\Delta}_{\varphi}\mathbf{H} = \frac{\epsilon(m+3)+3(m-1)}{4}\,\mathbf{H} \tag{4.5}$$

which is equivalent to

$$\left\{ \begin{array}{l} \displaystyle\sum_{i=1}^{m}(\nabla_{e_i}A_{\mathbf{H}})(e_i) + \sum_{i=1}^{m}A_{\nabla^{\perp}_{e_i}\mathbf{H}}(e_i) = 0, \\[4mm] \displaystyle\Delta^{\perp}\mathbf{H} + \sum_{i=1}^{m}B(A_{\mathbf{H}}(e_i),e_i) - \frac{\epsilon(m+3)+3(m-1)}{4}\,\mathbf{H} = 0. \end{array} \right. \tag{4.6}$$

Now, let us consider a Legendrian submanifold M^m of the $(2m+1)$-dimensional unit sphere $S^{2m+1}(1)$ with the standard metric ds_{std}^2 of constant sectional curvature 1. Then, we have, due to Theorem 3.3, and $J_{\overline{\varphi}} = \overline{\overline{\Delta}}$ which follows from that $R^{C(N)} = 0$ because of $(C(N), \overline{h}) = (\mathbb{C}^{m+1}, ds^2)$:

THEOREM 4.3. *Let $\varphi : (M^m, g) \to (S^{2m+1}(1), ds_{\text{std}}^2)$ be a Legendrian submanifold of $(S^{2m+1}(1), ds_{\text{std}}^2)$, and $\overline{\varphi} : (C(M), \overline{g}) \to (\mathbb{C}^{m+1}, ds^2)$, the corresponding Lagrangian cone submanifold of the standard complex space $(\mathbb{C}^{m+1}, ds^2)$. Then, it holds that $\varphi : (M^m, g) \to (S^{2m+1}(1), ds_{\text{std}}^2)$ is proper biharmonic if and only if $\tau(\overline{\varphi}) = \frac{1}{r^2}\tau(\varphi) = \frac{m}{r^2}\mathbf{H}$ is a non-zero eigen-section of the rough Laplacian $\overline{\overline{\Delta}}_{\overline{\varphi}}$ acting on $\Gamma(\overline{\varphi}^{-1}T\mathbb{C}^{m+1})$ with the eigenvalue $m = \dim M$: $\overline{\overline{\Delta}}_{\overline{\varphi}}\tau(\overline{\varphi}) = m\,\tau(\overline{\varphi})$.*

This Theorem 4.4 could be regarded as a biharmonic map version of the following T. Takahashi's theorem ([**143**]). For Takahashi-type theorem for harmonic maps into Grassmannian manifolds, see pp. 42 and 46 in [**106**]:

THEOREM 4.4. *(T. Takahashi) Let (M^m, g) be a compact Riemannian manifold, and let $\varphi : (M^m, g) \to (S^n, ds_{\text{std}}^2)$ be an isometric immersion. We write $\varphi = (\varphi_1, \cdots, \varphi_{n+1})$ where $\varphi_i \in C^\infty(M)$ $(1 \leq i \leq n+1)$ via the canonical embedding $S^n \hookrightarrow \mathbb{R}^{n+1}$. Then, $\varphi : (M, g) \to (S^n, ds_{\text{std}}^2)$ is minimal if and only if $\Delta_g \varphi_i = m\varphi_i$, $(1 \leq i \leq n+1)$. Here, Δ_g is the positive Laplacian acting on $C^\infty(M)$.*

Certain classification theorems about proper biharmonic Legendrian immersions into the unit sphere $(S^{2m+1}(1), ds_{\text{std}}^2)$ were obtained by T. Sasahara ([**131**], [**132**], [**133**]).

CHAPTER 13

Biharmonic Lagrangian Submanifolds in Kähler Manifolds

1

ABSTRACT. In this chapter, we give the necessary and sufficient conditions for Lagrangian submanifolds in Kähler manifolds to be biharmonic. We classify biharmonic PNMC Lagrangian H-umbilical submanifolds in the complex space forms. Finally, we classify biharmonic PNMC Lagrangian surfaces in the 2-dimensional complex space forms.

1. Introduction

Theory of harmonic maps has been applied into various fields in Differential geometry. Harmonic maps between two Riemannian manifolds are critical points of the energy functional $E(\phi) = \frac{1}{2} \int_M |d\phi|^2 v_g$ for smooth maps $\phi : M \to N$. The Euler-Lagrange equation is $\tau(\phi) = 0$, where $\tau(\phi) = \text{trace} \nabla d\phi$ is the *tension field* of ϕ.

On the other hand, in 1983, J. Eells and L. Lemaire [40] proposed the problem to consider *polyharmonic maps of order k*: they are critical points of the functional

$$E_k(\phi) = \int_M e_k(\phi) v_g, \quad (k = 1, 2, \dots),$$

where $e_k(\phi) = \frac{1}{2} |(d + \delta)^k \phi|^2$ for smooth maps $\phi : M \to N$, where δ is the codifferentiation. G. Y. Jiang [74] studied the first variational formula of the bi-energy E_2 ($k = 2$) which is written as

$$E_2(\phi) = \frac{1}{2} \int_M |\tau(\phi)|^2 v_g, \tag{1.1}$$

and the critical points of E_2 are called *biharmonic maps*. There have been extensive studies on biharmonic maps. Harmonic maps are always biharmonic maps by definition. By this, one of our central problem is to find non-harmonic biharmonic maps. Recently, T. Sasahara [132]

---

[1]This chapter is due to [95]: S. Maeta and H. Urakawa, *Biharmonic Lagrangian submanifolds in Kähler manifolds*, Glasgow Math. J. **55** (2013), 465–480.

classified the 2-dimensional biharmonic Lagrangian submanifolds in the two-dimensional complex space forms.

In this paper, we first obtain the biharmonic equations for a Lagrangian submanifold M^m in a Kähler manifold $(N^m, J, \langle \cdot, \cdot \rangle)$ of complex m dimension (cf. Theorem 3.2). We next give the necessary and sufficient conditions for Lagrangian submanifolds in the complex space forms to be biharmonic (cf. Proposition 4.1). In Section 5, we classify biharmonic Lagrangian H-umbilical submanifolds in the complex space forms $(N^m(4\varepsilon), J, \langle \cdot, \cdot \rangle)$ which have parallel normalized mean curvature vector field (cf. Definition 5.3 and Theorem 5.8). Finally, we classify biharmonic PNMC Lagrangian surfaces in the two-dimensional complex space forms (cf. Theorem 6.2).

2. Preliminaries

In this section, we give necessary notations on biharmonic maps for later use.

Let $\phi : (M, g) \to (N, h = \langle \cdot, \cdot \rangle)$ be a smooth map from an m dimensional Riemannian manifold (M, g) into an n dimensional Riemannian manifold (N, h). The second fundamental form of ϕ is a covariant differentiation $\widetilde{\nabla} d\phi$ of 1-form $d\phi$, which is a section of $\odot^2 T^* M \otimes \phi^{-1} TN$. For every vector fields X, Y on M,

$$
\begin{aligned}
(\widetilde{\nabla} d\phi)(X, Y) &= (\widetilde{\nabla}_X d\phi)(Y) = \overline{\nabla}_X d\phi(Y) - d\phi(\nabla_X Y) \\
&= \nabla^N_{d\phi(X)} d\phi(Y) - d\phi(\nabla_X Y).
\end{aligned}
$$

Here, $\nabla, \nabla^N, \overline{\nabla}$ and $\widetilde{\nabla}$ are the connections on the bundles TM, TN, $\phi^{-1} TN$ and $T^* M \otimes \phi^{-1} TN$ respectively.

We consider critical points of the energy functional

$$
E(\phi) = \int_M e(\phi) v_g,
$$

where $e(\phi) = \frac{1}{2} |d\phi|^2 = \frac{1}{2} \sum_{i=1}^m \langle d\phi(e_i), d\phi(e_i) \rangle$ is the *energy density* of ϕ, $\{e_i\}_{i=1}^m$ is a locally defined orthonormal frame field on (M, g). Here, $\langle \cdot, \cdot \rangle$ is an induced metric $\phi^* h$. The tension field $\tau(\phi)$ of ϕ is defined by

$$
\tau(\phi) = \sum_{i=1}^m (\widetilde{\nabla} d\phi)(e_i, e_i) = \sum_{i=1}^m (\widetilde{\nabla}_{e_i} d\phi)(e_i).
$$

Then, ϕ is a *harmonic map* if and only if $\tau(\phi) = 0$.

For the bi-energy E_2, G. Y. Jiang found in [74] the first variational formula. The map ϕ is called *biharmonic maps* if *bitension field* $\tau_2(\phi)$

vanishes, that is,

$$\tau_2(\phi) = \overline{\Delta}\tau(\phi) - \sum_{i=1}^{m} R^N(\tau(\phi), d\phi(e_i))d\phi(e_i) = 0, \qquad (2.1)$$

where R^N is the curvature tensor field, i.e.,

$$R^N(U,V)W = \nabla_U^N \nabla_V^N W - \nabla_V^N \nabla_U^N W - \nabla_{[U,V]}^N W, \qquad (U,V,W \in \mathfrak{X}(N)),$$

and $\overline{\Delta} = \overline{\nabla}^* \overline{\nabla} = -\sum_{k=1}^{m}(\overline{\nabla}_{e_k}\overline{\nabla}_{e_k} - \overline{\nabla}_{\nabla_{e_k} e_k})$ is the *rough Laplacian*.

The Gauss formula is

$$\nabla_X^N Y = d\phi(\nabla_X Y) + B(X,Y), \qquad X,Y \in \mathfrak{X}(M), \qquad (2.2)$$

where $\mathfrak{X}(M)$ is the space of vector field on M, and B denotes the second fundamental form. The Weingarten formula is

$$\nabla_X^N \xi = -A_\xi X + \nabla_X^\perp \xi, \ X \in \mathfrak{X}(M), \ \xi \in \Gamma(TM^\perp), \qquad (2.3)$$

where A_ξ is the shape operator for a normal vector field ξ on M and $\nabla^\perp$ stands for the normal connection of the normal bundle on M in N. It is well known that the second fundamental form and the shape operator are related by

$$\langle B(X,Y), \xi \rangle = \langle A_\xi X, Y \rangle. \qquad (2.4)$$

The equations of Gauss and Codazzi are given by

$$\langle R^N(X,Y)Z, W \rangle = \langle R(X,Y)Z, W \rangle + \langle A_{B(X,Z)}Y, W \rangle - \langle A_{B(Y,Z)}X, W \rangle, \qquad (2.5)$$

$$(\nabla_X^\perp B)(Y,Z) = (\nabla_Y^\perp B)(X,Z), \qquad (2.6)$$

where $\nabla^\perp B$ is given by

$$(\nabla_X^\perp B)(Y,Z) = \nabla_X^\perp(B(Y,Z)) - B(\nabla_X Y, Z) - B(Y, \nabla_X Z).$$

If $\phi : (M,g) \to (N,h)$ is a biharmonic isometric immersion, then M is called a *biharmonic submanifold*. In this case, the tension field satisfies $\tau(\phi) = m\,\mathbf{H}$, where $\mathbf{H}$ is the harmonic mean curvature vector field along ϕ. The bitension field $\tau_2(\phi)$ is rewritten as

$$\tau_2(\phi) = m\left\{\overline{\Delta}\,\mathbf{H} - \sum_{i=1}^{m} R^N(\mathbf{H}, d\phi(e_i))d\phi(e_i)\right\}, \qquad (2.7)$$

and ϕ is biharmonic if and only if

$$\overline{\Delta}\,\mathbf{H} - \sum_{i=1}^{m} R^N(\mathbf{H}, d\phi(e_i))d\phi(e_i) = 0. \qquad (2.8)$$

3. The necessary and sufficient conditions for biharmonic Lagrangian submanifolds in Kähler manifolds

In this section, we give the necessary and sufficient conditions for a Lagrangian submanifold in a Kähler manifold to be biharmonic.

Let us recall some fundamental facts on Lagrangian submanifolds in Kähler manifolds following Chen and Ogiue [**30**].

Let $(N^m, J, \langle \cdot, \cdot \rangle)$ be a Kähler manifold of complex dimension m, where J is the complex structure and $\langle \cdot, \cdot \rangle$ denotes the Kähler metric, which satisfies that $\langle JU, JV \rangle = \langle U, V \rangle$ and $d\Phi = 0$, where $\Phi(U, V) = \langle U, JV \rangle$, $(U, V \in \mathfrak{X}(N))$ is the fundamental 2-form. Let (M^m, g) be a Lagrangian submanifold in $(N^m, J, \langle \cdot, \cdot \rangle)$, that is, for all $x \in M$, $J(T_x M) \subset T_x M^\perp$, where $T_x M^\perp$ is the normal space at x. Then, it is well known that the following three equations hold:

$$\nabla_X^\perp JY = J(\nabla_X Y), \tag{3.1}$$

$$R^N(JX, JY) = R^N(X, Y), \tag{3.2}$$

for all $X, Y \in \mathfrak{X}(M)$, and

$$R^N(U, V) \cdot J = J \cdot R^N(U, V), \tag{3.3}$$

for all $U, V \in \mathfrak{X}(N)$.

To obtain the biharmonic equations for a Lagrangian submanifold in a Kähler manifold, we need the following lemma.

LEMMA 3.1. *Let* $\phi : (M, g) \to (N, \langle \cdot, \cdot \rangle)$ *be an isometric immersion of* (M, g) *into* $(N, \langle \cdot, \cdot \rangle)$. *Then it is biharmonic if and only if*

$$\mathrm{trace}_g \left(\nabla A_\mathbf{H} \right) + \mathrm{trace}_g \left(A_{\nabla_\bullet^\perp \mathbf{H}}(\bullet) \right) - \left(\sum_{i=1}^m R^N(\mathbf{H}, e_i) e_i \right)^T = 0, \tag{3.4}$$

$$\Delta^\perp \mathbf{H} + \mathrm{trace}_g B(A_\mathbf{H}(\bullet), \bullet) - \left(\sum_{i=1}^m R^N(\mathbf{H}, e_i) e_i \right)^\perp = 0, \tag{3.5}$$

where $(\cdot)^T$ *is the tangential part and* $(\cdot)^\perp$ *is the normal part.*

Proof. Due to (2.3), we have

$$\overline{\nabla}_X \mathbf{H} = -A_\mathbf{H}(X) + \nabla_X^\perp \mathbf{H},$$

and

$$\overline{\nabla}_Y \overline{\nabla}_X \mathbf{H} = -\nabla_Y A_\mathbf{H}(X) - B(Y, A_\mathbf{H}(X)) + \nabla_Y^\perp \nabla_X^\perp \mathbf{H} - A_{\nabla_X^\perp \mathbf{H}} Y,$$

for all $X, Y \in \mathfrak{X}(M)$.

Thus, we have

$$\overline{\Delta}\,\mathbf{H} = -\sum_{i=1}^{m}\Big\{-\nabla_{e_i}A_{\mathbf{H}}(e_i) - B(e_i, A_{\mathbf{H}}(e_i)) + \nabla_{e_i}^{\perp}\nabla_{e_i}^{\perp}\mathbf{H}$$

$$-A_{\nabla_{e_i}^{\perp}\mathbf{H}}(e_i) + A_{\mathbf{H}}(\nabla_{e_i}e_i) - \nabla_{\nabla_{e_i}e_i}^{\perp}\mathbf{H}\Big\}.$$

Dividing this into the tangential and the normal parts separately, we obtain Lemma 3.1.   □

By using Lemma 3.1, we obtain the following theorem.

THEOREM 3.1. *Let* $(N^m, J, \langle\cdot,\cdot\rangle)$ *be a Kähler manifold of complex dimension* m. *Assume that* $\phi : (M^m, g) \to (N^m, J, \langle\cdot,\cdot\rangle)$ *is a Lagrangian submanifold. Then* ϕ *is biharmonic if and only if the following two equations hold:*

$$\mathrm{trace}_g\,(\nabla A_{\mathbf{H}}) + \mathrm{trace}_g\left(A_{\nabla_{\bullet}^{\perp}\mathbf{H}}(\bullet)\right)$$

$$-\sum_{i=1}^{m}\Big\langle \mathrm{trace}_g\left(\nabla_{e_i}^{\perp}B\right) - \mathrm{trace}_g\left(\nabla_{\bullet}^{\perp}B\right)(e_i, \bullet), \mathbf{H}\Big\rangle e_i = 0,$$
$$(3.6)$$

$$\Delta^{\perp}\mathbf{H} + \mathrm{trace}_g B\left(A_{\mathbf{H}}(\bullet), \bullet\right) + \sum_{i=1}^{m}\mathrm{Ric}^N(J\mathbf{H}, e_i)Je_i$$

$$-\sum_{i=1}^{m}\mathrm{Ric}(J\mathbf{H}, e_i)Je_i - J\,\mathrm{trace}_g A_{B(J\mathbf{H},\bullet)}(\bullet) + mJA_{\mathbf{H}}(J\mathbf{H}) = 0,$$
$$(3.7)$$

where Ric *and* Ric^N *are the Ricci tensor of* (M^m, g) *and* $(N^m, \langle\cdot,\cdot\rangle)$ *respectively.*

Here, the trace, $\mathrm{trace}_g(A_{\nabla_{\bullet}^{\perp}\mathbf{H}(\bullet)})$ stands for $\sum_{i=1}^{m} A_{\nabla_{\bullet}^{\perp}\mathbf{H}(\bullet)}$, and so on.

Proof First note that due to (3.1), the harmonic mean curvature vector field $\mathbf{H}$ can be written as $\mathbf{H} = JZ$ for some vector field Z on M. By using (3.2) and (3.3), we obtain

$$\sum_{i=1}^{m}\Big\langle R^N(\mathbf{H}, e_i)e_i, JX\Big\rangle = \sum_{i=1}^{m}\Big\langle R^N(Z, Je_i)Je_i, X\Big\rangle,$$

which implies that

$$\sum_{i=1}^{m}\Big\langle R^N(\mathbf{H}, e_i)e_i, JX\Big\rangle + \sum_{i=1}^{m}\Big\langle R^N(Z, e_i)e_i, X\Big\rangle = \mathrm{Ric}^N(Z, X).$$

By the Gauss equation (2.5), we have

$$\sum_{i=1}^{m} \left\langle R^N(\mathbf{H}, e_i)e_i, JX \right\rangle$$

$$= \mathrm{Ric}^N(Z, X)$$

$$- \left\{ \sum_{i=1}^{m} \langle R(Z, e_i)e_i, X \rangle + \sum_{i=1}^{m} \langle B(Z, e_i), B(e_i, X) \rangle \right.$$

$$\left. - \sum_{i=1}^{m} \langle B(e_i, e_i), B(Z, X) \rangle \right\}$$

$$= \mathrm{Ric}^N(Z, X) - \mathrm{Ric}(Z, X) - \sum_{i=1}^{m} \langle B(Z, e_i), B(e_i, X) \rangle$$

$$+ m \langle \mathbf{H}, B(J\mathbf{H}, X) \rangle$$

$$= - \mathrm{Ric}^N(J\mathbf{H}, X) + \mathrm{Ric}(J\mathbf{H}, X)$$

$$+ \sum_{i=1}^{m} \langle B(J\mathbf{H}, e_i), B(e_i, X) \rangle - m \langle \mathbf{H}, B(J\mathbf{H}, X) \rangle. \qquad (3.8)$$

From (3.8), we have

$$\left(\sum_{i=1}^{m} R^N(\mathbf{H}, e_i)e_i \right)^{\perp}$$

$$= \sum_{j=1}^{m} \left\langle \sum_{i=1}^{m} R^N(\mathbf{H}, e_i)e_i, Je_j \right\rangle Je_j$$

$$= - \sum_{j=1}^{m} Ric^N(J\mathbf{H}, e_j)Je_j + \sum_{j=1}^{m} Ric(J\mathbf{H}, e_j)Je_j$$

$$+ \sum_{i,j=1}^{m} \langle B(J\mathbf{H}, e_i), B(e_i, e_j) \rangle Je_j - \sum_{j=1}^{m} m \langle \mathbf{H}, B(J\mathbf{H}, e_j) \rangle Je_j. \qquad (3.9)$$

By (2.4), we have

$$\sum_{i,j=1}^{m} \langle B(J\mathbf{H}, e_i), B(e_i, e_j) \rangle Je_j = \sum_{i,j=1}^{m} \left\langle A_{B(J\mathbf{H}, e_i)}(e_i), e_j \right\rangle Je_j$$

$$= \sum_{i,j=1}^{m} \left\langle JA_{B(J\mathbf{H}, e_i)}(e_i), Je_j \right\rangle Je_j$$

$$= \sum_{i=1}^{m} JA_{B(J\mathbf{H}, e_i)}(e_i)$$

$$= J \, \mathrm{trace}_g A_{B(J\mathbf{H}, \bullet)}(\bullet), \qquad (3.10)$$

and

$$\sum_{j=1}^{m}\langle \mathbf{H}, B(J\mathbf{H}, e_j)\rangle Je_j = \sum_{j=1}^{m}\langle A_{\mathbf{H}}(J\mathbf{H}), e_j\rangle Je_j$$

$$= \sum_{j=1}^{m}\langle JA_{\mathbf{H}}(J\mathbf{H}), Je_j\rangle Je_j$$

$$= JA_{\mathbf{H}}(J\mathbf{H}). \tag{3.11}$$

Combining (3.9)–(3.11), we obtain

$$\left(\sum_{i=1}^{m} R^N(\mathbf{H}, e_i)e_i\right)^{\perp} = -\sum_{j=1}^{m} Ric^N(J\mathbf{H}, e_j)Je_j + \sum_{j=1}^{m} Ric(J\mathbf{H}, e_j)Je_j$$

$$+ J \operatorname{trace}_g A_{B(J\mathbf{H},\bullet)}(\bullet) - mJA_{\mathbf{H}}(J\mathbf{H}). \tag{3.12}$$

By (2.5), we have

$$\left(\sum_{i=1}^{m} R^N(\mathbf{H}, e_i)e_i\right)^{T} = \sum_{i,j=1}^{m} \left\langle R^N(\mathbf{H}, e_i)e_i, e_j\right\rangle e_j$$

$$= \sum_{i,j=1}^{m} \left\langle \left(\nabla_{e_j}^{\perp} B\right)(e_i, e_i) - \left(\nabla_{e_i}^{\perp} B\right)(e_j, e_i), \mathbf{H}\right\rangle e_j. \tag{3.13}$$

Applying Lemma 3.1, we obtain the theorem.   □

4. Biharmonic Lagrangian submanifolds in complex space forms

In this section, we give the necessary and sufficient conditions for Lagrangian submanifolds in complex space forms to be biharmonic.

Let $N = N^m(4\varepsilon)$ be the simply connected complex m-dimensional complex space form of constant holomorphic sectional curvature 4ε. The curvature tensor R^N of $N^m(4\varepsilon)$ is given by

$$R^N(U, V)W$$
$$= \varepsilon\{\langle V, W\rangle U - \langle U, W\rangle V + \langle W, JV\rangle JU - \langle W, JU\rangle JV + 2\langle U, JV\rangle JW\}, \tag{4.1}$$

for U, V, $W \in \mathfrak{X}(N)$, where $\langle \cdot, \cdot \rangle$ is the Riemannian metric on $N^m(4\varepsilon)$ and J is the complex structure of $N^m(4\varepsilon)$. The complex space from $N^m(4\varepsilon)$ is the complex projective space $\mathbb{CP}^m(4\varepsilon)$, the complex Euclidean space $\mathbb{C}^m$ or the complex hyperbolic space $\mathbb{CH}^m(4\epsilon)$ according to $\varepsilon > 0, \varepsilon = 0$ or $\varepsilon < 0$.

By using Lemma 3.1, we obtain the following proposition which will be used in the next section.

PROPOSITION 4.1. *Let $(N^m(4\varepsilon), J, \langle \cdot, \cdot \rangle)$ be a complex space form of complex dimension m. Assume that $\phi : (M^m, g) \to (N^m(4\varepsilon), J, \langle \cdot, \cdot \rangle)$ is a Lagrangian submanifold. Then ϕ is biharmonic if and only if*

$$\operatorname{trace}_g (\nabla A_{\mathbf{H}}) + \operatorname{trace}_g \left(A_{\nabla^{\perp}_{\bullet} \mathbf{H}}(\bullet) \right) = 0, \qquad (4.2)$$

$$\Delta^{\perp} \mathbf{H} + \operatorname{trace}_g B \left(A_{\mathbf{H}}(\bullet), \bullet \right) - (m + 3)\varepsilon \mathbf{H} = 0. \qquad (4.3)$$

Proof By (4.1), we have

$$\sum_{i=1}^{m} R^N (\mathbf{H}, d\phi(e_i)) d\phi(e_i) = \varepsilon \sum_{i=1}^{m} \{ \langle d\phi(e_i), d\phi(e_i) \rangle \mathbf{H} - \langle d\phi(e_i), \mathbf{H} \rangle d\phi(e_i)$$

$$+ \langle d\phi(e_i), J d\phi(e_i) \rangle J\mathbf{H} - \langle d\phi(e_i), J\mathbf{H} \rangle J d\phi(e_i)$$

$$+ 2 \langle \mathbf{H}, J d\phi(e_i) \rangle J\phi(e_i) \}$$

$$= \left\{ m\mathbf{H} + \sum_{i=1}^{m} \langle \mathbf{H}, J d\phi(e_i) \rangle J d\phi(e_i) + 2\mathbf{H} \right\}$$

$$= (m + 3)\varepsilon \mathbf{H}.$$

By using this and Lemma 3.1, we conclude the proof. $\square$

REMARK 4.1. *Proposition 4.1 was previously obtained in* [**49**].

5. Biharmonic Lagrangian H-umbilical submanifolds in complex space forms

In this section, we classify biharmonic PNMC (see Definition 5.3) Lagrangian H-umbilical submanifolds in complex space forms.

Chen introduced the notion of Lagrangian H-umbilical submanifolds ([**25**]):

DEFINITION 5.1. ([**25**]) *If the second fundamental form of a Lagrangian submanifold M in a Kähler manifold takes the following form:*

$$\begin{cases} B(e_1, e_1) = \lambda J e_1, & B(e_i, e_i) = \mu J e_1, \\ B(e_1, e_i) = \mu J e_i, & B(e_i, e_j) = 0, \ (i \neq j), \quad i, j = 2, \cdots, m, \end{cases}$$

$$(5.1)$$

for suitable functions λ and μ with respect to a suitable orthonormal frame field $\{e_1, \cdots, e_m\}$ on M, then M is called a Lagrangian H-umbilical submanifold.

Lagrangian H-umbilical submanifolds are the simplest examples of Lagrangian submanifolds next to totally geodesic submanifolds. Since it is known that there are no totally umbilical Lagrangian submanifolds in the complex space forms $N^m(4\varepsilon)$ with $m \geq 2$, we should consider H-umbilical Lagrangian submanifolds.

In this case, the harmonic mean curvature vector $\mathbf{H}$ can be denoted by

$$\mathbf{H} = \frac{\lambda + (m-1)\mu}{m} J e_1.$$

Hereafter, we put $a = \frac{\lambda + (m-1)\mu}{m}$.

REMARK 5.1. *The class of Lagrangian H-umbilical submanifolds in the complex space forms includes the following interesting examples:*
(1) the Whitney's spheres in the complex Euclidean spaces (cf. [24]),
(2) twistor holomorphic Lagrangian surfaces in the complex projective planes (cf. [24]).
Furthermore, all Lagrangian H-umbilical submanifolds in the complex space forms were classified (cf. [24]–[26]).

B.Y. Chen also introduced PNMC submanifolds (cf. [6], [27]):

DEFINITION 5.2. *A submanifold M in a Riemannian manifold is said to have* parallel normalized mean curvature vector field *(PNMC) if it has nowhere zero mean curvature and the unit vector field in the direction of the mean curvature vector field is parallel in the normal bundle, i.e.,*

$$\nabla^\perp \left(\frac{\mathbf{H}}{|\mathbf{H}|} \right) = 0. \tag{5.2}$$

We denote as $\nabla_{e_i} e_j = \sum_{l=1}^{m} \omega_j^l(e_i) e_l \; (i, j = 1, \cdots, m)$. Then we obtain the following lemma.

LEMMA 5.1. ([25], [136]) *Let M^m be an m-dimensional Lagrangian H-umbilical submanifold in a complex space form. For an orthonormal* *frame field* $\{e_i\}_{i=1}^m$, *we have*

$$e_j\lambda = (2\mu - \lambda)\omega_j^1(e_1), \quad j > 1, \tag{5.3}$$

$$e_1\mu = (\lambda - 2\mu)\omega_1^l(e_l), \quad \textit{for all } l = 2, \cdots m, \tag{5.4}$$

$$(\lambda - 2\mu)\omega_1^i(e_j) = 0, \quad i \neq j > 1, \tag{5.5}$$

$$e_j\mu = 0, \quad j > 1, \tag{5.6}$$

$$\mu\,\omega_1^j(e_1) = 0 \tag{5.7}$$

$$\mu\,\omega_1^2(e_2) = \cdots = \mu\,\omega_1^m(e_m), \tag{5.8}$$

$$\mu\,\omega_1^i(e_j) = 0, \quad i \neq j > 1. \tag{5.9}$$

Proof By $(\nabla_{e_j}^\perp B)(e_1, e_1) = (\nabla_{e_1}^\perp B)(e_j, e_1)$ and (5.1), we obtain $(5.3) - (5.5)$. By $(\nabla_{e_1}^\perp B)(e_j, e_j) = (\nabla_{e_j}^\perp B)(e_1, e_j)$ and (5.1), we obtain (5.6) and (5.7). By $(\nabla_{e_i}^\perp B)(e_j, e_j) = (\nabla_{e_j}^\perp B)(e_i, e_i)$, $(i \neq j > 1)$, and (5.1), we obtain (5.8) and (5.9). □

By using Lemma 5.4, T. Sasahara showed the following (cf. [**136**]):

THEOREM 5.1. *Let* $(N^m(4\varepsilon), J, \langle\cdot, \cdot\rangle)$ *be a complex space form of complex dimension* m, *where* $\varepsilon \in \{-1, 0, 1\}$. *Assume that* $\phi : (M^m, g) \to (N^m(4\varepsilon), J, \langle\cdot, \cdot\rangle)$ *is a biharmonic Lagrangian* H-*umbilical submanifold. Then, the mean curvature of* M^m *is non-zero constant if and only if* $\varepsilon = 1$ *and* $\phi(M)$ *is congruent to an* m-*dimensional submanifold of* $\mathbb{CP}^m(4)$ *given by*

$$\pi\left(\sqrt{\frac{\mu^2}{\mu^2 + 1}}e^{-\frac{i}{\mu}x}, \sqrt{\frac{1}{\mu^2 + 1}}e^{i\mu x}y_1, \cdots, \sqrt{\frac{1}{\mu^2 + 1}}e^{i\mu x}y_m\right) \subset \mathbb{CP}^m(4), \tag{5.10}$$

where $x, y_1, \cdots, y_m$ *are real numbers satisfying* $y_1^2 + \cdots + y_m^2 = 1$ *and* $\mu = \pm\sqrt{\frac{m+5\pm\sqrt{m^2+6m+25}}{2m}}$.

Due to this theorem, we shall classify biharmonic PNMC Lagrangian H-umbilical submanifolds in complex space forms. We shall show the necessary and sufficient conditions for Lagrangian H-umbilical submanifolds in the complex space forms to be biharmonic.

PROPOSITION 5.1. *Let* (M^m, g) *be a Lagrangian* H-*umbilical submanifold in the complex space form* $(N^m(4\varepsilon), J, \langle\cdot, \cdot\rangle)$. *Then,* M^m *is biharmonic if and only if*

$$2\lambda(e_1 a) + a(e_1\lambda) + \lambda a\sum_{l=2}^m \omega_1^l(e_l) = 0, \tag{5.11}$$

$$2\mu(e_j a) + a\lambda\omega_1^j(e_1) = 0, \quad j > 1, \tag{5.12}$$

$$-\sum_{i=1}^{m} e_i(e_i a) + a \sum_{i,j=1}^{m} \omega_1^j(e_i)^2 + \sum_{i,j=1}^{m} (e_j a)\, \omega_i^j(e_i)$$

$$+ a\left\{\lambda^2 + (m-1)\mu^2 - \varepsilon(m+3)\right\} = 0, \tag{5.13}$$

$$-2\sum_{i=1}^{m} (e_i a)\omega_1^j(e_i) - a\sum_{i=1}^{m} e_i\left(\omega_1^j(e_i)\right)$$

$$- a\sum_{i,l=1}^{m} \omega_1^l(e_i)\omega_i^j(e_i) + a\sum_{i,l=1}^{m} \omega_i^l(e_i)\omega_i^j(e_l) = 0, \quad j > 1. \tag{5.14}$$

Proof   We shall calculate the tangential part (4.2). By using Lemma 5.4, we have

$$\operatorname{trace}_g\left(A_{\nabla_\bullet^\perp \mathbf{H}}(\bullet)\right) = \sum_{i=1}^{m} A_{\nabla_{e_i}^\perp a J e_1} e_i$$

$$= \sum_{i=1}^{m} (e_i a) A_{J e_1} e_i + a \sum_{i,l=1}^{m} \omega_1^l(e_i) A_{J e_l} e_i$$

$$= \lambda \sum_{i=1}^{m} (e_i a) e_1 + \mu \sum_{i=2}^{m} (e_i a) e_i$$

$$+ a\mu \sum_{l=2}^{m} \omega_1^l(e_1) e_l + a\mu \sum_{l=2}^{m} \omega_1^l(e_l) e_1$$

$$= \lambda \sum_{i=1}^{m} (e_i a) e_1 + \mu \sum_{i=2}^{m} (e_i a) e_i + a\mu \sum_{l=2}^{m} \omega_1^l(e_l) e_1, \tag{5.15}$$

and

$$\operatorname{trace}_g\left(\nabla A_{\mathbf{H}}\right) = \sum_{i=1}^{m} \nabla_{e_i}\left(A_{\mathbf{H}} e_i\right) - \sum_{i=1}^{m} A_{\mathbf{H}}\left(\nabla_{e_i} e_i\right)$$

$$= \sum_{i=1}^{m} \nabla_{e_i}(A_{a J e_1} e_i) - a \sum_{i,l=1}^{m} A_{J e_1}\left(\omega_i^l(e_i) e_l\right)$$

$$= \sum_{i=1}^{m} \left\{(e_i a) A_{J e_1} e_i + a\nabla_{e_i}(A_{J e_1} e_i)\right\} - a \sum_{i,l=1}^{m} \omega_i^l(e_i) A_{J e_1} e_l$$

$$= \lambda(e_1 a) e_1 + a(e_1 \lambda) e_1 + a\lambda \sum_{l=1}^{m} \omega_1^l(e_1) e_l$$

$$+ \sum_{i=2}^{m} \left\{\mu(e_i a) e_i + a(e_i \mu) e_i + a\mu \sum_{l=1}^{m} \omega_i^l(e_i) e_l\right\}$$

$$- a\lambda \sum_{l=1}^{m} \omega_l^1(e_l) e_1 - a\mu \sum_{l=1}^{m}\sum_{i=2}^{m} \omega_l^i(e_l) e_i. \tag{5.16}$$

By (5.15) and (5.16), we obtain

$$
\text{trace}_g \left(\nabla A_{\mathbf{H}} \right) + \text{trace}_g \left(A_{\nabla_\bullet^\perp}(\bullet) \right) = \left\{ 2\lambda(e_1 a) + a(e_1 \lambda) + a\lambda \sum_{l=2}^{m} \omega_1^l(e_l) \right\} e_1
$$
$$
+ \sum_{j=2}^{m} \left\{ 2\mu(e_j a) + a\lambda \omega_1^j(e_1) \right\} e_j, \tag{5.17}
$$

which yields (5.11) and (5.12).

We shall calculate the normal part (4.3). By using Lemma 5.4, we have

$$
\Delta^\perp \mathbf{H} = - \sum_{i=1}^{m} \nabla_{e_i}^\perp \nabla_{e_i}^\perp (aJe_1) + \sum_{i=1}^{m} \nabla_{\nabla_{e_i} e_i}^\perp (aJe_1)
$$
$$
= - \sum_{i=1}^{m} (e_i e_i a) Je_1 - \sum_{i,j=1}^{m} \left\{ 2(e_i a)\omega_1^j(e_i)Je_j + ae_i \left(\omega_1^j(e_j)Je_l \right) \right\}
$$
$$
- a \sum_{i,j,l=1}^{m} \omega_1^j(e_i)\omega_j^l(e_i)Je_l + \sum_{i,j=1}^{m} \omega_1^j(e_i)(e_j a)Je_1
$$
$$
+ a \sum_{i,j,l=1}^{m} \omega_i^j(e_i)\omega_1^l(e_j)Je_l, \tag{5.18}
$$

and

$$
\text{trace}_g B \left(A_{\mathbf{H}}(\bullet), \bullet \right) = a \left\{ \lambda^2 + (m-1)\mu^2 \right\} Je_1. \tag{5.19}
$$

By (5.18) and (5.19), we obtain

$$
\Delta^\perp \mathbf{H} + \text{trace}_g B \left(A_{\mathbf{H}}(\bullet), \bullet \right) - (m+3)\varepsilon \mathbf{H}
$$
$$
= \left\{ - \sum_{i=1}^{m} e_i e_i a + a \sum_{i,j=1}^{m} \omega_1^j(e_i)^2 + \sum_{i,j=1}^{m} (e_j a)\omega_i^j(e_i) \right.
$$
$$
\left. + a \left\{ \lambda^2 + (m-1)\mu^2 - \varepsilon(m+3) \right\} \right\} Je_1
$$
$$
+ \sum_{j=2}^{m} \left\{ - 2\sum_{i=1}^{m} (e_i a)\omega_1^j(e_i) - a \sum_{i=1}^{m} e_i \left(\omega_1^j(e_i) \right) \right.
$$
$$
\left. - a \sum_{i,l=1}^{m} \omega_1^l(e_i)\omega_l^j(e_i) + a \sum_{i,l=1}^{m} \omega_i^l(e_i)\omega_1^j(e_l) \right\} Je_j,
$$

which yields (5.13) and (5.14). □

From Proposition 5.6, we obtain the following proposition.

PROPOSITION 5.2. *Let (M^m, g) be a Lagrangian H-umbilical submanifold in the complex space form $(N^m(4\varepsilon), J, \langle \cdot, \cdot \rangle)$. Then, M^m is (non-harmonic) biharmonic if and only if $\mu \neq 0$ and*

$$2\lambda (e_1 a) + a (e_1 \lambda) + a\lambda(m-1)k = 0, \tag{5.20}$$

$$e_j a = 0, \quad j > 1, \tag{5.21}$$

$$- e_1(e_1 a) + a(m-1)k^2 - (e_1 a)(m-1)k$$
$$+ a\left\{\lambda^2 + (m-1)\mu^2 - \varepsilon(m+3)\right\} = 0, \tag{5.22}$$

$$e_j k = 0, \quad j > 1, \tag{5.23}$$

where, $k = \omega_1^2(e_2) = \cdots = \omega_1^m(e_m)$.

Proof   We shall prove that $\mu \neq 0$. If $\mu = 0$, then $a = \frac{1}{m}\lambda \neq 0$. By Lemma 5.4, we have

$$\omega_1^i(e_j) = 0, \quad j = 2, \cdots, m. \tag{5.24}$$

From (5.11) and (5.24), $e_1 a = 0$. From (5.11), we obtain

$$\omega_1^j(e_1) = 0, \quad j = 1, \cdots, m. \tag{5.25}$$

Combining (5.24) and (5.25), we have

$$\omega_1^i(e_j) = 0, \quad i, j = 1, \cdots, m. \tag{5.26}$$

It follows that $\langle R(e_1, e_i)e_i, e_1 \rangle = 0$. Thus, by (2.5), we have $\varepsilon = 0$. By (5.3), we have $e_j a = 0$, $(j > 1)$. From these and (5.13), we obtain $a = 0$, which contradicts the assumption.

We only have to consider the case of $\mu \neq 0$. Then, we have

$$\omega_1^i(e_j) = 0, \quad i \neq j, \tag{5.27}$$

$$\omega_1^2(e_2) = \cdots = \omega_1^m(e_m). \tag{5.28}$$

We put $k = \omega_1^2(e_2) = \cdots = \omega_1^m(e_m)$.

By using (5.28), we see that equation (5.11) is (5.20). Putting (5.27) into (5.12), we obtain (5.21). From (5.21) and (5.13), we have (5.22). Putting (5.21) into (5.14), we have

$$-a\sum_{i=1}^{m} e_i\left(\omega_i^j(e_i)\right) - a\sum_{i,l=1}^{m} \omega_1^l(e_i)\omega_l^j(e_i) + a\sum_{i,l=1}^{m} \omega_i^l(e_i)\omega_1^j(e_l) = 0.$$

From this and (5.27), we obtain (5.23).   □

By using Theorem 5.5, we shall classify all the biharmonic PNMC Lagrangian H-umbilical submanifolds in the complex space forms.

THEOREM 5.2. *Let* $(N^m(4\varepsilon), J, \langle \cdot, \cdot \rangle)$ *be a complex space form of complex dimension* m, *where* $\varepsilon \in \{-1, 0, 1\}$. *Assume that* $\phi : (M^m, g) \to (N^m(4\varepsilon), J, \langle \cdot, \cdot \rangle)$ *is a Lagrangian H-umbilical submanifold which has PNMC. Then,* ϕ *is biharmonic if and only if* $\varepsilon = 1$ *and* $\phi(M)$ *is congruent to an* m-*dimensional submanifold of* $\mathbb{CP}^m(4)$ *given by*

$$\pi \left(\sqrt{\frac{\mu^2}{\mu^2+1}} e^{-\frac{i}{\mu}x}, \sqrt{\frac{1}{\mu^2+1}} e^{i\mu x} y_1, \cdots, \sqrt{\frac{1}{\mu^2+1}} e^{i\mu x} y_m \right) \subset \mathbb{CP}^m(4), \tag{5.29}$$

where $x, y_1, \cdots, y_m$ *are real numbers satisfying* $y_1{}^2 + \cdots + y_m{}^2 = 1$ *and* $\mu = \pm\sqrt{\frac{m+5\pm\sqrt{m^2+6m+25}}{2m}}$.

REMARK 5.2. *The biharmonic immersion in* $\mathbb{CP}^m$, *given by* (5.2) *has parallel mean curvature vector field i.e.,* $\nabla^\perp \mathbf{H} = 0$.

Proof By the assumption

$$\nabla^\perp \left(\frac{\mathbf{H}}{|\mathbf{H}|} \right) = \nabla^\perp \left(\frac{aJe_1}{|a|} \right) = 0,$$

and $a \neq 0$, we have

$$J(\nabla e_1) = \nabla^\perp Je_1 = 0. \tag{5.30}$$

Thus, we obtain

$$0 = \nabla_{e_i} e_1 = \sum_{l=1}^m \omega_1^l(e_i) e_l, \quad (i = 1, \cdots, m), \tag{5.31}$$

which implies that

$$\omega_1^l(e_i) = 0, \quad (i, l = 1, \cdots, m). \tag{5.32}$$

In particular, we have

$$k = \omega_1^2(e_2) = \cdots = \omega_1^m(e_m) = 0. \tag{5.33}$$

By (4.3), we obtain

$$e_1\mu = 0. \tag{5.34}$$

Combining this and (4.5), μ is a constant. Since $\langle R(e_i, e_1)e_1, e_i \rangle = 0$, we have

$$\mu^2 - \lambda\mu = \varepsilon. \tag{5.35}$$

Thus, λ is a constant. Therefore, $a = \frac{\lambda+(m-1)\mu}{m}$ is a non-zero constant.

By using Theorem 5.5, we obtain (5.29).

Conversely, by a direct computation, it turns out that the immersion (5.29) is a biharmonic PNMC Lagrangian immersion. □

REMARK 5.3. (1) *We can not answer whether the same conclusion of Theorem 5.8 holds without the assumption PNMC.*

(2) *If $\mu_+ = \mu_0$, and $\mu_- = -\mu_0$, where $\mu_0 = \sqrt{\frac{m+5\pm\sqrt{m^2+6m+25}}{2m}}$, then it seems that the corresponding submanifolds to μ_+ and μ_- are isometric each other.*

6. Biharmonic PNMC surfaces

In this section, we classify all the biharmonic PNMC Lagrangian surfaces in a two-dimensional complex space forms $(N^2(4\varepsilon), J, \langle \cdot, \cdot \rangle)$.

Let $\phi : M^2 \to (N^2(4\varepsilon), J, \langle \cdot, \cdot \rangle)$ be a Lagrangian surface. Let $\{e_1, e_2\}$ be an orthonormal frame field on M^2 such that Je_1 is parallel to $\mathbf{H}$. Then, the second fundamental form takes the form:

$$
\begin{aligned}
B(e_1, e_1) &= (a - b)Je_1 + cJe_2, \\
B(e_1, e_2) &= cJe_1 + bJe_2, \\
B(e_2, e_2) &= bJe_1 - cJe_2,
\end{aligned}
\tag{6.1}
$$

for some functions $a(\neq 0)$, b and c. We put $\nabla_{e_1} e_1 = \alpha e_2$ and $\nabla_{e_2} e_1 = \beta e_2$ (then, we have $\nabla_{e_1} e_2 = -\alpha e_1$ and $\nabla_{e_2} e_2 = -\beta e_1$).

From these, we have

$$
(\nabla^\perp_{e_1} B)(e_2, e_2) = (e_1 b + 3c\alpha)Je_1 - (e_1 c - 3b\alpha)Je_2,
$$
$$
(\nabla^\perp_{e_2} B)(e_1, e_2) = \{e_2 c + (a - 3b)\beta\}Je_1 + (e_2 b + 3c\beta)Je_2,
$$
$$
(\nabla^\perp_{e_1} B)(e_1, e_2) = \{e_1 c + (a - 3b)\alpha\}Je_1 + (e_1 b + 3c\alpha)Je_2,
$$
$$
(\nabla^\perp_{e_2} B)(e_1, e_1) = \{e_2(a - b) - 3c\beta\}Je_1 + \{e_2 c + (a - 3b)\beta\}Je_2.
$$

By using (2.6), we obtain

$$
e_1 b + 3c\alpha = e_2 c + (a - 3b)\beta, 4.2.5 \tag{6.2}
$$
$$
-e_1 c + 3b\alpha = e_2 b + 3c\beta, 4.3.5 \tag{6.3}
$$
$$
e_2(a - b) - 3c\beta = e_1 c + (a - 3b)\alpha. 4.4.5 \tag{6.4}
$$

Combining (6.3) and (6.4) leads to

$$
e_2 a = a\alpha. \tag{6.5}
$$

From (2.5), we have

$$
ab - 2b^2 - 2c^2 + \varepsilon = -\alpha^2 - \beta^2 + e_2\alpha - e_1\beta. \tag{6.6}
$$

By using these results, we obtain the following proposition.

PROPOSITION 6.1. *Let* $(N^2(4\varepsilon), J, \langle \cdot, \cdot \rangle)$ *be a two-dimensional complex space form. Assume that* $\phi : M^2 \to (N^2(4\varepsilon), J, \langle \cdot, \cdot \rangle)$ *is a Lagrangian surface. Then* ϕ *is biharmonic if and only if*

$$3(e_1 a)a - 2(e_1 a)b + 4ac\alpha + 2ab\beta = 0, \tag{6.7}$$

$$2(e_1 a)c + 4a\alpha b + a^2\alpha - 2ac\beta = 0, \tag{6.8}$$

$$-e_1 e_1 a - \beta(e_1 a) + a(-5\epsilon + (a-b)^2 + b^2 + 2c^2 + \alpha^2 + \beta^2 - e_2\alpha) = 0, \tag{6.9}$$

$$2(e_1 a)\alpha + a(2\alpha\beta + e_1\alpha + e_2\beta - ac) = 0. \tag{6.10}$$

Proof We shall calculate the tangential part (4.2). Since $A_{Je_1}e_1 = (a-b)e_1 + ce_2$, $A_{Je_1}e_2 = A_{Je_2}e_1 = ce_1 + be_2$ and $A_{Je_2}e_2 = be_1 - ce_2$, we have

$$\text{trace}_g(A_{\nabla_\bullet^\perp \mathbf{H}}(\bullet)) = \sum_{i=1}^{2} A_{\nabla_{e_i}^\perp \mathbf{H}}(e_i)$$

$$= \frac{1}{2}\{(e_1 a)(a - b) + (e_2 a)c + a\alpha c + a\beta b\}e_1$$

$$+ \frac{1}{2}\{(e_1 a)c + (e_2 a)b + a\alpha b - a\beta c\}e_2,$$

and

$$\text{trace}_g(\nabla A_{\mathbf{H}}) = \sum_{i=1}^{2}(\nabla_{e_i} A_{\mathbf{H}})e_i = \sum_{i=1}^{2}\{\nabla_{e_i}(A_{\mathbf{H}}e_i) - A_{\mathbf{H}}(\nabla_{e_i}e_i)\}$$

$$= \frac{1}{2}\{(e_1 a)(a - b) + a(e_1 a) - a(e_1 b) - a\alpha c$$

$$+ (e_2 a)c + a(e_2 c) - ab\beta - a\alpha c + a\beta(a - b)\}e_1$$

$$+ \frac{1}{2}\{(e_1 a)c + a(a-b)\alpha + a(e_1 c) + (e_2 a)b + a(e_2 b) + ac\beta - a\alpha b + a\beta c\}e_2.$$

By using (6.2)–(6.5), we obtain (6.7) and (6.8).

We shall calculate the normal part (4.3). We have

$$\Delta^\perp \mathbf{H} = -\frac{1}{2}\left\{e_1(e_1 a)Je_1 + 2(e_1 a)\alpha Je_2 + a(e_1\alpha)Je_2 - a\alpha^2 Je_1 - \alpha(e_2 a)Je_1\right\}$$

$$- \frac{1}{2}\left\{e_2(e_2 a)Je_1 + 2(e_2 a)\beta Je_2 + a(e_2\beta)Je_2 - a\beta^2 Je_1 + \beta(e_1 a)Je_1\right\},$$

and

$$\sum_{i=1}^{2} B(A_{\mathbf{H}}e_i, e_i) = \frac{1}{2}a\left\{(a-b)^2 Je_1 + (a-b)cJe_2 + 2c^2 Je_1 + bcJe_2 + b^2 Je_1\right\}.$$

By using (6.5), we obtain (6.9) and (6.10).   □

We shall classify all the biharmonic PNMC Lagrangian surfaces in the 2-dimensional complex space forms.

THEOREM 6.1. *Let* $(N^2(4\varepsilon), J, \langle \cdot, \cdot \rangle)$ *be a 2-dimensional complex space form. Assume that* $\phi : M^2 \to (N^2(4\varepsilon), J, \langle \cdot, \cdot \rangle)$ *is a biharmonic Lagrangian surface. Then, the following properties are equivalent.*
(1) *the mean curvature is a non-zero constant.*
(2) M^2 *has PNMC.*
Moreover, if the biharmonic Lagrangian surface satisfies (1) *or* (2), *then we have* $\epsilon > 0$ *and* M^2 *is a H-umbilical surface. If* $\varepsilon = 1$, ϕ *is locally given by*

$\phi(x, y) =$

$$\pi \left(\sqrt{\frac{b^2}{b^2 + 1}} e^{-\frac{i}{b}x}, \sqrt{\frac{1}{b^2 + 1}} e^{ibx} \cos \sqrt{b^2 + 1}y, \sqrt{\frac{1}{b^2 + 1}} e^{ibx} \sin \sqrt{b^2 + 1}y \right),$$

(6.11)

where $b = \pm \frac{\sqrt{7 \pm \sqrt{41}}}{2}$.

Proof We shall show (1) $\Rightarrow$ (2). Since the mean curvature a is a non-zero constant, from (6.5), we have $\alpha = 0$. From these, we have that (6.7) is $b\beta = 0$, and (6.8) is $c\beta = 0$. If $\beta \neq 0$, then we have $b = c = 0$. But, from (6.2), we have $\beta = 0$. This contradicts our assumption $\beta \neq 0$. Therefore, we obtain $\beta = 0$, which means that M^2 has PNMC.

We shall show that (2) $\Rightarrow$ (1). By the assumption

$$\nabla^{\perp} \left(\frac{\mathbf{H}}{|\mathbf{H}|} \right) = \nabla^{\perp} \left(\frac{aJe_1}{|a|} \right) = 0,$$

and since $a \neq 0$, we have

$$J(\nabla e_1) = \nabla^{\perp} Je_1 = 0.$$

(6.12)

Thus, we obtain

$$0 = \nabla_{e_i} e_1 = \begin{cases} \alpha e_2 & (i = 1) \\ \beta e_2 & (i = 2), \end{cases}$$

(6.13)

which implies that $\alpha = 0$ and $\beta = 0$. From (6.7), we have that $e_1 a = 0$ or $3a - 2b = 0$.

The case $e_1 a = 0$. By (6.5), a is constant.

The case $3a - 2b = 0$. From (6.10), we have $c = 0$. It follows from $c = 0$, (6.2) and (6.3) that b is constant. By combining (6.6) and (6.9), we have $a = 0$. This contradicts our assumption $a \neq 0$.

Therefore a is constant.

If M^2 satisfies the condition (1) or (2), from (6.10), we have $c = 0$, i.e., M^2 is a $H-$umbilical surface. By using Theorem 5.8, we obtain the theorem. $\square$

Part 4

Further Developments on Biharmonic Maps

CHAPTER 14

Rigidity of Transversally Biharmonic Maps between Foliated Riemannian Manifolds

ABSTRACT. On a smooth foliated map from a complete, possibly non-compact, foliated Riemannian manifold into another foliated Riemannian manifold of which transversal sectional curvature is non-positive, we will show that, if it is transversally biharmonic and has the finite energy and finite bienergy, then it is transversally harmonic.

1. Introduction

Transversally biharmonic maps between two foliated Riemannian manifolds introduced by Chiang and Wolak (cf. [31]) are generalizations of transversally harmonic maps introduced by Konderak and Wolak (cf. [86], [87]).

Among smooth foliated maps φ between two Riemannian foliated manifolds, one can define the transversal energy and derive the Euler-Lagrange equation, and transversally harmonic map as its critical points, which are by definition the transversal tension field vanishes, $\tau_b(\varphi) \equiv 0$. The transverse bienergy can be also defined as $E_2(\varphi) = \frac{1}{2} \int_M |\tau_b(\varphi)|^2 \, v_g$ whose Euler-Lagrange equation is that the transversal bitension field $\tau_{2,b}(\varphi)$ vanishes and the transversally biharmonic maps which are, by definition, vanishing of the transverse bitension field.

Recently, S.D. Jung studied extensively the transversally harmonic maps and the transversally biharmonic maps on compact Riemannian foliated manifolds (cf. [76], [77], [79], [80]).

In this paper, we study transversally biharmonic maps of a complete (possibly non-compact) Riemannian foliated manifold $(M, g, \mathcal{F})$ into another Riemannian foliated manifold $(M', g', \mathcal{F}')$ of which transversal sectional curvature is non-positive. Then, we will show that:

---

[1]This chapter is due to [117]: S. Ohno, T. Sakai and H. Urakawa, *Rigidity of transversally biharmonic maps between foliated Riemannian manifolds*, accepted in Hokkaido Math. J., 2017.

THEOREM 1.1. *(cf. Theorem 2.11)   Let $(M, g, \mathcal{F})$ and $(M', g', \mathcal{F}')$ be two foliated Riemannian manifolds. Assume that the foliation $\mathcal{F}$ is transversally volume preserving (cf. Definition 2.1) and the transversal sectional curvature of $(M', g', \mathcal{F}')$ is non-positive. Let $\varphi : (M, g, \mathcal{F}) \to (M', g', \mathcal{F}')$ be a C^∞ foliated map satisfying the conservation law. If φ is transversally biharmonic with the finite transversal energy $E(\varphi) < \infty$ and finite transversal bienergy $E_2(\varphi) < \infty$, then it is transversally harmonic.*

This theorem can be regarded a natural analogue of B.Y. Chen's conjecture and the generalized Chen's conjecture (cf. [21], [73]).

B. Y. Chen's conjecture: *Every biharmonic submanifolds of the Euclidean space $\mathbb{R}^n$ must be harmonic (minimal).*

The generalized B. Y. Chen's conjecture: *Every biharmonic submanifolds of a Riemannian manifold of non-positive curvature must be harmonic (minimal).*

Several authors has contributed to give partial answers to solve these problems (cf. [1], [34], [57], [67], [63], [64], [108], [109], [111]). For the first and second variational formula of the bienergy, see [74]. For the CR analogue of biharmonic maps, see also [11], [37], [148].

Acknowledgement.   The last author would like to express his gratitude to Professor Seoung Dal Jung who invited him at Jeju National University at January, 2016, and noticed to the authors the errors in the manuscripts of [79] and [80]. This work has been started during the period of this period. Finally not the least, the authors express their thanks to the referee who pointed several errors in the first draft.

2. Preliminaries

We prepare the materials for the first and second variational formulas for the transversal energy of a smooth foliated map between two foliated Riemannian manifolds following [79], [80] and [167].

2.1. The Green's formula on a foliated Riemannian manifold.

Let $(M, g, \mathcal{F})$ be an $n(= p + q)$-dimensional foliated Riemannian manifold with foliation $\mathcal{F}$ of codimension q and a bundle-like Riemannian metric g with respect to $\mathcal{F}$ (cf. [145], [146]). Let TM be the tangent bundle of M, L, the tangent bundle of $\mathcal{F}$, and $Q = TM/L$, the corresponding normal bundle of $\mathcal{F}$. We denote g_Q the induced Riemannian metric on the normal bundle Q, and ∇^Q, the transversal

Levi-Civita connection on Q, R^Q, the transversal curvature tensor, and K^Q, the transversal sectional curvature, respectively. Notice that the bundle projection $\pi : TM \to Q$ is an element of the space $\Omega^1(M, Q)$ of Q-valued 1-forms on M. Then, one can obtain the Q-valued bilinear form α on M, called the second fundamental form of $\mathcal{F}$, defined by

$$\alpha(X, Y) = -(D_X \pi)(Y) = \pi(\nabla_X^Q Y), \qquad (X, Y \in \Gamma(L)),$$

where D is the torsion free connection on the bundle Q (cf. [**167**], Page 240, Proposition 1. See also Definition of α, (6) in Page 241 of [**167**]). The trace τ of α, called the *tension field* of $\mathcal{F}$ is defined by

$$\tau = \sum_{i,j=1}^{p} g^{ij} \, \alpha(X_i, X_j),$$

where $\{X_i\}_{i=1}^p$ spanns $\Gamma(L|U)$ on a neighborhood U on M. The Green's theorem, due to Yorozu and Tanemura([**167**]), of a foliated Riemannian manifold $(M, g, \mathcal{F})$ says that

$$\int_M \operatorname{div}_D(\nu) \, v_g = \int_M g_Q(\tau, \nu) \, v_g \quad (\nu \in \Gamma(Q)), \qquad (2.1)$$

where $\operatorname{div}_D(\nu)$ denotes the *transversal divergence* of ν with respect to ∇^Q given by $\operatorname{div}_D(\nu) := \sum_{a,b=1}^q g^{ab} \, g_Q(D_{X_a}\nu, \pi(X_b))$. Here $\{X_a\}_{a=1}^q$ spanns $\Gamma(L^\perp|U)$ where $L^\perp$ is the orthogonal complement bundle of L with a natural identification $\sigma : Q \overset{\cong}{\to} L^\perp$.

DEFINITION 2.1. *A foliation $\mathcal{F}$ is* transversally volume preserving *if $div(\tau) = 0$.*

Let us recall Gaffney's theorem ([**52**], [**111**]):

THEOREM 2.1. *Let (M, g) be a non-compact complete Riemannian manifold without boundary, If a C^1 vector field X on M satisfies that*

$$\int_M |X| \, v_g < \infty \quad and \quad \int_M div(X) \, v_g < \infty. \qquad (2.2)$$

Then, it holds that

$$\int_M div(X) \, v_g = 0. \qquad (2.3)$$

Furthermore, if $f \in C^1(M)$ and a C^1 vector field X on M satisfy $\operatorname{div}(X) = 0$, $\int_M Xf \, v_g < \infty$, $\int_M |f|^2 \, v_g < \infty$ and $\int_M |X|^2 \, v_g < \infty$, then it holds that

$$\int_M Xf \, v_g = 0. \qquad (2.4)$$

For the sake of completeness, we give a proof of Theorem 2.2 in the appendix.

If $\mathcal{F}$ is transversally volume preserving, it holds by definition that

$$\int_M \tau\, g_Q(\nu, \nu)\, v_g = 0 \quad (\nu \in \Gamma(Q) \text{ with compact support}).$$
(2.5)

2.2. The first and second variational formulas. Let $(M, g, \mathcal{F})$, and $(M', g', \mathcal{F}')$ be two compact foliated Riemannian manifolds. The *transversal energy* $E(\varphi)$ among the totality of smooth foliated maps from $(M, g, \mathcal{F})$ into $(M', g', \mathcal{F}')$ by

$$E(\varphi) = \frac{1}{2} \int_M |d_T\varphi|^2\, v_g.$$
(2.6)

Here, a smooth map φ is a foliated map is, by definition, for every leaf ℓ of $\mathcal{F}$, there exists a leaf ℓ' of $\mathcal{F}'$ satisfying $\varphi(\ell) \subset \ell'$. Then, $d_T\varphi := \pi' \circ d\varphi \circ \sigma$; $Q \to Q'$ can be regarded as a section of $Q^* \otimes \varphi^{-1}Q'$ where Q^* is a subspace of the cotangent bundle T^*M. Here, π, π' are the projections of $TM \to Q = TM/L$ and $TM' \to Q' = TM'/L'$. Notice that our definition of the transversal energy is the same as the one of Jung's definition (cf. [**80**], p. 11, (3.4)).

The first variational formula is given (cf. [**79**], the case $f = 1$ in Theorem 4.1, (4.2)), for every smooth foliated variation $\{\varphi_t\}$ with $\varphi_0 = \varphi$ and $\frac{d\varphi_t}{dt}\big|_{t=0} = V$ in which V being a section $\varphi^{-1}Q'$,

$$\frac{d}{dt}\bigg|_{t=0} E(\varphi_t) = -\int_M \langle V, \tau_b(\varphi) - d_T\varphi(\tau)\rangle\, v_g.$$
(2.7)

Here, $\tau_b(\varphi)$ is the *transversal tension field* defined by

$$\tau_b(\varphi) = \sum_{a=1}^{q} (\widetilde{\nabla}_{E_a} d_T\varphi)(E_a),$$
(2.8)

where $\widetilde{\nabla}$ is the induced connection in $Q^* \otimes \varphi^{-1}Q'$ from the Levi-Civita connection of (M', g'), and $\{E_a\}_{a=1}^{q}$ is a locally defined orthonormal frame field on Q.

DEFINITION 2.2. *A smooth foliated map* $\varphi : (M, g, \mathcal{F}) \to (M', g', \mathcal{F}')$ *is said to be* transversally harmonic *if* $\tau_b(\varphi) \equiv 0$.

Then, for a transversally harmonic map $\varphi : (M, g, \mathcal{F}) \to (M', g', \mathcal{F}')$, the second variation formula of the transversal energy $E(\varphi)$ is given as follows (cf. [**80**], p. 13, the case $f = 1$ in Theorem 4.1, (4.2)) : let

$\varphi_{s,t} : M \to M'$ $(-\epsilon < s, t < \epsilon)$ be any two parameter smooth foliated variation of φ with $V = \frac{\partial \varphi_{s,t}}{\partial s}\big|_{(s,t)=(0,0)}$, $W = \frac{\partial \varphi_{s,t}}{\partial t}\big|_{(s,t)=(0,0)}$ and $\varphi_{0,0} = \varphi$,

$$
\begin{aligned}
\mathrm{Hess}(E)_\varphi(V, W) &:= \frac{\partial^2}{\partial s \partial t}\bigg|_{(s,t)=(0,0)} E(\varphi_{s,t}) \\
&= \int_M \langle J_{b,\varphi}(V), W \rangle \, v_g + \int_M V \langle W, d_T\varphi(\tau) \rangle \, v_g,
\end{aligned}
\tag{2.9}
$$

where $J_{b,\varphi}$ is a second order semi-elliptic differential operator acting on the space $\Gamma(\varphi^{-1}Q')$ of sections of $\varphi^{-1}Q'$ which is of the form:

$$
\begin{aligned}
J_{b,\varphi}(V) &:= \widetilde{\nabla}^*\widetilde{\nabla}V - \widetilde{\nabla}_\tau V - \mathrm{trace}_Q R^{Q'}(V, d_T\varphi)d_T\varphi \\
&= -\sum_{a=1}^q (\widetilde{\nabla}_{E_a}\widetilde{\nabla}_{E_a} - \widetilde{\nabla}_{\nabla_{E_a}E_a})V \\
&\quad - \sum_{a=1}^q R^{Q'}(V, d_T\varphi(E_a))d_T\varphi(E_a)
\end{aligned}
\tag{2.10}
$$

for $V \in \Gamma(\varphi^{-1}Q')$. Here, ∇ is the Levi-Civita connection of (M, g), and recall also that:

$$
\widetilde{\nabla}^*\widetilde{\nabla}V = -\sum_{a=1}^q (\widetilde{\nabla}_{E_a}\widetilde{\nabla}_{E_a} - \widetilde{\nabla}_{\nabla_{E_a}E_a})V + \widetilde{\nabla}_\tau V,
\tag{2.11}
$$

$$
\mathrm{trace}_Q R^{Q'}(V, d_T\varphi)d_T\varphi := \sum_{a=1}^q R^{Q'}(V, d_T\varphi(E_a))d_T\varphi(E_a).
\tag{2.12}
$$

Here, $\widetilde{\nabla}^*$ is the adjoint of the connection $\widetilde{\nabla}$ which satisfies (cf. [76], Proposition 3.1) that

$$
\int_M \langle \widetilde{\nabla}^*V, W \rangle \, v_g = \int_M \langle V, \widetilde{\nabla}W \rangle \, v_g \qquad (V, W \in \Gamma(\varphi^{-1}Q')),
$$

and for all $V, W \in \Gamma(\varphi^{-1}Q')$, it holds that

$$
\int_M \langle \widetilde{\nabla}^*\widetilde{\nabla}V, W \rangle \, v_g = \int_M \langle \widetilde{\nabla}V, \widetilde{\nabla}W \rangle \, v_g = \int_M \langle V, \widetilde{\nabla}^*\widetilde{\nabla}W \rangle \, v_g.
$$

DEFINITION 2.3. *The* transversal bitension field $\tau_{2,b}(\varphi)$ *of a smooth foliated map φ is defined by*

$$
\tau_{2,b}(\varphi) := J_{b,\varphi}(\tau_b(\varphi)).
\tag{2.13}
$$

DEFINITION 2.4. *The* transversal bienergy E_2 *of a smooth foliated map φ is defined by*

$$
E_2(\varphi) := \frac{1}{2} \int_M |\tau_b(\varphi)|^2 \, v_g.
\tag{2.14}
$$

Remark that this definition of the transversal bienergy is also the same as the one of Jung (cf. Jung [**80**], p. 16, the case $f = 1$ in Definition 6.1, (6.2)) because $\tau_b(\varphi) = \sum_{a=1}^{q}(\widetilde{\nabla}_{E_a}d_T\varphi)(E_a) = -\tilde{\delta}d_T\varphi$ (cf. Jung [**80**], p. 11, the case $f = 1$ in (3.3)). On the first variation formula of the transversal bienergy is given as follows. For a smooth foliated map φ and a smooth foliated variation $\{\varphi_t\}$ of φ, it holds (cf. [**80**], p. 16, the case $f = 1$ in (6.3)) that

$$\frac{d}{dt}\bigg|_{t=0} E_2(\varphi_t) = -\int_M \left\{ \langle V, \tau_{2,b}(\varphi)\rangle + \langle\widetilde{\nabla}_\tau V, \tau_b(\varphi)\rangle - \langle V, \widetilde{\nabla}_\tau \tau_b(\varphi)\rangle \right\} v_g. \tag{2.15}$$

DEFINITION 2.5. *A smooth foliated map* $\varphi : (M, g, \mathcal{F}) \to (M', g', \mathcal{F}')$ *is said to be* transversally biharmonic *if* $\tau_{2,b}(\varphi) \equiv 0$.

Let us recall that

DEFINITION 2.6. *A smooth foliated map* $\varphi : (M, g, \mathcal{F}) \to (M', g', \mathcal{F}')$ satisfies the conservation law *if*

$$\text{div}_{\widetilde{\nabla}} S(\varphi)(X) = 0 \qquad (\forall X \in \Gamma(Q)). \tag{2.16}$$

Here, $\text{div}_{\widetilde{\nabla}} S(\varphi)(X)$ is defined by

$$\text{div}_{\widetilde{\nabla}} S(\varphi)(X) := \sum_{a=1}^{q}(\widetilde{\nabla}_{E_a}S(\varphi))(E_a, X), \qquad (X \in \Gamma(Q)), \tag{2.17}$$

and recall icf. [**80**], Page 11) the *transversal stress-energy tensor* $S(\varphi) := \frac{1}{2}|d_T\varphi|^2 g_Q - \varphi^* g_{Q'}$, and Jung showed (cf. Jung, [**80**], Page 11, Proposition 3.4) that:

PROPOSITION 2.1. *For every a* C^∞ *foliated map* $\varphi : (M, g, \mathcal{F}) \to (M', g', \mathcal{F}')$, *it holds that*

$$\text{div}_{\widetilde{\nabla}} S(\varphi)(X) = -\langle \tau_b(\varphi), d_T\varphi(X)\rangle, \qquad (X \in \Gamma(Q)). \tag{2.18}$$

Then, one can ask the following generalized B.Y. Chen's conjecture:

The generalized Chen's conjecture:
Let φ *be a transversally biharmonic map from a foliated Riemannian manifold* $(M, g, \mathcal{F})$ *into another foliated Riemannian manifold* $(M', g', \mathcal{F}')$ *whose transversal sectional curvature* $K^{Q'}$ *is non-positive. Then,* φ *must be transversally harmonic.*

To this conjecture, Jung showed (cf. [**80**], Page 19) that

THEOREM 2.2. *(Jung) Assume that $(M, g, \mathcal{F})$ is a compact foliated Riemannian manifold whose transversal Ricci curvature is non-negative and positive at some point, and $(M', g', \mathcal{F}')$ has a positive constant transversal sectional curvature: $K^{Q'} = C > 0$. Then, every transversally stable, transversally biharmonic map $\varphi : (M, g, \mathcal{F}) \to (M', g', \mathcal{F}')$ which satisfies the conservation law must be transversally harmonic.*

Jung also showed (cf. [**80**], Page 5, Theorem 6.5) that

THEOREM 2.3. *(Jung) Assume that $(M, g, \mathcal{F})$ is a compact foliated Riemannian manifold whose transversal Ricci curvature is non-negative and positive at some point, and $(M', g', \mathcal{F}')$ has non-positive transversal sectional curvature $K^{Q'} \leq 0$. Then, every transversally biharmonic map $\varphi : (M, g, \mathcal{F}) \to (M', g', \mathcal{F}')$ must be transversally harmonic.*

Then, we can state our main theorem which gives an affirmative partial answer to the above generalized Chen's conjecture under the additional assumption that φ has both the finite transversal energy and the finite transversal bienergy:

THEOREM 2.4. *Let $\varphi : (M, g, \mathcal{F}) \to (M', g', \mathcal{F}')$ a smooth foliated map satisfying the conservation law. Assume that (M, g) is complete (possibly non-compact), $\mathcal{F}$ is transversally volume preserving, i.e., $\mathrm{div}(\tau) = 0$, and the transversal sectional curvature $K^{Q'}$ of $(M', g', \mathcal{F}')$ is non-positive: $K^{Q'} \leq 0$.*

If φ is transversally biharmonic having both the finite transversal energy $E(\varphi) < \infty$ and the finite transversal bienergy $E_2(\varphi)$, then it is transversally harmonic.

Remark that in the case that M is compact, Theorem 2.11 is true due to Jung's work (cf. [**80**], p.17, Theorem 6.5).

3. Proof of main theorem

In this section, we give a proof of Theorem 2.11.

(The first step) First, let us take a cut off function η from a fixed point $x_0 \in M$ on (M, g), i.e.,

$$\begin{cases} 0 \leq \eta(x) \leq 1 & (x \in M), \\ \eta(x) = 1 & (x \in B_r(x_0)), \\ \eta(x) = 0 & (x \notin B_{2r}(x_0), \\ |\nabla^g \eta| \leq \dfrac{2}{r} & (x \in M), \end{cases}$$

where $B_r(x_0) := \{x \in M | r(x) < r\}$, $r(x)$ is a distance function from x_0 on (M, g), ∇^g is the Levi-Civita connection of (M, g), respectively.

Assume that φ is a transversally biharmonic map of $(M, g, \mathcal{F})$ into $(M', g', \mathcal{F}')$, i.e.,

$$
\begin{aligned}
\tau_{2,b}(\varphi) &= J_{b,\varphi}(\tau_b(\varphi)) \\
&= \widetilde{\nabla}^* \widetilde{\nabla} \tau_b(\varphi) - \widetilde{\nabla}_\tau \tau_b(\varphi) - \operatorname{trace}_Q R^{Q'}(\tau_b(\varphi), d_T\varphi) d_T\varphi) \\
&= 0,
\end{aligned}
\tag{3.1}
$$

where recall $\widetilde{\nabla}$ is the induced connection on $\varphi^{-1} Q' \otimes T^* M$.

(The second step) Since $\tau_b(\varphi) \in \Gamma(Q)$ satisfies that $\int_M |\tau_b(\varphi)|^2 \, v_g < \infty$, it holds that as $r \to \infty$,

$$
\begin{aligned}
\int_M \langle \widetilde{\nabla}_\tau \tau_b(\varphi), \eta^2 \, \tau_b(\varphi) \rangle \, v_g &= \frac{1}{2} \int_M \tau \langle \tau_b(\varphi), \tau_b(\varphi) \rangle \, \eta^2 \, v_g \\
&\longrightarrow \frac{1}{2} \int_M \tau \langle \tau_b(\varphi), \tau_b(\varphi) \rangle \, v_g = 0
\end{aligned}
\tag{3.2}
$$

due to the completeness of (M, g), $\operatorname{div}(\tau) = 0$, $\int_M |\widetilde{\nabla}_\tau \tau_b(\varphi)|^2 \, v_g < \infty$ and Gaffney's theorem (cf. Theorem 2.2).

Furthermore, by (3.1), we obtain that

$$
\begin{aligned}
\int_M \langle \widetilde{\nabla}^* \widetilde{\nabla} \tau_b(\varphi), \eta^2 \, \tau_b(\varphi) \rangle \, v_g &= \int_M \eta^2 \, \langle \operatorname{trace}_Q R^{Q'}(\tau_b(\varphi), d_T\varphi) d_T\varphi, \tau_b(\varphi) \rangle \, v_g \\
&= \int_M \eta^2 \sum_{a=1}^q \langle R^{Q'}(\tau_b(\varphi), d_T\varphi(E_a)) d_T\varphi(E_a), \tau_b(\varphi) \rangle \, v_g \\
&= \int_M \eta^2 \sum_{a=1}^q K^{Q'}(\Pi_{\varphi,a}) \, v_g \\
&\leq 0,
\end{aligned}
\tag{3.3}
$$

where the sectional curvature $K^{Q'}(\Pi_{\varphi,a})$ of $(M', g', \mathcal{F}')$ corresponding to the plane spanned by $\tau_b(\varphi)$ and $d_T\varphi(E_a)$ is non-positive.

(The third step) On the other hand, by the properties of the adjoint $\widetilde{\nabla}^*$ of $\widetilde{\nabla}$, the left hand side of (3.3) is equal to

$$
\begin{aligned}
\int_M \langle \widetilde{\nabla} \tau_b(\varphi), \widetilde{\nabla}(\eta^2 \, \tau_b(\varphi)) \rangle \, v_g &= \int_M \sum_{a=1}^q \langle \widetilde{\nabla}_{E_a} \tau_b(\varphi), \widetilde{\nabla}_{E_a}(\eta^2 \, \tau_b(\varphi)) \rangle \, v_g \\
&= \int_M \eta^2 \sum_{a=1}^q |\widetilde{\nabla}_{E_a} \tau_b(\varphi)|^2 \, v_g + 2 \int_M \sum_{a=1}^q \langle \eta \, \widetilde{\nabla}_{E_a} \tau_b(\varphi), (E_a \eta) \, \tau_b(\varphi) \rangle \, v_g
\end{aligned}
\tag{3.4}
$$

since

$$\widetilde{\nabla}_{E_a}(\eta^2\, \tau_b(\varphi)) = \eta^2\, \widetilde{\nabla}_{E_a}\tau_b(\varphi) + 2\,\eta\,(E_a\eta)\,\tau_b(\varphi).$$

Together (3.3) and (3.4), we obtain

$$\int_M \eta^2 \sum_{a=1}^q \left|\widetilde{\nabla}\tau_b(\varphi)\right|^2 v_g \leq -2 \int_M \sum_{a=1}^q \langle \eta\, \widetilde{\nabla}_{E_a}\tau_b(\varphi), (E_a\eta)\,\tau_b(\varphi)\rangle\, v_g$$

$$\leq \frac{1}{2}\int_M \eta^2 \sum_{a=1}^q \left|\widetilde{\nabla}_{E_a}\tau_b(\varphi)\right|^2 v_g + 2\int_M \sum_{a=1}^q |E_a\eta|^2\, |\tau_b(\varphi)|^2\, v_g.$$

$$(3.5)$$

Because, putting $V_a := \eta\, \widetilde{\nabla}_{E_a}\tau_b(\varphi)$, $W_a := (E_a\eta)\,\tau_b(\eta)$ $(a = 1, \cdots, q)$, we have

$$0 \leq \left|\sqrt{\epsilon}\, V_a \pm \frac{1}{\sqrt{\epsilon}}\, W_a\right|^2 = \epsilon\, |V_a|^2 \pm 2\, \langle V_a, W_a\rangle + \frac{1}{\epsilon}\, |W_a|^2$$

which is

$$\mp 2\, \langle V_a, W_a\rangle \leq \epsilon\, |V_a|^2 + \frac{1}{\epsilon}\, |W_a|^2. \qquad (3.6)$$

If we put $\epsilon = \frac{1}{2}$ in (3.6), then we obtain

$$\mp 2\, \langle V_a, W_a\rangle \leq \frac{1}{2}\, |V_a|^2 + 2\, |W_a|^2 \qquad (a = 1, \cdots, q). \qquad (3.7)$$

By (3.7), we have the second inequality of (3.5).

   (The fourth step) Noticing that $\eta = 1$ on $B_r(x_0)$ and $|E_a\eta|^2 \leq \frac{2}{r}$ in the inequality (3.5), we obtain

$$\int_{B_r(x_0)} \sum_{a=1}^q |\widetilde{\nabla}_{E_a}\tau_b(\varphi)|^2\, v_g = \int_{B_r(x_0)} \eta^2 \sum_{a=1}^q \left|\widetilde{\nabla}_{E_a}\tau_b(\varphi)\right|^2 v_g$$

$$\leq \int_M \eta^2 \sum_{a=1}^q \left|\widetilde{\nabla}_{E_a}\tau_b(\varphi)\right|^2 v_g$$

$$\leq 4 \int_M \sum_{a=1}^q |E_a\eta|^2\, |\tau_b(\varphi)|^2\, v_g$$

$$\leq \frac{16}{r^2} \int_M |\tau_b(\varphi)|^2\, v_g. \qquad (3.8)$$

   Letting $r \to \infty$, the right hand side of (3.8) converges to zero since $E_2(\varphi) = \frac{1}{2}\int_M |\tau_b(\varphi)|^2\, v_g < \infty$. But due to (3.8), the left hand side of (3.8) must converge to $\int_M \sum_{a=1}^q |\widetilde{\nabla}_{E_a}\tau_b(\varphi)|^2\, v_g$ since $B_r(X_0)$ tends to

M because (M, g) is complete. Therefore, we obtain that

$$0 \le \int_M \sum_{a=1}^q \left| \widetilde{\nabla}_{E_a} \tau_b(\varphi) \right|^2 v_g \le 0,$$

which implies that

$$\widetilde{\nabla}_{E_a} \tau_b(\varphi) = 0 \ (a = 1, \cdots, q), \text{ i.e., } \widetilde{\nabla}_X \tau_b(\varphi) = 0 \ (\forall X \in \Gamma(Q)). \tag{3.9}$$

(The fifth step) Let us define a 1-form α on M by

$$\alpha(X) := \langle d\varphi(\pi(X)), \tau_b(\varphi) \rangle, \quad (X \in \mathfrak{X}(M)), \tag{3.10}$$

and a canonical dual vector field $\alpha^\# \in \mathfrak{X}(M)$ on M by $\langle \alpha^\#, Y \rangle := \alpha(Y), (Y \in \mathfrak{X}(M))$. Then, its divergence $\operatorname{div}(\alpha^\#)$ written as $\operatorname{div}(\alpha^\#) = \sum_{i=1}^p g(\nabla^g_{E_i} \alpha^\#, E_i) + \sum_{a=1}^q g(\nabla^g_{E_a} \alpha^\#, E_a)$, can be given as follows. Here, $\{E_i\}_{i=1}^p$ and $\{E_a\}_{a=1}^q$ are locally defined orthonormal frame fields on leaves L of $\mathcal{F}$ and Q, respectively, (dim $L_x = p$, dim $Q_x = q$, $x \in M$). Then, we can calculate $\operatorname{div}(\alpha^\#)$ as follows:

$$\operatorname{div}(\alpha^\#) = \sum_{i=1}^p \left\{ E_i(\alpha(E_i)) - \alpha(\nabla^g_{E_i} E_i) \right\}$$

$$+ \sum_{a=1}^q \left\{ E_a(\alpha(E_a)) - \alpha(\nabla^g_{E_a} E_a) \right\}$$

$$= \left\langle d\varphi \left(\pi \left(-\sum_{i=1}^p \nabla^g_{E_i} E_i \right) \right), \tau_b(\varphi) \right\rangle$$

$$+ \sum_{a=1}^q \left\{ E_a \langle d\varphi(E_a), \tau_b(\varphi) \rangle - \langle d\varphi (\pi (\nabla^g_{E_a} E_a)), \tau_b(\varphi) \rangle \right\}$$

$$= \left\langle d\varphi \left(\pi \left(-\sum_{i=1}^p \nabla^g_{E_i} E_i \right) \right), \tau_b(\varphi) \right\rangle$$

$$+ \sum_{a=1}^q \left\{ \left\langle \widetilde{\nabla}_{E_a} (d\varphi(E_a)), \tau_b(\varphi) \right\rangle + \left\langle d\varphi(E_a), \widetilde{\nabla}_{E_a} \tau_b(\varphi) \right\rangle \right.$$

$$\left. - \langle d\varphi (\pi (\nabla^g_{E_a} E_a)), \tau_b(\varphi) \rangle \right\}$$

$$= \left\langle d\varphi \left(\pi \left(-\sum_{i=1}^p \nabla^g_{E_i} E_i \right) \right) \right.$$

$$\left. + \sum_{a=1}^q \left\{ \widetilde{\nabla}_{E_a} (d\varphi(E_a)) - d\varphi (\pi (\nabla^g_{E_a} E_a)) \right\}, \tau_b(\varphi) \right\rangle. \tag{3.11}$$

since $\widetilde{\nabla}_{E_a} \tau_b(\varphi) = 0$ in the last equality of (3.11). Integrating the both hands of (3.11) over M, we have

$$\int_M \left\langle d\varphi \left(\pi \left(\sum_{i=1}^p \nabla^g{}_{E_i} E_i \right) \right), \tau_b(\varphi) \right\rangle v_g$$

$$= \int_M \left\langle \sum_{a=1}^q \left\{ \widetilde{\nabla}_{E_a}(d\varphi(E_a)) - d\varphi \left(\pi \left(\nabla^g{}_{E_a} E_a \right) \right) \right\}, \tau_b(\varphi) \right\rangle v_g.$$

$$(3.12)$$

because of $\int_M \operatorname{div}(\alpha^\#) \, v_g = 0$. Notice that the both hands in (3.12) are well defined because of $E(\varphi) < \infty$ and $E_2(\varphi) < \infty$.

Since $\kappa^\# := \pi(\sum_{i=1}^p \nabla^g{}_{E_i} E_i)$ is the second fundamental form of each leaf L in (M, g) and

$$\tau_b(\varphi) = \sum_{a=1}^q \left\{ \widetilde{\nabla}_{E_a}(d\varphi(E_a)) - d\varphi \left(\nabla^g{}_{E_a} E_a \right) \right\}$$

$$= \sum_{a=1}^q \left\{ \widetilde{\nabla}_{E_a}(d\varphi(E_a)) - d\varphi \left(\pi \left(\nabla^g{}_{E_a} E_a \right) \right) \right\} - d\varphi \left(\left(\sum_{a=1}^q \nabla^g{}_{E_a} E_a \right)^\perp \right),$$

$$(3.13)$$

the right hand side of (3.12) coincides with

$$\int_M \left\langle \tau_b(\varphi) + d\varphi \left(\left(\sum_{a=1}^q \nabla^g{}_{E_a} E_a \right)^\perp \right), \tau_b(\varphi) \right\rangle v_g,$$

$$(3.14)$$

(3.12) is equivalent to that

$$\int_M \left\langle d\varphi(\kappa^\#), \tau_b(\varphi) \right\rangle v_g = \int_M \left\langle \tau_b(\varphi), \tau_b(\varphi) \right\rangle v_g$$

$$+ \int_M \left\langle d\varphi \left(\left(\sum_{a=1}^q \nabla^g{}_{E_a} E_a \right)^\perp \right), \tau_b(\varphi) \right\rangle v_g.$$

$$(3.15)$$

Finally, $\varphi : (M, g) \to (M', g')$ satisfies the conservation law, then it holds due to Proposition 2.6 that $\langle d\varphi_x(Q_x), \tau_b(\varphi) \rangle = 0$. Furthermore, recall that $X^\perp$ ($X \in \mathfrak{X}(M)$) is the Q-component of $X \in \mathfrak{X}(M)$ relative to the decomposition $TM = L \oplus Q$ of the bundles. Therefore, these imply that both the left hand side and the second term of the right hand side of (3.15) must vanish. That is, we obtain that $\int_M \langle \tau_b(\varphi), \tau_b(\varphi) \rangle v_g = 0$. Therefore $\tau_b(\varphi) \equiv 0$. We have Theorem 2.11. $\qquad\square$

4. Appendix

Here, we give a proof of Theorem 2.2. For the first part of the proof, see Appendix, Page 271 in [**111**]. We give a proof of the latter half.

THEOREM 4.1. *(cf. Theorem 2.2) Let (M, g) be a non-compact complete Riemannian manifold without boundary, If a C^1 vector field X on M satisfies that*

$$\int_M |X| \, v_g < \infty \quad and \quad \int_M div(X) \, v_g < \infty. \tag{4.1}$$

Then, it holds that

$$\int_M div(X) \, v_g = 0. \tag{4.2}$$

Furthermore, if $f \in C^1(M)$ and a C^1 vector field X on M satisfy $\mathrm{div}(X) = 0$, $\int_M Xf \, v_g < \infty$, $\int_M |f|^2 \, v_g < \infty$ and $\int_M |X|^2 \, v_g < \infty$, then it holds that

$$\int_M Xf \, v_g = 0. \tag{4.3}$$

Proof. (The first step)   For $f \in C_c^2(M)$ ($f \in C^2(M)$ with compact suport) and a C^1 vector field X on M satisfying $\mathrm{div}(X) = 0$, let us define m-form $\omega = f \, v_g$, $(m = \dim M)$. Then, the Lie derivative $L_X \omega$ of ω by X is calculated as follows:

$$\begin{cases} L_X \omega = Xf \, v_g + f \, L_X v_g = Xf \, v_g + f \, \mathrm{div}(X) \, v_g = Xf \, v_g, \\ L_X \omega = i_X \, d\omega + d \, i_X \, \omega = d \, i_X \omega \end{cases} \tag{4.4}$$

due to $\mathrm{div}(X) = 0$, the H. Cartan's identity and $d\omega = 0$, where $i_X K$ is the interior product of a tensor field K by X. By (4.1), we have

$$\int_M Xf \, v_g = \int_M L_X \omega = \int_M d \, i_X \omega = \int_{\partial M} i_X \omega = 0 \tag{4.5}$$

because each integral is finite due to $f \in C_c^2(M)$, and $\partial M = \emptyset$.

(The second step)   Let us take $f \in C^1(M)$ and a C^1 vector field X on M satisfying $\mathrm{div}(X) = 0$ and $\int_M Xf \, v_g < \infty$. Then there exists a sequence $f_n \in C^2(M)$ $(n = 1, 2, \cdots)$ such that $f_n \to f$ in the C^1 topology in a Riemannian manifold (M, g). Then, it holds that

$$\int_M Xf_n \, v_g \to \int_M Xf \, v_g \tag{4.6}$$

in the C^0 topology in (M, g).

(The third step)   Let us take a cutoff function μ from a fixed point $x_0 \in M$ on (M, g) as in the first step of the proof of Theorem 2.11 in Section Three.

Applying the first step to the functions $f_n \, \mu \in C_c^2(M)$, it holds that

$$\int_M X \, (f_n \, \mu) \, v_g = 0. \tag{4.7}$$

But, we have

$$\int_M X \, (f_n \, \mu) \, v_g = \int_M (X \, f_n) \, \mu \, v_g + \int_M f_n \, (X \mu) \, v_g. \tag{4.8}$$

By (4.4) and (4.5), we have,

$$\left| \int_M (X \, f_n) \, \mu \, v_g \right| = \left| - \int_M f_n \, (X \, \mu) \, v_g \right| \leq \int_M |f_n| \, |X \, \mu| \, v_g$$

$$\leq \int_M |f_n| \, |X| |\nabla \mu| \, v_g$$

$$\leq \frac{2}{r} \int_M |f_n| \, |X| \, v_g$$

$$\leq \frac{2C}{r} \int_M |f| \, |X| \, v_g \leq \frac{2C}{r} \|f\| \, \|X\| \tag{4.9}$$

with $\|f\|^2 = \int_M |f|^2 \, v_g < \infty$ and $\|X\|^2 = \int_M |X|^2 \, v_g < \infty$ for a certain positive constant $C > 0$. Tending $r \to \infty$ in (4.9), since the right hand side of (4.9) goes to zero,

$$\int_M (X f_n) \, \mu \, v_g \longrightarrow 0 \qquad (\text{as } r \to \infty). \tag{4.10}$$

On the other hand, as $r \to \infty$,

$$\int_M (X f_n) \, \mu \, v_g \longrightarrow \int_M X f_n \, v_g \tag{4.11}$$

which implies that

$$\int_M X f_n \, v_g = 0. \tag{4.12}$$

Due to (4.6), as $n \to \infty$, we have

$$\int_M X f \, v_g = 0 \tag{4.13}$$

which is the desired. $\qquad\qquad\qquad\qquad\qquad\qquad\qquad\qquad\qquad\qquad$ $\square$

CHAPTER 15

CR-Rigidity of Pseudo Harmonic Maps and Pseudo Biharmonic Maps

1

ABSTRACT. The CR analogue of B.-Y. Chen's conjecture on pseudo biharmonic maps will be shown. Pseudo biharmonic, but not pseudo harmonic, isometric immersions with parallel pseudo mean curvature vector fields, will be characterized.

1. Introduction

Harmonic maps play a central role in geometry; they are critical points of the energy functional $E(\varphi) = \frac{1}{2} \int_M |d\varphi|^2 \, v_g$ for smooth maps φ of (M, g) into (N, h). The Euler-Lagrange equations are given by the vanishing of the tension filed $\tau(\varphi)$. In 1983, Eells and Lemaire [40] extended the notion of harmonic map to biharmonic map, which are critical points of the bienergy functional $E_2(\varphi) = \frac{1}{2} \int_M |\tau(\varphi)|^2 \, v_g$. After Jiang [74] studied the first and second variation formulas of E_2, extensive studies in this area have been done (for instance, see [16], [90], [102], [63], [64], [73]). Every harmonic maps is always biharmonic by definition. Chen raised ([21]) famous Chen's conjecture and later, Caddeo, Montaldo, Piu and Oniciuc raised ([16]) the generalized Chen's conjecture.

B.Y. Chen's conjecture:
Every biharmonic submanifold of the Euclidean space $\mathbb{R}^n$ must be harmonic (minimal).

The generalized B.Y. Chen's conjecture:
Every biharmonic submanifold of a Riemannian manifold of non-positive curvature must be harmonic (minimal).

For the generalized Chen's conjecture, Ou and Tang gave ([123]) a counter example in a Riemannian manifold of negative curvature. For Chen's conjecture, some affirmative answers were known for surfaces in

---

[1]This chapter is due to [158]: H. Urakawa, *CR rigidity of pseudo harmonic maps and pseudo biharmonic maps*, Hokkaido Math. J., **46** (2017), 141–187.

the three dimensional Euclidean space ([21]), and hypersurfaces of the four dimensional Euclidean space ([57], [34]). Akutagawa and Maeta showed ([1]) that any properly immersed biharmonic submanifold of the Euclidean space $\mathbb{R}^n$ is harmonic (minimal).

To the generalized Chen's conjecture, we showed ([111]) that: for a complete Riemannian manifold (M, g), a Riemannian manifold (N, h) of non-positive curvature, then, every biharmonic map $\varphi : (M, g) \to (N, h)$ with finite energy and finite bienergy is harmonic. In the case $\mathrm{Vol}(M, g) = \infty$, every biharmonic map $\varphi : (M, g) \to (N, h)$ with finite bienergy is harmonic. This gave ([108], [109], [111]) affirmative answers to the generalized Chen's conjecture under the L^2-condition and the completeness of (M, g).

In 1970's, Chern and Moser initiated ([29]) the geometry and analysis of strictly convex CR manifolds, and many mathematicians works on CR manifolds (cf. [38]). Recently, Barletta, Dragomir and Urakawa gave ([11]) the notion of pseudo harmonic map, and also Dragomir and Montaldo settled ([37]) the one of pseudo biharmonic map.

In this paper, we raise

The CR analogue of the generalized Chen's conjecture:

Let (M, g_θ) be a complete strictly pseudoconvex CR manifold, and assume that (N, h) is a Riemannian manifold of non-positive curvature. Then, every pseudo biharmonic isometric immersion $\varphi : (M, g_\theta) \to (N, h)$ must be pseudo harmonic.

We will show this conjecture holds under some L^2 condition on a complete strongly pseudoconvex CR manifold (cf. Theorem 3.2), and will give characterization theorems on pseudo biharmonic immersions from CR manifolds into the unit sphere or the complex projective space (cf. Theorems 6.2 and 7.1). More precisely, we will show

THEOREM 1.1. *(cf. Theorem 3.2) Let φ be a pseudo biharmonic map of a complete CR manifold (M, g_θ) into a Riemannian manifold (N, h) of non-positive curvature. Then,*

If the pseudo energy $E_b(\varphi)$ and the pseudo bienergy $E_{b,2}(\varphi)$ are finite, then φ is pseudo harmonic.

For isometric immersions of a CR manifold (M^{2n+1}, g_θ) into the unit sphere $S^{2n+2}(1)$ of curvature 1, we have

THEOREM 1.2. *(cf. Theorem 6.2) For such immersion, assume that the pseudo mean curvature is parallel, but not pseudo harmonic.*

Then, φ is pseudo biharmonic if and only if the restriction of the second fundamental form B_φ to the holomorphic subspace $H_x(M)$ of

T_xM $(x \in M)$ *satisfies that*

$$\| B_\varphi|_{H(M) \times H(M)} \|^2 = 2n.$$

For isometric immersions of a CR manifold (M^{2n+1}, g_θ) into the complex projective space $(\mathbb{P}^{n+1}(c), h, J)$ of holomorphic sectional curvature $c > 0$, we have

THEOREM 1.3. *(cf. Theorem 7.1) For such immersion, assume that the pseudo mean curvature is parallel, but not pseudo harmonic. Then, φ is pseudo biharmonic if and only if one of the following holds:*
(1) $J(d\varphi(T))$ *is tangent to* $\varphi(M)$ *and*

$$\| B_\varphi|_{H(M) \times H(M)} \|^2 = \frac{c}{4}(2n + 3).$$

(2) $J(d\varphi(T))$ *is normal to* $\varphi(M)$ *and*

$$\| B_\varphi|_{H(M) \times H(M)} \|^2 = \frac{c}{4}(2n) = \frac{n}{2} c.$$

Here, T is the charactersitic vector field of (M, g_θ), $H_x(M) \oplus \mathbb{R}T_x = T_x(M)$, and $B_\varphi|_{H(M) \times H(M)}$ is the restriction of the second fundamental form B_φ to $H_x(M)$ $(x \in M)$.

Several examples of pseudo biharmonic immersions of (M, g_θ) into the unit sphere or complex projective space will be given.

Acknowledgement. This work was finished during the stay at the University of Basilicata, Potenza, Italy, September of 2014. The author was invited by Professor Sorin Dragomir to the University of Basilicata, Italy. The author would like to express his sincere gratitude to Professor Sorin Dragomir and Professor Elisabetta Barletta for their kind hospitality and helpful discussions.

2. Preliminaries

2.1. We prepare the materials for the first and second variational formulas for the bienergy functional and biharmonic maps. Let us recall the definition of a harmonic map $\varphi : (M, g) \to (N, h)$, of a compact Riemannian manifold (M, g) into another Riemannian manifold (N, h), which is an extremal of the *energy functional* defined by

$$E(\varphi) = \int_M e(\varphi) \, v_g,$$

where $e(\varphi) := \frac{1}{2}|d\varphi|^2$ is called the energy density of φ. That is, for any variation $\{\varphi_t\}$ of φ with $\varphi_0 = \varphi$,

$$\frac{d}{dt}\Big|_{t=0} E(\varphi_t) = -\int_M h(\tau(\varphi), V)v_g = 0, \tag{2.1}$$

where $V \in \Gamma(\varphi^{-1}TN)$ is a variation vector field along φ which is given by $V(x) = \frac{d}{dt}\Big|_{t=0}\varphi_t(x) \in T_{\varphi(x)}N$, $(x \in M)$, and the *tension field* is given by $\tau(\varphi) = \sum_{i=1}^m B_\varphi(e_i, e_i) \in \Gamma(\varphi^{-1}TN)$, where $\{e_i\}_{i=1}^m$ is a locally defined orthonormal frame field on (M, g), and B_φ is the second fundamental form of φ defined by

$$\begin{aligned} B_\varphi(X, Y) &= (\widetilde{\nabla} d\varphi)(X, Y) \\ &= (\widetilde{\nabla}_X d\varphi)(Y) \\ &= \overline{\nabla}_X(d\varphi(Y)) - d\varphi(\nabla^g_X Y), \end{aligned} \tag{2.2}$$

for all vector fields $X, Y \in \mathfrak{X}(M)$. Here, ∇^g, and ∇^h, are Levi-Civita connections on TM, TN of (M, g), (N, h), respectively, and $\overline{\nabla}$, and $\widetilde{\nabla}$ are the induced ones on $\varphi^{-1}TN$, and $T^*M \otimes \varphi^{-1}TN$, respectively. By (2.1), φ is *harmonic* if and only if $\tau(\varphi) = 0$.

The second variation formula is given as follows. Assume that φ is harmonic. Then,

$$\frac{d^2}{dt^2}\Big|_{t=0} E(\varphi_t) = \int_M h(J(V), V)v_g, \tag{2.3}$$

where J is an elliptic differential operator, called the *Jacobi operator* acting on $\Gamma(\varphi^{-1}TN)$ given by

$$J(V) = \overline{\Delta}V - \mathcal{R}(V), \tag{2.4}$$

where $\overline{\Delta}V = \overline{\nabla}^*\overline{\nabla}V = -\sum_{i=1}^m \{\overline{\nabla}_{e_i}\overline{\nabla}_{e_i}V - \overline{\nabla}_{\nabla^g_{e_i}e_i}V\}$ is the *rough Laplacian* and $\mathcal{R}$ is a linear operator on $\Gamma(\varphi^{-1}TN)$ given by $\mathcal{R}(V) = \sum_{i=1}^m R^h(V, d\varphi(e_i))d\varphi(e_i)$, and R^h is the curvature tensor of (N, h) given by $R^h(U, V) = \nabla^h_U\nabla^h_V - \nabla^h_V\nabla^h_U - \nabla^h_{[U,V]}$ for $U, V \in \mathfrak{X}(N)$.

J. Eells and L. Lemaire [40] proposed polyharmonic (k-harmonic) maps and Jiang [74] studied the first and second variation formulas of biharmonic maps. Let us consider the *bienergy functional* defined by

$$E_2(\varphi) = \frac{1}{2}\int_M |\tau(\varphi)|^2 v_g, \tag{2.5}$$

where $|V|^2 = h(V, V)$, $V \in \Gamma(\varphi^{-1}TN)$.

The first variation formula of the bienergy functional is given by

$$\frac{d}{dt}\Big|_{t=0} E_2(\varphi_t) = -\int_M h(\tau_2(\varphi), V) v_g. \tag{2.6}$$

Here,

$$\tau_2(\varphi) := J(\tau(\varphi)) = \overline{\Delta}(\tau(\varphi)) - \mathcal{R}(\tau(\varphi)), \tag{2.7}$$

which is called the *bitension field* of φ, and J is given in (2.4).

A smooth map φ of (M, g) into (N, h) is said to be *biharmonic* if $\tau_2(\varphi) = 0$. By definition, every harmonic map is biharmonic. For an isometric immersion, it is minimal if and only if it is harmonic.

2.2.   Following Dragomir and Montaldo [37], and also Barletta, Dragomir and Urakawa [11], we will prepare the materials on pseudo harmonic maps and pseudo biharmonic maps.

Let M be a strictly pseudoconvex CR manifold of $(2n+1)$-dimension, T, the characteristic vector field on M, J is the complex structure of the subspace $H_x(M)$ of $T_x(M)$ $(x \in M)$, and g_θ, the Webster Riemannian metric on M defined for $X, Y \in H(M)$ by

$$g_\theta(X, Y) = (d\theta)(X, JY), \ g_\theta(X, T) = 0, \ g_\theta(T, T) = 1.$$

Let us recall for a C^∞ map φ of (M, g_θ) into another Riemannian manifold (N, h), the *pseudo energy* $E_b(\varphi)$ is defined ([11]) by

$$E_b(\varphi) = \frac{1}{2} \int_M \sum_{i=1}^{2n} (\varphi^* h)(X_i, X_i) \, \theta \wedge (d\theta)^n, \tag{2.8}$$

where $\{X_i\}_{i=1}^{2n}$ is an orthonormal frame field on $(H(M), g_\theta)$. Then, the first variational formula of $E_b(\varphi)$ is as follows ([11]). For every variation $\{\varphi_t\}$ of φ with $\varphi_0 = \varphi$,

$$\frac{d}{dt}\Big|_{t=0} E_b(\varphi_t) = -\int_M h(\tau_b(\varphi), V) \, d\theta \wedge (d\theta)^n = 0, \tag{2.9}$$

where $V \in \Gamma(\varphi^{-1} TN)$ is defined by $V(x) = \frac{d}{dt}\big|_{t=0}\varphi_t(x) \in T_{\varphi(x)}N$, $(x \in M)$. Here, $\tau_b(\varphi)$ is the *pseudo tension field* which is given by

$$\tau_b(\varphi) = \sum_{i=1}^{2n} B_\varphi(X_i, X_i), \tag{2.10}$$

where $B_\varphi(X, Y)$ $(X, Y \in \mathfrak{X}(M))$ is the second fundamental form (2.2) for a C^∞ map of (M, g_θ) into (N, h). Then, φ is *pseudo harmonic* if $\tau_b(\varphi) = 0$.

The second variational formula of E_b is given as follows ([**11**], p.733):

$$\frac{d^2}{dt^2}\bigg|_{t=0} E_b(\varphi_t) = \int_M h(J_b(V), V)\,\theta \wedge (d\theta)^n, \qquad (2.11)$$

where J_b is a subelliptic operator acting on $\Gamma(\varphi^{-1}TN)$ given by

$$J_b(V) = \Delta_b V - \mathcal{R}_b(V). \qquad (2.12)$$

Here, for $V \in \Gamma(\varphi^{-1}TN))$,

$$\begin{cases} \Delta_b V = (\overline{\nabla}^H)^* \overline{\nabla}^H V = -\sum_{i=1}^{2n} \left\{ \overline{\nabla}_{X_i}(\overline{\nabla}_{X_i} V) - \overline{\nabla}_{\nabla_{X_i} X_i} V \right\}, \\ \mathcal{R}_b(V) = \sum_{i=1}^{2n} R^h(V, d\varphi(X_i))d\varphi(X_i), \end{cases} \qquad (2.13)$$

where ∇ is the Tanaka-Webster connection, and $\overline{\nabla}$, the induced connection on $\phi^{-1}TN$ induced from the Levi-Civita connection ∇^h, and $\{X_i\}_{i=1}^{2n}$, a local orthonormal frame field on $(H(M), g_\theta)$, respectively. Here, $(\overline{\nabla}^H)_X V := \overline{\nabla}_{X^H} V$ $(X \in \mathfrak{X}(M), V \in \Gamma(\phi^{-1}TN))$, corresponding to the decomposition $X = X^H + g_\theta(X, T)\,T$ $(X^H \in H(M))$, and define $\pi_H(X) = X^H$ $(X \in T_x(M))$, and $(\overline{\nabla}^H)^*$ is the formal adjoint of $\overline{\nabla}^H$.

Dragomir and Montaldo [**37**] introduced the *pseudo bienergy* given by

$$E_{b,2}(\varphi) = \frac{1}{2}\int_M h(\tau_b(\varphi), \tau_b(\varphi))\,\theta \wedge (d\theta)^n, \qquad (2.14)$$

where $\tau_b(\varphi)$ is the *pseudo tension field* of φ. They gave the first variational formula of $E_{b,2}$ as follows ([**37**], p.227):

$$\frac{d}{dt}\bigg|_{t=0} E_{b,2}(\varphi_t) = -\int_M h(\tau_{b,2}(\varphi), V)\,\theta \wedge (d\theta)^n, \qquad (2.15)$$

where $\tau_{b,2}(\varphi)$ is called the *pseudo bitension field* given by

$$\tau_{b,2}(\varphi) = \Delta_b\big(\tau_b(\varphi)\big) - \sum_{i=1}^{2n} R^h(\tau_b(\varphi), d\varphi(X_i))\,d\varphi(X_i). \qquad (2.16)$$

Then, a smooth map φ of (M, g_θ) into (N, h) is said to be *pseudo biharmonic* if $\tau_{b,2}(\varphi) = 0$. By definition, a pseudo harmonic map is always pseudo biharmonic.

3. Generalized Chen's conjecture for pseudo biharmonic maps

3.1 First, let us recall the usual Weitzenbeck formula for a C^∞ map from a Riemannian manifod (M, g) of $(2n + 1)$ dimension into a Riemannian manifold (N, h):

LEMMA 3.1. *(The Weitzenbeck formula)* *For every C^∞ map φ of (M, g) of $(2n+1)$-dimension into a Riemannian manifold (N, h), the Hodge Laplacian Δ acting on the 1-form $d\varphi$, regarded as a $\varphi^{-1}TN$- valued 1 form, $d\varphi \in \Gamma(T^*M \otimes \varphi^{-1}TN)$, we have*

$$\Delta\, d\varphi = \widetilde{\nabla}^* \widetilde{\nabla}\, d\varphi + S. \tag{3.1}$$

Here, let us recall the rough Laplacian

$$\widetilde{\nabla}^* \widetilde{\nabla} := \sum_{k=1}^{2n+1} \left\{ \widetilde{\nabla}_{e_k} \widetilde{\nabla}_{e_k} - \widetilde{\nabla}_{\nabla^g_{e_k} e_k} \right\} \tag{3.2}$$

$$S(X) := -(\widetilde{R}(X, e_k)d\varphi)(e_k), \qquad (X \in \mathfrak{X}(M)). \tag{3.3}$$

Here, ∇^g, ∇^h are the Levi-Civita connections of (M, g), (N, h), and $\widetilde{\nabla}$ is the induced connection on $T^*M \otimes \varphi^{-1}TN$ defined by $(\widetilde{\nabla}_X d\varphi)(Y) = \overline{\nabla}_X d\varphi(Y) - d\varphi(\nabla^g_X Y)$, $\overline{\nabla}$ is the induced connection on $\varphi^{-1}TN$ given by $\overline{\nabla}_X d\varphi(Y) = \nabla^h_{d\varphi(X)} d\varphi(Y)$, $(X, Y \in \mathfrak{X}(M))$, and $\{e_k\}_{k=1}^{2n+1}$ is a locally defined orthonormal vector field on (M, g). The curvature tensor field $\widetilde{R}$ in (3.3) is defined by

$$(\widetilde{R}(X,Y)d\varphi)(Z) := \overline{R}(X,Y)\, d\varphi(Z) - d\varphi(R^g(X,Y)Z)$$
$$= R^h(d\varphi(X), d\varphi(Y))d\varphi(Z) - d\varphi(R^g(X,Y)Z),$$

for $X, Y, Z \in \mathfrak{X}(M)$, where $\overline{R}$, R^g, and R^h are the curvature tensors of the induced connection $\overline{\nabla}$, ∇^g and ∇^h, respectively.

Notice that for an isometric immersion $\varphi : (M, g) \to (N, h)$, it holds that

$$(\widetilde{\nabla}_X d\varphi)(Y) = B_\varphi(X, Y), \qquad (X, Y \in \mathfrak{X}(M)). \tag{3.4}$$

3.2 In this part, we first raise the CR analogue of the generalized Chen's conjecture, and settle it for pseudo biharmonic maps with finite pseudo energy and finite pseudo bienergy.

Let us recall a strictly pseudoconvex CR manifold (possibly non compact) (M, g_θ) of $(2n + 1)$-dimension, and the Webster Riemannian metric g_θ given by

$$g_\theta(X, Y) = (d\theta)(X, JY), \quad g_\theta(X, T) = 0, \quad g_\theta(T, T) = 1$$

for X, $Y \in H(M)$. Recall the material on the Levi-Civita connection ∇^{g_θ} of (M, g_θ). Due to Lemma 1.3, Page 38 in [**38**], it holds that,

$$\nabla^{g_\theta} = \nabla + (\Omega - A) \otimes T + \tau \otimes \theta + 2 \theta \odot J, \qquad (3.5)$$

where ∇ is the Tanaka-Webster connection, $\Omega = d\theta$, and $A(X,Y) = g_\theta(\tau X, Y)$, $\tau X = T_\nabla(T, X)$, and T_∇ is the torsion tensor of ∇. And also, $(\tau \otimes \theta)(X, Y) = \theta(Y) \tau X$, $(\theta \odot J)(X, Y) = \frac{1}{2} \{\theta(X) JY + \theta(Y) JX\}$ for all vector fields X, Y on M. Here, J is the complex structure on $H(M)$ and is extended as an endomorphism on (M) by $JT = 0$.

Then, we have

$$\nabla^{g_\theta}_{X_k} X_k = \nabla_{X_k} X_k - A(X_k, X_k) \, T, \qquad (3.6)$$
$$\nabla^{g_\theta}_T T = 0, \qquad (3.7)$$

where $\{X_k\}_{k=1}^{2n}$ is a locally defined orthonormal frame field on $H(M)$ with respect to g_θ, and T is the characteristic vector field of (M, g_θ). For (3.6), it follows from that $\Omega(X_k, X_k) = 0$, $(\tau \otimes \theta)(X_k, X_k) = 0$, and $(\theta \odot J)(X_k, X_k) = 0$ since $\theta(X_k) = 0$. For (3.7), notice that the Tanaka-Webster connection ∇ satisfies $\nabla_T T = 0$, and also $\tau T = 0$ and $JT = 0$, so that $\Omega(T, T) = 0$, $A(T, T) = 0$, $(\tau \otimes \theta)(T, T) = 0 \, (\theta \odot J)(T, T) = 0$ which imply (3.7).

For (3.2) in the Weitenbeck formula in Lemma 3.1, by taking $\{X_k \; (k = 1, \cdots, 2n), \; T\}$, as an orthonormal basis $\{e_k\}$ of our (M, g^θ), and due to (3.6) and (3.7), we have

$$(\widetilde{\nabla}^* \widetilde{\nabla} \, d\varphi)(X) = (\tilde{\Delta}_b \, d\varphi)(X)$$

$$= - \sum_{k=1}^{2n+1} \{\widetilde{\nabla}_{e_k} \widetilde{\nabla}_{e_k} - \widetilde{\nabla}_{\nabla^{g_\theta}_{e_k} e_k}\} \, d\varphi(X)$$

$$= - \sum_{k=1}^{2n} \{\widetilde{\nabla}_{X_k} \widetilde{\nabla}_{X_k} - \widetilde{\nabla}_{\nabla^{g_\theta}_{X_k} X_k}\} \, d\varphi(X)$$
$$\quad - \{\widetilde{\nabla}_T \widetilde{\nabla}_T - \widetilde{\nabla}_{\nabla^{g_\theta}_T T}\} \, d\varphi(X)$$

$$= - \sum_{k=1}^{2n} \{\widetilde{\nabla}_{X_k} \widetilde{\nabla}_{X_k} - \widetilde{\nabla}_{\nabla_{X_k} X_k}\} \, d\varphi(X)$$
$$\quad - \{\widetilde{\nabla}_T \widetilde{\nabla}_T + \sum_{k=1}^{2n} A(X_k, A_k) \widetilde{\nabla}_T\} \, d\varphi(X)$$

$$= - \sum_{k=1}^{2n} \{\widetilde{\nabla}_{X_k} \widetilde{\nabla}_{X_k} - \widetilde{\nabla}_{\nabla_{X_k} X_k}\} \, d\varphi(X) - \widetilde{\nabla}_T \widetilde{\nabla}_T \, d\varphi(X). \qquad (3.8)$$

since $\sum_{k=1}^{2n} A(X_k, X_k) = 0$ (cf. [**38**], p. 35).

For (3.3) in the Weitzenbeck formula in Lemma 3.1, we have

$$S(X) = -\sum_{k=1}^{2n+1} (\widetilde{R}(X, e_k)d\varphi)(e_k)$$

$$= -\sum_{k=1}^{2n} (\widetilde{R}(X, X_k)d\varphi)(X_k) - (\widetilde{R}(X, T)d\varphi)(T)$$

$$= -\sum_{k=1}^{2n} \left\{ R^h(d\varphi(X), d\varphi(X_k))d\varphi(X_k) - d\varphi(R^{g_\theta}(X, X_k)X_k) \right\}$$

$$- \left\{ R^h(d\varphi(X), d\varphi(T))d\varphi(T) - d\varphi(R^{g_\theta}(X, T)T) \right\}. \qquad (3.9)$$

And, we have the following formulas for (3.1) in our case,

$$\Delta\, d\varphi(X) = d\, d^*\, d\varphi(X)$$

$$= -d\,\tau(\varphi)(X)$$

$$= -\overline{\nabla}_X \tau(\varphi). \qquad (3.10)$$

Therefore, we have

$$-(\widetilde{\Delta}_b\, d\varphi)(X) = \sum_{k=1}^{2n} \left\{ \widetilde{\nabla}_{X_k} \widetilde{\nabla}_{X_k} - \widetilde{\nabla}_{\nabla_{X_k} X_k} \right\} d\varphi(X)$$

$$= -(\Delta\, d\varphi)(X) + S(X) - \widetilde{\nabla}_T \widetilde{\nabla}_T\, d\varphi(X)$$

$$= \overline{\nabla}_X \tau(\varphi) - \sum_{k=1}^{2n} \{ R^h(d\varphi(X), d\varphi(X_k))d\varphi(X_k)$$

$$- d\varphi(R^{g_\theta}(X, X_k)X_k) \}$$

$$- \{ R^h(d\varphi(X), d\varphi(T))d\varphi(T) - d\varphi(R^{g_\theta}(X, T)T) \}$$

$$- \widetilde{\nabla}_T \widetilde{\nabla}_T d\varphi(X). \qquad (3.11)$$

3.3 Let us consider the generalized B.-Y. Chen's conjecture for pseudo biharmonic maps which is CR analogue of the usual generalized Chen's conjecture for biharmonic maps:

The CR analogue of the generalized B.-Y. Chen's conjecture for pseudo biharmonic maps:

Let (M, g_θ) be a complete strictly pseudoconvex CR manifold, and assume that (N, h) is a Riemannian manifold of non-positive curvature.

Then, every pseudo biharmonic isometric immersion $\varphi : (M, g_\theta) \to (N, h)$ must be pseudo harmonic.

In this section, we want to show that the above conjecture is true under the finiteness of the pseudo energy and pseudo bienergy.

THEOREM 3.1. *Assume that φ is a pseudo biharmonic map of a strictly pseudoconvex complete CR manifold (M, g_θ) into another Riemannian manifold (N, h) of non positive curvature.*
If φ has finite pseudo bienergy $E_{b,2}(\varphi) < \infty$ and finite pseudo energy $E_b(\varphi) < \infty$, then it is pseudo harmonic, i.e., $\tau_b(\varphi) = 0$.

(*Proof of Theorem 3.2*)   The proof is divided into several steps.

(The first step)   For an arbitrarily fixed point $x_0 \in M$, let $B_r(x_0) = \{x \in M : r(x) < r\}$ where $r(x)$ is a distance function on (M, g_θ), and let us take a cut off function η on (M, g_θ), i.e.,

$$\begin{cases} 0 \leq \eta(x) \leq 1 & (x \in M), \\ \eta(x) = 1 & (x \in B_r(x_0)), \\ \eta(x) = 0 & (x \notin B_{2r}(x_0)), \\ |\nabla^{g_\theta} \eta| \leq \dfrac{2}{r} & (x \in M), \end{cases} \tag{3.12}$$

where r, ∇^{g_θ} are the distance function, the Levi-Civita connection of (M, g_θ), respectively. Assume that $\varphi : (M, g_\theta) \to (N, h)$ is a *pseudo biharmonic map*, i.e.,

$$\tau_{b,2}(\varphi) = J_b(\tau_b(\varphi))$$

$$= \Delta_b(\tau_b(\varphi)) - \sum_{j=1}^{2n} R^h(\tau_b(\varphi), d\varphi(X_j)) \, d\varphi(X_j)$$

$$= 0. \tag{3.13}$$

(The second step)   Then, we have

$$\int_M \langle \Delta_b(\tau_b(\varphi)), \eta^2 \, \tau_b(\varphi) \rangle \, \theta \wedge (d\theta)^n$$

$$= \int_M \eta^2 \sum_{j=1}^{2n} \langle R^h(\tau_b(\varphi), d\varphi(X_j)) \, d\varphi(X_j), \tau_b(\varphi) \rangle \, \theta \wedge (d\theta)^n$$

$$\leq 0 \tag{3.14}$$

since (N, h) has the non-positive sectional curvature. But, for the left hand side of (3.14), it holds that

$$\int_M \langle \Delta_b(\tau_b(\varphi)), \eta^2 \tau_b(\varphi) \rangle \, \theta \wedge (d\theta)^n$$

$$= \int_M \langle \overline{\nabla}^H \tau_b(\varphi), \overline{\nabla}^H (\eta^2 \tau_b(\varphi)) \rangle \, \theta \wedge (d\theta)^n$$

$$= \int_M \sum_{j=1}^{2n} \langle \overline{\nabla}_{X_j} \tau_b(\varphi), \overline{\nabla}_{X_j} (\eta^2 \tau_b(\varphi)) \rangle \, \theta \wedge (d\theta)^n.$$

$$(3.15)$$

Here, let us recall, for $V, W \in \Gamma(\varphi^{-1}TN))$,

$$\langle \overline{\nabla}^H V, \overline{\nabla}^H W \rangle = \sum_\alpha \langle \overline{\nabla}^H_{e_\alpha} V, \overline{\nabla}^H_{e_\alpha} W \rangle = \sum_{j=1}^{2n} \langle \overline{\nabla}_{X_i} V, \overline{\nabla}_{X_i} W \rangle,$$

where $\{e_\alpha\}$ is a locally defined orthonormal frame field of (M, g_θ) and $\overline{\nabla}^H_X W$ $(X \in \mathfrak{X}(M), W \in \Gamma(\varphi^{-1}TN))$ is defined by

$$\overline{\nabla}^H_X W = \sum_j \{(X^H f_j) V_j + f_j \overline{\nabla}_{X^H} V_j\}$$

for $W = \sum_j f_i V_j$ $(f_j \in C^\infty(M)$ and $V_j \in \Gamma(\varphi^{-1}TN)$. Here, X^H is the $H(M)$-component of X corresponding to the decomposition of $T_x(M) = H_x(M) \oplus \mathbb{R}T_x$ $(x \in M)$, and $\overline{\nabla}$ is the induced connection of $\varphi^{-1}TN$ from the Levi-Civita connection ∇^h of (N, h).

Since

$$\overline{\nabla}_{X_j}(\eta^2 \tau_b(\varphi)) = 2\eta X_j \eta \tau_b(\varphi) + \eta^2 \overline{\nabla}_{X_j} \tau_b(\varphi), \qquad (3.16)$$

the right hand side of (3.15) is equal to

$$\int_M \eta^2 \sum_{j=1}^{2n} \left| \overline{\nabla}_{X_j} \tau_b(\varphi) \right|^2 \theta \wedge (d\theta)^n$$

$$+ 2 \int_M \sum_{j=1}^{2n} \langle \eta \overline{\nabla}_{X_j} \tau_b(\varphi), (X_j \eta) \tau_b(\varphi) \rangle \, \theta \wedge (d\theta)^n.$$

$$(3.17)$$

Therefore, together with (3.14), we have

$$\int_M \eta^2 \sum_{j=1}^{2n} \left| \overline{\nabla}_{X_j} \tau_b(\varphi) \right|^2 \theta \wedge (d\theta)^n$$

$$\leq -2 \int_M \sum_{j=1}^{2n} \langle \eta \overline{\nabla}_{X_j} \tau_b(\varphi), (X_j \eta) \tau_b(\varphi) \rangle \, \theta \wedge (d\theta)^n$$

$$=: -2 \int_M \sum_{j=1}^{2n} \langle V_j, W_j \rangle \, \theta \wedge (d\theta)^n, \qquad (3.18)$$

where we define V_j, $W_j \in \Gamma(\varphi^{-1}TN)$ $(j = 1, \cdots, 2n)$ by

$$V_j := \eta \, \overline{\nabla}_{X_j} \tau_b(\varphi), \quad W_j := (X_j \eta) \, \tau_b(\varphi).$$

Then, since it holds that $0 \leq \left| \sqrt{\epsilon} \, V_i \pm \frac{1}{\sqrt{\epsilon}} W_i \right|^2$ for every $\epsilon > 0$, we have,

the right hand side of (3.18)

$$\leq \epsilon \int_M \sum_{j=1}^{2n} |V_j|^2 \, \theta \wedge (d\theta)^n + \frac{1}{\epsilon} \int_M \sum_{j=1}^{2n} |W_j|^2 \, \theta \wedge (d\theta)^n \tag{3.19}$$

foe every $\epsilon > 0$. By taking $\epsilon = \frac{1}{2}$, we obtain

$$\int_M \eta^2 \sum_{j=1}^{2n} \left| \overline{\nabla}_{X_j} \tau_b(\varphi) \right|^2 \theta \wedge (d\theta)^n$$

$$\leq \frac{1}{2} \int_M \sum_{j=1}^{2n} \eta^2 \left| \overline{\nabla}_{X_j} \tau_b(\varphi) \right|^2 \theta \wedge (d\theta)^n + 2 \int_M \sum_{j=1}^{2n} \left| X_j \eta \right|^2 \left| \tau_b(\varphi) \right|^2 \theta \wedge (d\theta)^n. \tag{3.20}$$

Therefore, we obtain, due to the properties that $\eta = 1$ on $B_r(x_0)$, and $\sum_{j=1}^{2n} |X_j \eta|^2 \leq |\nabla^{g_\theta} \eta|^2 \leq \left(\frac{2}{r} \right)^2$,

$$\int_{B_r(x_0)} \sum_{j=1}^{2n} \left| \overline{\nabla} \tau_b(\varphi) \right|^2 \theta \wedge (d\theta)^n \leq \int_M \eta^2 \sum_{j=1}^{2n} \left| \overline{\nabla}_{X_j} \tau_b(\varphi) \right|^2 \theta \wedge (d\theta)^n$$

$$\leq 4 \int_M \sum_{j=1}^{2n} \left| X_j \eta \right|^2 \left| \tau_b(\varphi) \right|^2 \theta \wedge (d\theta)^n$$

$$\leq \frac{16}{r^2} \int_M |\tau_b(\varphi)|^2 \theta \wedge (d\theta)^n. \tag{3.21}$$

(The third step)   By our assumption that $E_{b,2}(\varphi) = \frac{1}{2} \int_M |\tau_b(\varphi)|^2 \theta \wedge (d\theta)^n < \infty$ and (M, g_θ) is complete, if we let $r \to \infty$, then $B_r(x_0)$ goes to M, and the right hand side of (3.21) goes to zero. We have

$$\int_M \sum_{j=1}^{2n} \left| \overline{\nabla}_{X_j} \tau_b(\varphi) \right|^2 \theta \wedge (d\theta)^n = 0. \tag{3.22}$$

This implies that

$$\overline{\nabla}_X \tau_b(\varphi) = 0 \qquad \text{(for all } X \in H(M)\text{).} \tag{3.23}$$

(The fourth step)   Let us take a 1 form α on M defined by

$$\alpha(X) = \begin{cases} \langle d\varphi(X), \tau_b(\varphi) \rangle, & (X \in H(M)), \\ 0 & (X = T). \end{cases}$$

Then, we have

$$\int_M |\alpha| \, \theta \wedge (d\theta)^n = \int_M \left(\sum_{j=1}^{2n} \alpha(X_j)|^2 \right)^{\frac{1}{2}} \theta \wedge (d\theta)^n$$

$$\leq \left(|d_b\varphi|^2 \, \theta \wedge (d\theta)^n \right)^{\frac{1}{2}} \left(\int_M |\tau_b(\varphi)|^2 \, \theta \wedge (d\theta)^n \right)^{\frac{1}{2}}$$

$$= 2\sqrt{E_b(\varphi) \, E_{b,2}(\varphi)} < \infty, \tag{3.24}$$

where we put $d_b\varphi := \sum_{i=1}^{2n} d\varphi(X_i) \otimes X_i$,

$$|d_b\varphi|^2 = \sum_{i,j=1}^{2n} g_\theta(X_i, X_j) \, h(d\varphi(X_i), d\varphi(X_j)) = \sum_{i=1}^{2n} h(d\varphi(X_i), d\varphi(X_i)),$$

and

$$E_b(\varphi) = \frac{1}{2} \int_M |d_b\varphi|^2 \, \theta \wedge (d\theta)^n. \tag{3.25}$$

Furthermore, let us define a C^∞ function $\delta_b\alpha$ on M by

$$\delta_b\alpha = -\sum_{j=1}^{2n} (\nabla_{X_j}\alpha)(X_j) = -\sum_{j=1}^{2n} \left\{ X_j(\alpha(X_j)) - \alpha(\nabla_{X_j}X_j) \right\}, \tag{3.26}$$

where ∇ is the Tanaka-Webster connection. Notice that

$$\text{div}(\alpha) = \sum_{j=1}^{2n} (\nabla_{X_j}^{g_\theta}\alpha)(X_j) + (\nabla_T^{g_\theta}\alpha)(T)$$

$$= \sum_{j=1}^{2n} \left\{ X_j(\alpha \circ \pi_H(X_j)) - \alpha \circ \pi_H(\nabla_{X_j}^{g_\theta}X_j) \right\}$$

$$\quad + T(\alpha \circ \pi_H(T)) - \alpha \circ \pi_H(\nabla_T^{g_\theta}T)$$

$$= \sum_{j=1}^{2n} \left\{ X_j(\alpha(X_j)) - \alpha(\pi_H(\nabla_{X_j}^{g_\theta}X_j)) \right\}$$

$$= \sum_{j=1}^{2n} \left\{ X_j(\alpha(X_j)) - \alpha(\nabla_{X_j}X_j) \right\}$$

$$= -\delta_b\alpha, \tag{3.27}$$

where $\pi_H : T_x(M) \to H_x(M)$ is the natural projection. We used the facts that $\nabla_T^{g_\theta}T = 0$, and $\pi_H(\nabla_X^{g_\theta}Y) = \nabla_X Y$ $(X, Y \in H(M))$ ([**3**], p.37). Here, recall again ∇^{g_θ} is the Levi-Civita connection of g_θ, and ∇ is the Tanaka-Webster connection. Then, we have, for (3.26),

$$
\begin{aligned}
\delta_b \alpha &= -\sum_{j=1}^{2n} \left\{ X_j \langle d\varphi(X_j), \tau_b(\varphi) \rangle - \langle d\varphi(\nabla_{X_j} X_j), \tau_b(\varphi) \rangle \right\} \\
&= -\sum_{j=1}^{2n} \left\{ \begin{array}{l} \langle \overline{\nabla}_{X_j}(d\varphi(X_j)), \tau_b(\varphi) \rangle + \langle d\varphi(X_j), \overline{\nabla}_{X_j} \tau_b(\varphi) \rangle \\ - \langle d\varphi(\nabla_{X_j} X_j), \tau_b(\varphi) \rangle \end{array} \right\} \\
&= -\left\langle \sum_{j=1}^{2n} \left\{ \overline{\nabla}_{X_j}(d\varphi(X_j)) - d\varphi(\nabla_{X_j} X_j) \right\}, \tau_b(\varphi) \right\rangle \\
&= -|\tau_b(\varphi)|^2.
\end{aligned}
\tag{3.28}
$$

We used (3.23) $\overline{\nabla}_{X_j} \tau_b(\varphi) = 0$ to derive the last second equality of (3.28). Then, due to (3.28), we have for $E_{b,2}(\varphi)$,

$$
\begin{aligned}
E_{b,2}(\varphi) &= \frac{1}{2} \int_M |\tau_b(\varphi)|^2 \, \theta \wedge (d\theta)^n \\
&= -\frac{1}{2} \int_M \delta_b \alpha \, \theta \wedge (d\theta)^n \\
&= \frac{1}{2} \int_M \operatorname{div}(\alpha) \, \theta \wedge (d\theta)^n \\
&= 0.
\end{aligned}
\tag{3.29}
$$

In the last equality, we used Gaffney's theorem ([111], p. 271, [52]).

Therefore, we obtain $\tau_b(\varphi) \equiv 0$, i.e., φ is pseudo harmonic. □

4. Parallel pseudo biharmonic isometric immersion into rank one symmetric spaces

On the contrary of the Section Three, we consider isometric immersions into the unit sphere or the complex projective spaces which are pseudo biharmonic. One of the main theorem of this section is as follows:

THEOREM 4.1. Let $\varphi : (M, g_\theta) \to S^{2n+2}(1)$ be an isometric immersion of a CR manifold (M, g_θ) of $(2n + 1)$-dimension into the unit sphere $S^{2n+2}(1)$ of constant sectional curvature 1 and $(2n + 2)$-dimension. Assume that φ admits a parallel pseudo mean curvature vector field with non-zero pseudo mean curvature. The following equivalences hold: The immersion φ is pseudo biharmonic if and only if

$$
\sum_{i=1}^{2n} \lambda_i^2 = 2n
\tag{4.1}
$$

if and only if

$$\left\|B_\varphi\big|_{H(M)\times H(M)}\right\|^2 = 2n, \tag{4.2}$$

where λ_i ($1 \le i \le 2n+1$ are the principal curvatures of the immersion φ whose λ_{2n+1} corresponds to the characteristic vector field T of (M, g_θ), and $B_\varphi|_{H(M)\times H(M)}$ is the restriction of the second fundamental form od φ to the orthogonal complement $H(M)$ of T in the tangent space $(T_x(M), g_\theta)$.

As applications of this theorem, we will give pseudo biharmonic immersions into the unit sphere which are not pseudo harmonic.

The other case of rank one symmetric space is the complex projective space $\mathbb{P}^{n+1}(c)$. We obtain the following theorem:

THEOREM 4.2. *Let $\varphi : (M^{2n+1}, g_\theta \to \mathbb{P}^{n+1}(c)$ be an isometric immersion of CR manifold (M, g_θ) into the complex projective space $\mathbb{P}^{n+1}(c)$ of constant holomorphic sectional curvature c and complex $(n+1)$-dimension. Assume that φ has parallel pseudo-mean curvature vector filed with non-zero pseudo mean curvature. Then, the following equivalence relation holds: The immersion φ is pseudo-biharmonic if and only if the following hold:*
Either (1) $J(d\varphi(T))$ is tangent to $\varphi(M)$ and

$$\left\|B_\varphi\big|_{H(M)\times H(M)}\right\|^2 = \frac{c}{4}(2n+3), \tag{4.3}$$

or (2) $J(d\varphi(T))$ is normal to $\varphi(M)$ and

$$\left\|B_\varphi\big|_{H(M)\times H(M)}\right\|^2 = \frac{c}{4}(2n) = c\frac{n}{2}. \tag{4.4}$$

As applications of this theorem, we will give pseudo biharmonic, but not pseudo harmonic immersions (M, g_θ) into the complex projective space $\mathbb{P}^{n+1}(c)$.

5. Admissible immersions of strongly pseudoconvex CR manifolds

In this section, we introduce the notion of admissible isometric immersion of strongly pseudoconvex CR manifold (M, g_θ), and will show the following two lemmas related to $\Delta_b(\tau_b(\varphi))$ which are necessary to prove main theorems.

DEFINITION 5.1. *Let (M^{2n+1}, g_θ) be a strictly pseudoconvex CR manifold, and $T_x M = H_x(M) \oplus \mathbb{R}T_x$, $(x \in M)$, the orthogonal decomposition of the tangent space $T_x M$ $(x \in M)$, where T is the characteristic vector field of (M^{2n+1}, g_θ), $\varphi : (M^{2n+1}, g_\theta) \to (N, h)$ be an isometric immersion. The immersion φ is called to be* admissible *if the second fundamental form B_φ satisfies that*

$$B_\varphi(X, T) = 0 \qquad (5.1)$$

for all vector field X in $H(M)$.

The following clarifies the meaning of the admissibility condition:

PROPOSITION 5.1. *Let φ be an isometric immersion of a strongly pseudoconvex CR manifold (M^{2n+1}, g_θ) into another Riemannian manifold (N, h). Then, φ is admissible if and only if*
(1) $d\varphi(T_x)$ $(x \in M)$ is a principal curvature vector field along φ with some principal curvature $\lambda(x)$ $(x \in M)$.
    This is equivalent the following:
(2) The shape operator A_ξ of the immersion $\varphi : (M, g_\theta) \to (N, h)$ preserves $H_x(M)$ $(x \in M)$ invariantly for a normal vector field ξ.

(*Proof of Proposition 5.2*)   We first note for every normal vector field ξ of the isometric immersion $\varphi : (M, g_\theta) \to (N, h)$, it holds that

$$\langle B_\varphi(X, T), \xi \rangle = g_\theta(A_\xi X, T) = g_\theta(X, A_\xi T), \qquad (X \in H_x(M)). \quad (\#)$$

Thus, if φ is admissible, then the left hand side of ($\#$) vanishes, then we have immediately that

$$\begin{cases} A_\xi X \in H_x(M) & (X \in H_x(M)), \\ A_\xi T_x = \lambda(x)\, T_x & \text{(for some real number } \lambda(x)). \end{cases} \qquad (\flat)$$

Conversely, if one of the conditions of ($\flat$) holds, then it turns out immediately that φ is admissible. $\square$

The following two lemmas will be essential to us later.

LEMMA 5.1. *Let $\varphi : (M^{2n+1}, g_\theta) \to (N, h)$ be an admissible isometric immersion with parallel pseudo mean curvature vector field. Then, the pseudo tension field $\tau_b(\varphi)$ satisfies that*

$$-\Delta_b(\tau_b(\varphi)) = \langle -\Delta_b(\tau_b(\varphi)), d\varphi(X_i) \rangle \, d\varphi(X_i)$$
$$+ \left\langle \overline{\nabla}_{X_i} \tau_b(\varphi), d\varphi(X_j) \right\rangle \left(\widetilde{\nabla}_{X_i} d\varphi \right)(X_j),$$
$$(5.2)$$

where $\{X_j\}_{j=1}^{2n}$ is a local orthonormal frame field of $H(M)$ with respect to g_θ.

LEMMA 5.2. *Under the same assumptions of the above lemma, we have*

$$-\Delta_b(\tau_b(\varphi)) = \left\langle \tau_b(\varphi), R^h(d\varphi(X_j), d\varphi(X_k))d\varphi(X_k) \right\rangle d\varphi(X_j)$$
$$+ \left\langle \tau_b(\varphi), R^h(d\varphi(X_j), d\varphi(T))d\varphi(T) \right\rangle d\varphi(X_j)$$
$$- \left\langle \tau_b(\varphi), \left(\widetilde{\nabla}_{X_i}d\varphi\right)(X_j) \right\rangle \left(\widetilde{\nabla}_{X_i}d\varphi\right)(X_j), \qquad (5.3)$$

where $R^h(U,V)W$ *is the curvature tensor field of* (N,h) *defined by* $R^h(U,V)W = \nabla_U^h(\nabla_V^h W) - \nabla_V^h(\nabla_U^N W) - \nabla_{[U,V]}^h W$ *for vector fields* U, V, W *on* N, *and* ∇^h *is the Levi-Civita connection of* (N,h).

(*Proof of Lemma 5.3*)   The proof is divided into several steps.

(The first step)   Since we assume the pseudo mean curvature vector field $\tau_b(\varphi)$ is parallel, i.e., $\overline{\nabla}_X^\perp \tau_b(\varphi) = 0$ $(X \in \mathfrak{X}(M))$, the induced connection $\overline{\nabla}$ of the Levi-Civita connection ∇^h to the induced bundle $\varphi^{-1}TN$ satisfies that, for all $X \in \mathfrak{X}(M)$,

$$\overline{\nabla}_X \tau_b(\varphi) = \overline{\nabla}_X^\top \tau_b(\varphi) + \overline{\nabla}_X^\perp \tau_b(\varphi) = \overline{\nabla}_X^\top \tau_b(\varphi) \in \Gamma(\varphi_* TM).$$

Then. we have, for all $X \in H(M)$,

$$\overline{\nabla}_X \tau_b(\varphi) = \sum_{j=1}^{2n} \langle \overline{\nabla}_X \tau_b(\varphi), d\varphi(X_j) \rangle \, d\varphi(X_j) + \langle \overline{\nabla}_X \tau_b(\varphi), d\varphi(T) \rangle \, d\varphi(T)$$

$$= \sum_{j=1}^{2n} \langle \overline{\nabla}_X \tau_b(\varphi), d\varphi(X_j) \rangle \, d\varphi(X_j). \qquad (5.4)$$

Due to the assumption of the admissibility of φ, for all $X \in H(M)$,

$$\langle \overline{\nabla}_X \tau_b(\varphi), d\varphi(T) \rangle = X \langle \tau_b(\varphi), d\varphi(T) \rangle - \langle \tau_b(\varphi), \overline{\nabla}_X d\varphi(T) \rangle = 0. \qquad (5.5)$$

In fact, $\tau_b(\varphi) = \sum_{i=1}^{2n} B_\varphi(X_i, X_i)$ is orthogonal to $d\varphi(TM)$ with respect to $\langle \ , \ \rangle$, we have $\langle \tau_b(\varphi), d\varphi(T) \rangle = 0$. So, the first term of (5.5) vanishes. By the admissibility of φ, for all $X \in H(M)$,

$$0 = B_\varphi(X, T) = \overline{\nabla}_X d\varphi(T) - d\varphi(\nabla_X^{g_\theta} T), \qquad (5.6)$$

so that $\overline{\nabla}_X d\varphi(T)$ is tangential, which implies that

$$\langle \tau_b(\varphi), \overline{\nabla}_X d\varphi(T) \rangle = 0.$$

We have (5.5), and then (5.4).

(The second step)   We calculate $-\Delta_b(\tau_b(\varphi))$. We have by (5.4),

$$-\Delta_b(\tau_b(\varphi)) = \sum_{i=1}^{2n} \left\{ \overline{\nabla}_{X_i}(\overline{\nabla}_{X_i}\tau_b(\varphi)) - \overline{\nabla}_{\nabla_{X_i}X_i}\tau_b(\varphi) \right\}$$

$$= \sum_{i=1}^{2n} \left[\begin{array}{c} \sum_{j=1}^{2n} \overline{\nabla}_{X_i} \left\{ \langle \overline{\nabla}_X \tau_b(\varphi), d\varphi(X_j) \rangle \, d\varphi(X_j) \right\} \\ - \sum_{j=1}^{2n} \langle \overline{\nabla}_{\nabla_{X_i}X_i}\tau_b(\varphi), d\varphi(X_j) \rangle \, d\varphi(X_j) \end{array} \right]$$

$$= \sum_{i,j=1}^{2n} \left[\begin{array}{c} \langle \overline{\nabla}_{X_i}(\overline{\nabla}_{X_i}\tau_b(\varphi)), d\varphi(X_j) \rangle \, d\varphi(X_j) \\ + \langle \overline{\nabla}_{X_i}\tau_b(\varphi), \overline{\nabla}_{X_i}(d\varphi(X_j)) \rangle \, d\varphi(X_j) \\ + \langle \overline{\nabla}_{X_i}\tau_b(\varphi), d\varphi(X_j) \rangle \, \overline{\nabla}_{X_i}d\varphi(X_j) \\ - \langle \overline{\nabla}_{\nabla_{X_i}X_i}\tau_b(\varphi), d\varphi(X_j) \rangle \, d\varphi(X_j) \end{array} \right]$$

$$= \sum_{j=1}^{2n} \langle -\Delta_b(\tau_b(\varphi)), d\varphi(X_j) \rangle \, d\varphi(X_j)$$

$$+ \sum_{i,j=1}^{2n} \left[\begin{array}{c} \langle \overline{\nabla}_{X_i}\tau_b(\varphi), \overline{\nabla}_{X_i}(d\varphi(X_j)) \rangle \, d\varphi(X_j) \\ + \langle \overline{\nabla}_{X_i}\tau_b(\varphi), d\varphi(X_j) \rangle \, \overline{\nabla}_{X_i}d\varphi(X_j) \end{array} \right]. \quad (5.7)$$

(The third step)   Here, we have

$$\begin{cases} (\widetilde{\nabla}_{X_i}d\varphi)(X_j) = \overline{\nabla}_{X_i}d\varphi(X_j) - d\varphi(\nabla_{X_i}X_j) \in T^\perp M, \\ \overline{\nabla}_{X_i}\tau_b(\varphi) \in T^\perp M, \end{cases} \quad (5.8)$$

where ∇ is the Tanaka-Webster connection and $\nabla_{X_i}X_j \in H(M)$. Then, we have, in the first term of the second sum of (5.7),

$$\sum_{i,j=1}^{2n} \langle \overline{\nabla}_{X_i}\tau_b(\varphi), \overline{\nabla}_{X_i}d\varphi(X_j) \rangle \, d\varphi(X_j)$$

$$= \sum_{i,j=1}^{2n} \left\langle \overline{\nabla}_{X_i}\tau_b(\varphi), (\widetilde{\nabla}_{X_i}d\varphi)(X_j) + d\varphi(\nabla_{X_i}X_j) \right\rangle d\varphi(X_j)$$

$$= \sum_{i,j=1}^{2n} \langle \overline{\nabla}_{X_i}\tau_b(\varphi), d\varphi(\nabla_{X_i}X_j) \rangle \, d\varphi(X_j)$$

$$= \sum_{i,j=1}^{2n} \left\langle \overline{\nabla}_{X_i}\tau_b(\varphi), d\varphi \left(\sum_{k=1}^{2n} \langle \nabla_{X_i}X_j, X_k \rangle X_k \right) \right\rangle d\varphi(X_j), \quad (5.9)$$

because of $\nabla_{X_i} X_j \in H(M)$. Since the Tanaka-Webster connection ∇ satisfies $\nabla g_\theta = 0$, we have

$$\langle \nabla_{X_i} X_j, X_k \rangle = X_i \langle X_j, X_k \rangle - \langle X_j, \nabla_{X_i} X_k \rangle = -\langle X_j, \nabla_{X_i} X_k \rangle.$$

Thus, (5.9) turns to

$$\sum_{i,j=1}^{2n} \langle \overline{\nabla}_{X_i} \tau_b(\varphi), \overline{\nabla}_{X_i} d\varphi(X_j) \rangle \, d\varphi(X_j)$$

$$= \sum_{i,j,k=1}^{2n} \langle \nabla_{X_i} X_j, X_k \rangle \langle \overline{\nabla}_{X_i} \tau_b(\varphi), d\varphi(X_k) \rangle \, d\varphi(X_j)$$

$$= - \sum_{i,j,k=1}^{2n} \langle X_j, \nabla_{X_i} X_k \rangle \langle \overline{\nabla}_{X_i} \tau_b(\varphi), d\varphi(X_k) \rangle \, d\varphi(X_j)$$

$$= - \sum_{i,k=1}^{2n} \langle \overline{\nabla}_{X_i} \tau_b(\varphi), d\varphi(X_k) \rangle \, d\varphi(\nabla_{X_i} X_k). \tag{5.10}$$

(The fourth step)   By inserting (5.10) into (5.7), the second sum of (5.7) turns to

$$\sum_{i,j=1}^{2n} \left[\begin{array}{l} \langle \overline{\nabla}_{X_i} \tau_b(\varphi), \overline{\nabla}_{X_i}(d\varphi(X_j)) \rangle \, d\varphi(X_j) \\ + \langle \overline{\nabla}_{X_i} \tau_b(\varphi), d\varphi(X_j) \rangle \, \overline{\nabla}_{X_i} d\varphi(X_j) \end{array} \right]$$

$$= \sum_{i,j=1}^{2n} \left[\begin{array}{l} \langle \overline{\nabla}_{X_i} \tau_b(\varphi), d\varphi(X_j) \rangle \, \overline{\nabla}_{X_i} d\varphi(X_j) \\ - \langle \overline{\nabla}_{X_i} \tau_b(\varphi), d\varphi(X_j) \rangle \, d\varphi(\nabla_{X_i} X_j) \end{array} \right]$$

$$= \sum_{i,j=1}^{2n} \langle \overline{\nabla}_{X_i} \tau_b(\varphi), d\varphi(X_j) \rangle \left(\widetilde{\nabla}_{X_i} d\varphi \right)(X_j). \tag{5.11}$$

Thus, by (5.7) and (5.11), we obtain Lemma 5.3.   □

(*Proof of Lemma 5.4*)   We will calculate the right hand side of (5.2) in Lemma 5.3. The proof is divided into several steps.

(The first step)   We first note that

$$\langle \overline{\nabla}_{X_i} \tau_b(\varphi), d\varphi(X_j) \rangle + \langle \tau_b(\varphi), \overline{\nabla}_{X_i} d\varphi(X_j) \rangle = X_i \langle \tau_b(\varphi), d\varphi(X_j) \rangle$$

$$= 0. \tag{5.12}$$

Thus, by (5.12), we have

$$
\begin{aligned}
\langle \overline{\nabla}_{X_i} \tau_b(\varphi), d\varphi(X_j) \rangle &= -\langle \tau_b(\varphi), \overline{\nabla}_{X_i} d\varphi(X_j) \rangle \\
&= -\langle \tau_b(\varphi), \overline{\nabla}_{X_i} d\varphi(X_j) - d\varphi(\nabla_{X_i} X_j) \rangle \\
&= -\langle \tau_b(\varphi), \overline{\nabla}_{X_i} d\varphi(X_j) - d\varphi(\nabla^{g_\theta}_{X_i} X_j) \rangle \\
&\qquad\qquad (\text{by } \langle \tau_b(\varphi), d\varphi(T) \rangle = 0) \\
&= -\langle \tau_b(\varphi), \left(\widetilde{\nabla}_{X_i} d\varphi \right)(X_j) \rangle. \qquad (5.13)
\end{aligned}
$$

(The second step)   By differentiating (5.12), we have

$$
\left\langle \overline{\nabla}_{X_i} \left(\overline{\nabla}_{X_i} \tau_b(\varphi) \right), d\varphi(X_j) \right\rangle + 2 \left\langle \overline{\nabla}_{X_i} \tau_b(\varphi), \overline{\nabla}_{X_i} d\varphi(X_j) \right\rangle \\
+ \left\langle \tau_b(\varphi), \overline{\nabla}_{X_i} \left(\overline{\nabla}_{X_i} d\varphi(X_j) \right) \right\rangle = 0. \\
(5.14)
$$

And we have

$$
\left\langle \overline{\nabla}_{\nabla_{X_i} X_i} \tau_b(\varphi), d\varphi(X_j) \right\rangle + \left\langle \tau_b(\varphi), \overline{\nabla}_{\nabla_{X_i} X_i} d\varphi(X_j) \right\rangle \\
= \nabla_{X_i} X_i \langle \tau_b(\varphi), d\varphi(X_j) \rangle = 0. \\
(5.15)
$$

Thus, by (5.14) and (5.15), we have

$$
\langle -\Delta_b \left(\tau_b(\varphi) \right), d\varphi(X_j) \rangle + 2 \langle \overline{\nabla}_{X_i} \tau_b(\varphi), \overline{\nabla}_{X_i} d\varphi(X_j) \rangle \\
+ \langle \tau_b(\varphi), -\Delta_b(d\varphi(X_j)) \rangle = 0. \qquad (5.16)
$$

(The third step)   For the second term of the left hand side of (5.16), we have

$$
2 \langle \overline{\nabla}_{X_i} \tau_b(\varphi), \overline{\nabla}_{X_i} d\varphi(X_j) \rangle = -2 \left\langle \tau_b(\varphi), \left(\widetilde{\nabla}_{X_i} d\varphi \right) (\nabla_{X_i} X_j) \right\rangle. \\
(5.17)
$$

Because, the left hand side of (5.17) is

$$
\begin{aligned}
2 \langle \overline{\nabla}_{X_i} \tau_b(\varphi), \overline{\nabla}_{X_i} d\varphi(X_j) \rangle &= 2 \left\langle \overline{\nabla}_{X_i} \tau_b(\varphi), \left(\widetilde{\nabla}_{X_i} d\varphi \right)(X_j) + d\varphi \left(\nabla^{g_\theta}_{X_i} X_j \right) \right\rangle \\
&= 2 \left\langle \overline{\nabla}_{X_i} \tau_b(\varphi), \left(\widetilde{\nabla}_{X_i} d\varphi \right)(X_j) + d\varphi \left(\nabla_{X_i} X_j \right) \right\rangle \\
&= 2 \left\langle \overline{\nabla}_{X_i} \tau_b(\varphi), d\varphi \left(\nabla_{X_i} X_j \right) \right\rangle \\
&= -2 \left\langle \tau_b(\varphi), \overline{\nabla}_{X_i} d\varphi(\nabla_{X_i} X_j) \right\rangle \\
&\qquad\qquad (\text{by } \langle \tau_b(\varphi), d\varphi(\nabla_{X_i} X_j) \rangle = 0) \\
&= -2 \left\langle \tau_b(\varphi), \left(\widetilde{\nabla}_{X_i} d\varphi \right)(\nabla_{X_i} X_j) \right\rangle \qquad (5.18)
\end{aligned}
$$

which is the right hand side of (5.17). In the last step of (5.18), we used the equality $\langle \tau_b(\varphi), \overline{\nabla}_{X_i} d\varphi(T) \rangle = 0$ which follows from that $\overline{\nabla}_{X_i} d\varphi(T)$ is tangential.

(The fourth step)   For the third term of the left hand side of (5.16), we have

$$\langle \tau_b(\varphi), -\Delta_b(d\varphi(X_j))) \rangle = \langle \tau_b(\varphi), \left(-\tilde{\Delta}_b d\varphi\right)(X_j)$$
$$+ 2\sum_{k=1}^{2n}(\widetilde{\nabla}_{X_k} d\varphi)(\nabla_{X_k} X_j)\rangle. \qquad (5.19)$$

Because, by the definition of Δ_b, we have

$$\langle \tau_b(\varphi), -\Delta_b(d\varphi(X_j))) \rangle$$
$$= \langle \tau_b(\varphi), \sum_{k=1}^{2n}\left\{\overline{\nabla}_{X_k}\left(\widetilde{\nabla}_{X_k} d\varphi(X_j)\right) - \widetilde{\nabla}_{\nabla_{X_k} X_k} d\varphi(X_j)\right\}\rangle$$
$$= \langle \tau_b(\varphi), \sum_{k=1}^{2n}\left\{\overline{\nabla}_{X_k}\left((\widetilde{\nabla}_{X_k} d\varphi)(X_j) + d\varphi(\nabla_{X_k} X_j)\right)\right.$$
$$\left. - \left(\widetilde{\nabla}_{\nabla_{X_k} X_k} d\varphi\right)(X_j) - d\varphi\left(\nabla_{\nabla_{X_k} X_k} X_j\right)\right\}\rangle$$
$$= \langle \tau_b(\varphi), \sum_{k=1}^{2n}\left\{\left(\widetilde{\nabla}_{X_k}\widetilde{\nabla}_{X_k} d\varphi\right)(X_j) + \left(\widetilde{\nabla}_{X_k} d\varphi\right)(\nabla_{X_k} X_j)\right.$$
$$+ \left(\widetilde{\nabla}_{X_k} d\varphi\right)(\nabla_{X_k} X_j) + d\varphi(\nabla_{X_k}\nabla_{X_k} X_j)$$
$$\left. - \left(\widetilde{\nabla}_{\nabla_{X_k} X_k} d\varphi\right)(X_j)\right\}\rangle$$
$$= \langle \tau_b(\varphi), \left(-\tilde{\Delta}_b d\varphi\right)(X_j) + 2\sum_{k=1}^{2n}\left(\widetilde{\nabla}_{X_k} d\varphi\right)(\nabla_{X_k} X_j)\rangle,$$

which is (5.19). To get the last equality of the above, we used the following equations: for all $X \in H(M)$, it holds that

$$\langle \tau_b(\varphi), (\overline{\nabla}_X d\varphi)(T)\rangle = \langle \tau_b(\varphi), d\varphi(X)\rangle = \langle \tau_b(\varphi), (\widetilde{\nabla}_X d\varphi)(T)\rangle = 0.$$
$$(5.20)$$

To get (5.20), due to the admissibility of φ, we have

$$(\widetilde{\nabla}_X d\varphi)(T) = (\overline{\nabla}_X d\varphi)(T) - d\varphi(\nabla_X^{g_\theta} T) = B_\varphi(X, T) = 0,$$

and then, $(\overline{\nabla}_X d\varphi)(T)$ is tangential for all $X \in H(M)$. We have (5.20), and then (5.19).

(The fifth step)   Then, the right hand side of (5.19) is equal to

$$\langle \tau_b(\varphi), (-\tilde{\Delta}_b d\varphi)(X_j) + 2 \sum_{k=1}^{2n} (\tilde{\nabla}_{X_k} d\varphi)(\nabla_{X_k} X_j) \rangle$$

$$= \langle \tau_b(\varphi), \tilde{\nabla}_{X_j} \tau(\varphi) \rangle$$

$$- \sum_{k=1}^{2n} \left\{ R^h(d\varphi(X_j), d\varphi(X_k)) d\varphi(X_k) - d\varphi(R^{g_\theta}(X_j, X_k) X_k) \right\}$$

$$- \sum_{k=1}^{2n} \left\{ R^h(d\varphi(X_j), d\varphi(T)) d\varphi(T) - d\varphi((R^{g_\theta}(X_j, T) T \right\}$$

$$- \tilde{\nabla}_T \tilde{\nabla}_T d\varphi(X_j)$$

$$+ 2(\tilde{\nabla}_{X_k} d\varphi)(\nabla_{X_k} X_j) \rangle, \tag{5.21}$$

which follows from the formula (3.11).

Here, notice that

$$\langle \tau_b(\varphi), \tilde{\nabla}_{X_j} \tau(\varphi) \rangle = 0. \tag{5.22}$$

Because $\tau_b(\varphi)$ is normal, and $\tilde{\nabla}_{X_j} \tau(\varphi)$ is tangential. And also we have

$$\langle \tau_b(\varphi), \tilde{\nabla}_T (d\varphi(X_j)) \rangle = 0, \tag{5.23}$$

$$\langle \tau_b(\varphi), \tilde{\nabla}_T \tilde{\nabla}_T d\varphi(X_j) \rangle = 0. \tag{5.24}$$

To see (5.23), since we assume φ is an admissible isometric immersion, we have

$$\tilde{\nabla}_T d\varphi(X_j) = \nabla_T^h X_j = \nabla_T^{g_\theta} X_j + B_\varphi(T, X_j) = \nabla_T^{g_\theta} X_j \tag{5.25}$$

which is tangential, so that we have (5.23). Furthermore, to see (5.24), we have

$$\tilde{\nabla}_T \tilde{\nabla}_T d\varphi(X_j) = \tilde{\nabla}(\nabla_T^{g_\theta} X_j)$$

$$= \nabla_T^h (\nabla_T^{g_\theta} X_j)$$

$$= \nabla_T^{g_\theta} (\nabla_T^{g_\theta} X_j) + B(T, \nabla_T^{g_\theta} X_j). \tag{5.26}$$

Here, for every $X \in H(M)$,

$$\nabla_T^{g_\theta} X \in H(M).$$

Indeed, since $g_\theta(T, X) = 0$, and $\nabla_T^{g_\theta} T = 0$ (cf. [**38**], pp. 47, and 48),

$$g_\theta(T, \nabla_T^{g_\theta} X) = T(g_\theta(T, X)) - g_\theta(\nabla_T^{g_\theta} T, X) = 0,$$

which implies $\nabla_T^{g_\theta} X \in H(M)$. Thus, the admissibility implies that the second term of (5.26) vanishes. Thus, the right hand side of (5.26) is tangential, which implies (5.24).

Therefore, we obtain

$$\langle \tau_b(\varphi), (-\tilde{\Delta}_b d\varphi)(X_j) + 2\sum_{k=1}^{2n} (\widetilde{\nabla}_{X_k} d\varphi)(\nabla_{X_k} X_j)\rangle$$

$$= -\sum_{k=1}^{2n} \langle \tau_b(\varphi), R^h(d\varphi(X_j), d\varphi(X_k))d\varphi(X_k)\rangle$$

$$- \sum_{k=1}^{2n} \langle \tau_b(\varphi), R^h(d\varphi(X_j), d\varphi(T))d\varphi(T)\rangle$$

$$+ 2\sum_{k=1}^{2n} \langle \tau_b(\varphi), (\widetilde{\nabla}_{X_k} d\varphi)(\nabla_{X_k} X_j)\rangle. \tag{5.27}$$

(The sixth step)   Now, return to (5.16), by using (5.17), (5.19) and (5.27), we have

$$0 = \langle -\Delta_b(\tau_b(\varphi)), d\varphi(X_j)\rangle + 2\langle \overline{\nabla}_{X_i} \tau_b(\varphi), \overline{\nabla}_{X_i} d\varphi(X_j)\rangle$$

$$+ \langle \tau_b(\varphi), -\Delta_b(d\varphi(X_j))\rangle$$

$$= \langle -\Delta_b(\tau_b(\varphi)), d\varphi(X_j)\rangle - 2\langle \tau_b(\varphi), (\widetilde{\nabla}_{X_i} d\varphi)(\nabla_{X_i} X_j)\rangle$$

$$+ \langle \tau_b(\varphi), -\Delta_b(d\varphi(X_j))\rangle$$

$$= \langle -\Delta_b(\tau_b(\varphi)), d\varphi(X_j)\rangle - 2\langle \tau_b(\varphi), (\widetilde{\nabla}_{X_i} d\varphi)(\nabla_{X_i} X_j)\rangle$$

$$+ \langle \tau_b(\varphi), (-\tilde{\Delta}_b d\varphi)(X_j) + 2\sum_{k=1}^{2n} (\widetilde{\nabla}_{X_k} d\varphi)(\nabla_{X_k} X_j)\rangle$$

$$= \langle -\Delta_b(\tau_b(\varphi)), d\varphi(X_j)\rangle - 2\langle \tau_b(\varphi), (\widetilde{\nabla}_{X_i} d\varphi)(\nabla_{X_i} X_j)\rangle$$

$$+ \Big\langle \tau_b(\varphi), -\sum_{k=1}^{2n} R^h(d\varphi(X_j), d\varphi(X_k))d\varphi(X_k)$$

$$- R^h(d\varphi(X_j), d\varphi(T))d\varphi(T)$$

$$+ 2\sum_{k=1}^{2n} (\widetilde{\nabla}_{X_k} d\varphi)(\nabla_{X_k} X_j)\Big\rangle$$

$$= \langle -\Delta_b(\tau_b(\varphi)), d\varphi(X_j)\rangle$$

$$+ \Big\langle \tau_b(\varphi), -\sum_{k=1}^{2n} R^h(d\varphi(X_j), d\varphi(X_k))d\varphi(X_k)$$

$$- R^h(d\varphi(X_j), d\varphi(T))d\varphi(T)\Big\rangle. \tag{5.28}$$

(The seventh step)   Inserting (5.28) into (5.2) of Lemma 5.3, we obtain

$$-\Delta_b(\tau_b(\varphi)) = \langle \tau_b(\varphi), \sum_{k=1}^{2n} R^h(d\varphi(X_j), d\varphi(X_k))d\varphi(X_k) \rangle \, d\varphi(X_j)$$
$$+ \langle \tau_b(\varphi), R^h(d\varphi(X_j), d\varphi(T))d\varphi(T) \rangle \, d\varphi(X_j)$$
$$+ \langle \overline{\nabla}_{X_i} \tau_b(\varphi), d\varphi(X_j) \rangle \, (\widetilde{\nabla}_{X_i} d\varphi)(X_j). \qquad (5.29)$$

At last, for the third term of (5.29), we have

$$\langle \overline{\nabla}_{X_i} \tau_b(\varphi), d\varphi(X_j) \rangle = X_i \, \langle \tau_b(\varphi), d\varphi(X_j) \rangle - \langle \tau_b(\varphi), \overline{\nabla}_{X_i} d\varphi(X_j) \rangle$$
$$= -\langle \tau_b(\varphi), (\widetilde{\nabla}_{X_i} d\varphi)(X_j) \rangle. \qquad (5.30)$$

Together with (5.29) and (5.30), we have (5.3) of Lemma 5.4.   □

Due to Lemma 5.4 and the definition of biharmonicity, we obtain immediately

THEOREM 5.1. *Let φ be an admissible isometric immersion of a strongly pseudoconvex CR manifold (M, g_θ) into another Riemannian manifold (N, h) whose pseudo mean curvature vector field along φ is parallel. Then, φ is pseudo biharmonic if and only if*

$$\tau_{b,2}(\varphi) := \Delta_b\big(\tau_b(\varphi)\big) - \sum_{j=1}^{2n} R^h\big(\tau_b(\varphi), d\varphi(X_j)\big) d\varphi(X_j) = 0 \qquad (5.31)$$

if and only if

$$-\sum_{j,k=1}^{2n} h\big(\tau_b(\varphi), R^h(d\varphi(X_j), d\varphi(X_k))d\varphi(X_k)\big) \, d\varphi(X_j)$$
$$-\sum_{j=1}^{2n} h\big(\tau_b(\varphi), R^h(d\varphi(X_j), d\varphi(T))d\varphi(T)\big) \, d\varphi(X_j)$$
$$+\sum_{j,k=1}^{2n} h\big(\tau_b(\varphi), B_\varphi(X_j, X_k)\big) \, B_\varphi(X_j, X_k)$$
$$-\sum_{j=1}^{2m} R^h\big(\tau_b(\varphi), d\varphi(X_j)\big) d\varphi(X_j) = 0, \qquad (5.32)$$

where $\{X_j\}_{j=1}^{2n}$ is an orthonormal frame field of $(H(M), g_\theta)$.

REMARK 5.1. *Due to [**74**], p. 220, Lemma 10, and the definition of bi-tension field $\tau_2(\varphi)$ for an isometric immersion φ of a Riemannian manifold (M, g) into another Riemannian manifold (N, h), we can also obtain immediately the following useful theorem:*

Theorem *Let φ be an isometric immersion of a Riemannian manifold (M^m, g) into another Riemannian manifold (N^n, h) whose mean curvature vector field along φ is parallel. Let $\{e_j\}_{j=1}^m$ be an orthonormal frame field of (M, g). Then, φ is biharmonic if and only if*

$$\tau_2(\varphi) := \overline{\Delta}\big(\tau(\varphi)\big) - \sum_{j=1}^m R^h\big(\tau(\varphi), e_j\big)e_j = 0 \tag{5.33}$$

if and only if

$$-\sum_{j,k=1}^m h\big(\tau(\varphi), R^h(d\varphi(e_j), d\varphi(e_k))d\varphi(e_k)\big)\, d\varphi(e_j)$$

$$+\sum_{j,k=1}^m h\big(\tau(\varphi), B_\varphi(e_j, e_k)\big)\, B_\varphi(e_j, e_k)$$

$$-\sum_{j=1}^m R^h\big(\tau(\varphi), d\varphi(e_j)\big)d\varphi(e_j) = 0. \tag{5.34}$$

6. Isometric immersions into the unit sphere

In this section, we treat with admissible isometric immersions of (M^{2n+1}, g_θ) into the unit sphere $(N, h) = S^{2n+2}(1)$ with parallel pseudo mean curvature vector field with non-zero pseudo mean curvature.

The curvature tensor field R^h of the target space $(N, h) = S^{2n+2}(1)$ satifies that

$$R^h(X, Y)Z = h(Z, Y)\, X - h(Z, X)\, Y \tag{6.1}$$

for all vector fields X, Y, Z on N. Then, we have

$$\left(R^h(d\varphi(X_j), d\varphi(X_k))\, d\varphi(X_k)\right)^\perp = 0, \tag{6.2}$$

$$\left(R^h(d\varphi(X_j), d\varphi(T))d\varphi(T)\right)^\perp = 0, \tag{6.3}$$

for all $i, j = 1, \cdots, 2n$. Therefore, we obtain by (5.3) in Lemma 5.4,

$$-\Delta_b(\tau_b(\varphi)) = -\left\langle \tau_b(\varphi), \left(\widetilde{\nabla}_{X_i}d\varphi\right)(X_j)\right\rangle \left(\widetilde{\nabla}_{X_i}d\varphi\right)(X_j). \tag{6.4}$$

On the other hand, we have

$$\sum_{k=1}^{2n} R^h(\tau_b(\varphi), d\varphi(X_k))d\varphi(X_k)$$

$$= \sum_{k=1}^{2n} h(d\varphi(X_k), d\varphi(X_k))\, \tau_b(\varphi) - \sum_{k=1}^{2n} h(d\varphi(X_k), \tau_b(\varphi))\, \tau_b(\varphi)$$

$$= 2n\, \tau_b(\varphi). \tag{6.5}$$

Now, let us recall the pseudo biharmonicity of φ is equivalent to that

$$-\Delta_b(\tau_b(\varphi)) + \sum_{k=1}^{2n} R^h(\tau_b(\varphi), d\varphi(X_k))\, d\varphi(X_k) = 0 \tag{6.6}$$

which is equivalent to that

$$-\left\langle \tau_b(\varphi), \left(\widetilde{\nabla}_{X_i} d\varphi\right)(X_j)\right\rangle \left(\widetilde{\nabla}_{X_i} d\varphi\right)(X_j) + 2n\, \tau_b(\varphi) = 0. \tag{6.7}$$

For our immersion $\varphi : (M, g_\theta) \to S^{2n+2}(1)$, let ξ be the unit normal vector filed on M along φ, we have by definition,

$$\left(\widetilde{\nabla}_{X_i} d\varphi\right)(X_j) = B_\varphi(X_i, X_j) = H_{ij}\, \xi. \tag{6.8}$$

Then, we have by definition of $\tau_b(\varphi)$,

$$\tau_b(\varphi) = \sum_{i=1}^{2n} \left(\widetilde{\nabla}_{X_i} d\varphi\right)(X_i) = \left(\sum_{i=1}^{2n} H_{ii}\right)\xi. \tag{6.9}$$

Therefore, we have

$$\|\tau_b(\varphi)\|^2 = \left(\sum_{i=1}^{2n} H_{ii}\right)^2 \|\xi\|^2 = \left(\sum_{i=1}^{2n} H_{ii}\right)^2. \tag{6.10}$$

By the admissibility, we have

$$\|B_\varphi\|^2 = \sum_{i,j=1}^{2n} \|B_\varphi(X_i, X_j)\|^2 + 2\sum_{i=1}^{2n} \|B_\varphi(X_i, T)\|^2 + \|B_\varphi(T, T)\|^2$$

$$= \sum_{i,j=1}^{2n} \|H_{ij}\, \xi\|^2 + \|B_\varphi(T, T)\|^2$$

$$= \sum_{i,j=1}^{2n} H_{ij}{}^2 + \|B_\varphi(T, T)\|^2. \tag{6.11}$$

Due to (6.7), (6.8) and (6.9), the biharmonicity of φ is equivalent to that

$$0 = -\left\langle \left(\sum_{k=1}^{2n} H_{kk}\right)\xi, H_{ij}\,\xi\right\rangle H_{ij}\,\xi + 2n\left(\sum_{k=1}^{2n} H_{kk}\right)\xi \tag{6.12}$$

which is equivalent to that

$$0 = \left(\sum_{k=1}^{2n} H_{kk}\right)\left\{-\sum_{i,j=1}^{2n} H_{ij}{}^2 + 2n\right\}$$

$$= \|\tau_b(\varphi)\|\left\{-\|B_\varphi\|^2 + \|B_\varphi(T,T)\|^2 + 2n\right\} \tag{6.13}$$

by (6.11). By our assumption of non-zero pseudo mean curvature, $\|\tau_b(\varphi)\| \neq 0$ at every point, we obtain the following equivalence relation: φ is pseudo biharmonic if and only if

$$\|B_\varphi\|^2 = \|B_\varphi(T,T)\|^2 + 2n \tag{6.14}$$

at every point in M.

By summing up the above, we obtain the following theorem:

THEOREM 6.1. *Let φ be an sdmissible isometric immersion of a strictly pseudoconvex CR manifold (M, g_θ) into the unit sphere $(N, h) = S^{2n+2}(1)$. Assume that the pseudo mean curvature vector field is parallel with non-zero pseudo mean curvature. Then, φ is pseudo biharmonic if and only if*

$$\|B_\varphi\|^2 = \|B_\varphi(T,T)\|^2 + 2n. \tag{6.15}$$

The admissibility condition is that: $d\varphi(T)$ is the principal curvature vector field along φ with some principal curvature, say λ_{2n+1}. I.e.,

$$A_\xi T = \lambda_{2n+1}\,T.$$

Then, we have

$$\|B_\varphi\|^2 = \sum_{i=1}^{2n+1}\lambda_i{}^2, \quad \text{and } \|B_\varphi(T,T)\|^2 = \lambda_{2n+1}.$$

By Theorem 6.1, we have immediately Thus, we obtain

COROLLARY 6.1. *Let $\varphi : (M^{2n+1}, g_\theta) \to S^{2n+2}(1)$ be an isometric immersion whose the pseudo mean curvature vector field is parallel and has non-zero pseudo mean curvature. Then, φ is pseudo biharmonic if and only if it holds that*

$$\sum_{i=1}^{2n}\lambda_i{}^2 = 2n \tag{6.16}$$

which is equivalent to that

$$\left\| B_\varphi \Big|_{H(M) \times H(M)} \right\|^2 = 2n, \tag{6.17}$$

where $B_\varphi|_{H(M) \times H(M)}$ is the restriction of B_φ to the subspace $H(M)$ of the tangent space $T_x M$ ($x \in M$).

7. Isometric immersions to the complex projective space

In this section, we will consider admissible isometric immersions of (M^{2n+1}, g_θ) into the complex projective space $(N, h) = \mathbb{P}^{n+1}(c)$ $(c > 0)$ whose mean curvature vector field is parallel with non-zero pseudo mean curvature.

7.1.   Let us recall that the curvature tensor field $(N, h) = \mathbb{P}^{n+1}(c)$ is given by

$$R^h(U, V)W = \frac{c}{4} \Big\{ h(V, W) U - h(U, W) V$$
$$+ h(JV, W) JU - h(JU, W) JV + 2h(U, JV) JW \Big\}, \tag{7.1}$$

where J is the adapted complex structure of $\mathbb{P}^{n+1}(c)$, and U, V and W are vector fields on $\mathbb{P}^{n+1}(c)$, respectively. Therefore, we have

$$R^h(d\varphi(X_j), d\varphi(X_k))d\varphi(X_k)$$
$$= \frac{c}{4} \Big\{ h(d\varphi(X_k), d\varphi(X_k)) \, d\varphi(X_j) - h(d\varphi(X_j), d\varphi(X_k)) \, d\varphi(X_k)$$
$$+ h(J \, d\varphi(X_k), d\varphi(X_k)) \, Jd\varphi(X_j) - h(Jd\varphi(X_j), d\varphi(X_k)) \, Jd\varphi(X_k)$$
$$+ 2h(d\varphi(X_j), Jd\varphi(X_k)) \, Jd\varphi(X_k) \Big\}$$
$$= \frac{c}{4} \Big\{ d\varphi(X_j) - \delta_{jk} \, d\varphi(X_k) + 3 \, h(d\varphi(X_j), Jd\varphi(X_k)) \, Jd\varphi(X_k) \Big\}. \tag{7.2}$$

We show first

$$\sum_{j,k=1}^{2n} h\Big(\tau_b(\varphi), R^h(d\varphi(X_j), d\varphi(X_k)) \, d\varphi(X_k) \Big) \, d\varphi(X_j)$$
$$= -\frac{3c}{4} h(\tau_b(\varphi), J \, d\varphi(T)) \left(J \, d\varphi(T) \right)^\top$$
$$- \frac{3c}{4} \sum_{j=1}^{2n} h\Big(d\varphi(X_j), J \left(J \, \tau_b(\varphi) \right)^\top \Big) \, d\varphi(X_j). \tag{7.3}$$

Recall here that the tangential part of $Z \in T_{\varphi(x)} N$ $(x \in M)$ is given by

$$Z^\top = \sum_{i=1}^{2n} h(Z, d\varphi(X_i))\, d\varphi(X_i) + h(Z, d\varphi(T))\, d\varphi(T).$$

$$(7.4)$$

Since $h(\tau_b(\varphi), d\varphi(X_j)) = 0$ $(j = 1, \cdots, 2n)$, and (7.2), one can calculate the left hand side of (7.3) as follows:

$$\sum_{j,k=1}^{2n} h\Big(\tau_b(\varphi), R^h(d\varphi(X_j), d\varphi(X_k))d\varphi(X_k)\Big)\, d\varphi(X_j)$$

$$= \frac{3c}{4} \sum_{j,k=1}^{2n} h(\tau_b(\varphi), h(d\varphi(X_j), J\, d\varphi(X_k))\, h(\tau_b(\varphi), J\, d\varphi(X_k))\, d\varphi(X_j)$$

$$= \frac{3c}{4} \sum_{j,k=1}^{2n} h(J\, d\varphi(X_j), d\varphi(X_k))\, h(J\, \tau_b(\varphi), d\varphi(X_k))\, d\varphi(X_j)$$

$$= \frac{3c}{4} \sum_{j=1}^{2n} h\Big(J\, d\varphi(X_j), \sum_{k=1}^{2n} h(J\, \tau_b(\varphi), d\varphi(X_k))\, d\varphi(X_k))\Big)\, d\varphi(X_j)$$

$$= \frac{3c}{4} \sum_{j=1}^{2n} h\Big(J\, d\varphi(X_j), (\, J\tau_b(\varphi))^\top - h(J\, \tau_b(\varphi), d\varphi(T))\, d\varphi(T)\Big)\, d\varphi(X_j)$$

$$= \frac{3c}{4} \sum_{j=1}^{2n} h(J\, d\varphi(X_j), (J\, \tau_b(\varphi))^\top)\, d\varphi(X_j)$$

$$+ \frac{3c}{4} h(J\tau_b(\varphi), d\varphi(T)) \sum_{j=1}^{2n} h(d\varphi(X_j), J\, d\varphi(T))\, d\varphi(X_j)$$

$$= -\frac{3c}{4} \sum_{j=1}^{2n} h(d\varphi(X_j), J\, (J\tau_b(\varphi))^\top)\, d\varphi(X_j)$$

$$- \frac{3c}{4} h(\tau_b(\varphi), J\, d\varphi(T))\, (J\, d\varphi(T))^\top.$$

$$(7.5)$$

Then, (7.5) is just (7.3).

Second, by a similar way,

$$\sum_{j=1}^{2n} \langle \tau_b(\varphi), R^h(d\varphi(X_j), d\varphi(T))\, d\varphi(T)\rangle\, d\varphi(X_j)$$

$$= \sum_{j=1}^{2n} \langle \tau_b(\varphi), \frac{3c}{4} h(d\varphi(X_j), J\, d\varphi(T))\, J\, d\varphi(T)\rangle\, d\varphi(X_j)$$

$$= \frac{3c}{4} h(\tau_b(\varphi), J\, d\varphi(T))\, (J\, d\varphi(T))^\top$$

$$(7.6)$$

in the last equality of (7.6) we used that $h(d\varphi(T), J\, d\varphi(T)) = 0$.

Thus, we have

$$\Delta_b(\tau_b(\varphi)) = \frac{3c}{4} h(\tau_b(\varphi), Jd\varphi(T)) (Jd\varphi(T))^\top$$

$$+ \frac{3c}{4} \sum_{j=1}^{2n} h(d\varphi(X_j), J(J\tau_b(\varphi))^\top) d\varphi(X_j)$$

$$- \frac{3c}{4} h(\tau_b(\varphi), J d\varphi(T)) (J d\varphi(T))^\top$$

$$+ \langle \tau_b(\varphi), B_\varphi(X_i, X_j) \rangle B_\varphi(X_i, X_j)$$

$$= \frac{3c}{4} \sum_{j=1}^{2n} h(d\varphi(X_j), J((J\tau_b(\varphi))^\top) d\varphi(X_j)$$

$$+ \langle (\tau_b(\varphi), B_\varphi(X_i, X_j) \rangle B_\varphi(X_i, X_j). \tag{7.7}$$

Therefore, an isometric immersion φ is pseudo biharmonic if and only if the pseudo biharmonic map equation folds:

$$\Delta_b(\tau_b(\varphi)) - \sum_{k=1}^{2n} R^h(\tau_b(\varphi), d\varphi(X_k) d\varphi(X_k) = 0. \tag{7.8}$$

By (7.7) and (7.1), (7.8) is equivalent to that the following (7.9) holds:

$$\frac{3c}{4} \sum_{j=1}^{2n} h(d\varphi(X_j), J((J\tau_b(\varphi)^\top)) d\varphi(X_j)$$

$$+ \langle \tau_b(\varphi), B_\varphi(X_i, X_j) \rangle B_\varphi(X_i, X_j)$$

$$- \frac{2nc}{4} \tau_b(\varphi) + \frac{3c}{4} \sum_{k=1}^{2n} h(d\varphi(X_j), J\tau_b(\varphi)) J d\varphi(X_k)$$

$$= 0. \tag{7.9}$$

7.2.   Let ξ be the unit normal vector field along the admissible isometric immersion $\varphi : (M, g_\theta) \to \mathbb{P}^{n+1}(c)$ $(c > 0)$.

We have immediately

$$\begin{cases} B_\varphi(X_i, X_i) = (\widetilde{\nabla}_{X_i} d\varphi)(X_j) = H_{ij} \, \xi, \\ \tau_b(\varphi) = \sum_{k=1}^{2n} (\widetilde{\nabla}_{X_k} d\varphi)(X_k) = \left(\sum_{k=1}^{2n} H_{kk} \right) \xi, \\ J \, \tau_b(\varphi) = \left(\sum_{k=1}^{2n} H_{kk} \right) J\xi, \end{cases} \tag{7.10}$$

and then, we have

$$h(\xi, J\xi) = 0. \tag{7.11}$$

Indeed, we have

$$- h(J\xi, \xi) = h(J\xi, J(J\xi)) = h(\xi, J\xi),$$

which implies that $h(\xi, J\xi) = 0$.

Due to (7.11), $J\xi$ is tangential. By (7.10), $J\tau_b(\varphi)$ is also tangential. Therefore, we have

$$(J\tau_b(\varphi))^\top = J\tau_b(\varphi). \tag{7.12}$$

In particular, we have

$$\sum_{j=1}^{2n} h(d\varphi(X_j), J(J\tau_b(\varphi))^\top)) d\varphi(X_j)$$

$$= \sum_{j=1}^{2n} h(d\varphi(X_j), J(J\tau_b(\varphi)))) d\varphi(X_j)$$

$$= -\sum_{j=1}^{2n} h(d\varphi(X_j), \tau_b(\varphi)) d\varphi(X_j)$$

$$= 0 \tag{7.13}$$

by using (7.12) and $\tau_b(\varphi)$ is a normal vector field along φ.

Since $J\tau_b(\varphi)$ is tangential, we can write as

$$J\tau_b(\varphi) = \sum_{k=1}^{2n} h(d\varphi(X_k), J\tau_b(\varphi)) d\varphi(X_k) + h(d\varphi(T), J\tau_b(\varphi)) d\varphi(T),$$

which implies that

$$\sum_{k=1}^{2n} h(d\varphi(X_k), J\tau_b(\varphi)) d\varphi(X_k)$$

$$= J\tau_b(\varphi) - h(d\varphi(T), J\tau_b(\varphi)) d\varphi(T). \tag{7.14}$$

Therefore, applying J to (7.14), we have

$$\sum_{k=1}^{2n} h(d\varphi(X_k), J\tau_b(\varphi)) J d\varphi(X_k)$$

$$= J^2 \tau_b(\varphi) - h(d\varphi(T), J\tau_b(\varphi)) J d\varphi(T)$$

$$= -\tau_b(\varphi) - h(d\varphi(T), J\tau_b(\varphi)) J d\varphi(T). \tag{7.15}$$

Inserting (7.15) into (7.9), the left hand side of (7.9) is equal to

$$\frac{3c}{4} \sum_{j=1}^{2n} h(d\varphi(X_j), J(\,(J\,\tau_b(\varphi))^\top)\,)\,d\varphi(X_j)$$

$$+ \sum_{i,j=1}^{2n} h(\tau_b(\varphi), B_\varphi(X_i, X_j)\,B_\varphi(X_i, X_j)$$

$$- \frac{2nc}{4}\,\tau_b(\varphi) + \frac{3c}{4}\left\{\,-\tau_b(\varphi) - h(d\varphi(T), J\,\tau_b(\varphi))\,Jd\varphi(T)\right\}$$

$$= \sum_{i,j=1}^{2n} h(\tau_b(\varphi), B_\varphi(X_i, X_j))\,B_\varphi(X_i, X_j)$$

$$- \frac{c(2n+3)}{4}\,\tau_b(\varphi) - \frac{3c}{4}h(d\varphi(T), J\,\tau_b(\varphi))\,J\,d\varphi(T),$$
$$\tag{7.16}$$

where we used (7.13) for vanishing the first term of the left hand side of (7.16).

Due to (7.9) and (7.16), we obtain the equivalence relation that φ is biharmonic if and only if both the equations

(1) $$h(d\varphi(T), J\,\tau_b(\varphi))\,(J\,d\varphi(T))^\top = 0,$$
$$\tag{7.17}$$

and

(2) $$\sum_{i,j=1}^{2n} h(\tau_b(\varphi), B_\varphi(X_i, X_j))\,B_\varphi(X_i, X_j) - \frac{c(2n+3)}{4}\,\tau_b(\varphi)$$

$$- \frac{3c}{4}h(d\varphi(T), J\,\tau_b(\varphi))\,(J\,d\varphi(T))^\perp = 0,$$
$$\tag{7.18}$$

hold.

7.3   For the first equation (1) (7.17) is equivalent to that

$$h(d\varphi(T), J\,\tau_b(\varphi)) = 0 \quad \text{or} \quad (J\,d\varphi(T))^\top = 0. \tag{7.19}$$

But, by (7.10), we have

$$h(d\varphi(T), J\,\tau_b(\varphi)) = \Big(\sum_{k=1}^{2n} H_{kk}\Big)\,h(d\varphi(T), J\,\xi)$$

$$= -\Big(\sum_{k=1}^{2n} H_{kk}\Big)\,h(J\,d\varphi(T), \xi). \tag{7.20}$$

By our assumption that the pseudo mean curvature $\sum_{k=1}^{2n} H_{kk} \neq 0$, to hold that $h(d\varphi(T), J\,\tau_b(\varphi)) = 0$ is equivalent to that

$$h(J\,d\varphi(T), \xi) = 0. \tag{7.21}$$

And to hold that $(J\, d\varphi(T))^{\top} = 0$ is equivalent to that

$$J\, d\varphi(T) = h(J\, d\varphi(T), \xi)\, \xi. \tag{7.22}$$

Thus, (1) (7.17) holds if and only if

$$(7.21) \qquad h(J\, d\varphi(T), \xi) = 0, \qquad\qquad \text{or}$$

$$(7.22) \qquad J\, d\varphi(T) = h(J\, d\varphi(T), \xi)\, \xi.$$

In the case (7.21) holds, we have

$$h(d\varphi(T), J\, \tau_b(\varphi))\, (J\, d\varphi(T))^{\top} = 0, \tag{7.23}$$

which implies that (2) (7.18) turns out that

$$\sum_{i,j=1}^{2n} h(\tau_b(\varphi), B_\varphi(X_i, X_j)\, B_\varphi(X_i, X_j) - \frac{c(2n+3)}{4}\, \tau_b(\varphi) = 0. \tag{7.24}$$

In the case that (7.22) holds, we have that

$$h(d\varphi(T), J\, \tau_b(\varphi))\, (J\, d\varphi(T))^{\top}$$
$$= h(d\varphi(T), J\, \tau_b(\varphi))\, h(J\, d\varphi(T), \xi)\, \xi$$
$$= \Big(\sum_{k=1}^{2n} H_{kk} \Big)\, h(d\varphi(T), J\, \xi)\, h(J\, d\varphi(T), \xi)\, \xi \qquad \text{(by (7.10)}$$
$$= -\Big(\sum_{k=1}^{2n} H_{kk} \Big)\, h(J\, d\varphi(T), \xi)^2\, \xi. \tag{7.25}$$

In the case that (7.21) holds, (2) (7.18) turns out that

$$\sum_{i,j=1}^{2n} h(\tau_b(\varphi), B_\varphi(X_i, X_j)) - \frac{c(2n+3)}{4}\, \tau_b(\varphi)$$
$$+ \frac{3c}{4} \Big(\sum_{k=1}^{2n} H_{kk} \Big)\, h(J\, d\varphi(T), \xi)^2\, \xi = 0. \tag{7.26}$$

By inserting (7.10) into (7.24), the left hand side of (7.24) is equal to

$$\sum_{i,j=1}^{2n} h\Big(\sum_{k=1}^{2n} H_{kk}\, \xi, H_{ij}\, \xi \Big)\, H_{ij}\, \xi - \frac{c(2n+3)}{4} \sum_{k=1}^{2n} H_{kk}\, \xi$$
$$= \Big(\sum_{k=1}^{2n} H_{kk} \Big) \Big\{ \sum_{i,j=1}^{2n} H_{ij}{}^2 - \frac{c(2n+3)}{4} \Big\}\, \xi. \tag{7.27}$$

(2) (7.18) is equivalent to that

$$\sum_{i,j=1}^{2n} H_{ij}{}^2 = \frac{c(2n+3)}{4} \tag{7.28}$$

by our assumption that $\sum_{k=1}^{2n} H_{kk} \neq 0$.

In the case that (7.22) holds, by inserting (7.10) into (7.26), the left hand side of (7.26) is equal to

$$\sum_{i,j=1}^{2n} h\left(\sum_{k=1}^{2n} H_{kk}\,\xi,\, H_{ij}\,\xi \right) H_{ij}\,\xi - \frac{c(2n+3)}{4} \sum_{k=1}^{2n} H_{kk}\,\xi$$

$$+ \frac{3c}{4}\, h(J\,d\varphi(X),\,\xi)^2 \left(\sum_{k=1}^{2n} H_{kk} \right)\xi$$

$$= \left(\sum_{k=1}^{2n} H_{kk} \right)\left\{ \sum_{i,j=1}^{2n} H_{ij}{}^2 - \frac{c(2n+3)}{4} + \frac{3c}{4}\, h(J\,d\varphi(T),\,\xi)^2 \right\}\xi.$$

$$(7.29)$$

Since (7.22) $J\,d\varphi(T) = h(J\,d\varphi(T),\,\xi)\,\xi$, we have

$$h(J\,d\varphi(T),\xi)^2 = h(J\,d\varphi(T),d\,\varphi(T))$$
$$= h(d\varphi(T),d\varphi(T)) = g_\theta(T,T) = 1$$

which implies again by our assumption $\sum_{k=1}^{2n} H_{kk} \neq 0$, that (2) (7.18) is equivalent to that

$$\sum_{i,j=1}^{2n} H_{ij}{}^2 - \frac{c(2n+3)}{4} + \frac{3c}{4} = \sum_{i,j=1}^{2n} H_{ij}{}^2 - \frac{n}{2}\,c = 0.$$

$$(7.30)$$

Therefore, we obtain

THEOREM 7.1.   *Assume that $\varphi : (M,g_\theta) \to \mathbb{P}^{n+1}(c) = (N,h)$ $(c>0)$ is an admissible isometric immersion whose pseudo mean curvature vector filed along φ is parallel with non-zero pseudo mean curvature. Then, φ is biharmonic if and only if one of the following two cases occurs:*

   (1) $h(J\,d\varphi(T),\,\xi) = 0$ *and*

$$\left\| B_\varphi \right|_{H(M)\times H(M)} \Big\|^2 = \frac{c(2n+3)}{4},$$

$$(7.31)$$

   (2) $J\,d\varphi(T) = h(J\,d\varphi(T),\xi)\,\xi$ *and*

$$\left\| B_\varphi \right|_{H(M)\times H(M)} \Big\|^2 = \frac{n}{2}\,c.$$

$$(7.32)$$

8. Examples of pseudo harmonic maps and pseudo biharmonic maps

In this section, we give some examples of pseudo biharmonic maps.

Example 8.1.   Let $(M^{2n+1}, g_0) = S^{2n+1}(r)$ be the sphere of radius r $(0 < r < 1)$ which is embedded in the unit sphere $S^{2n+2}(1)$, i.e., the natural embedding $\varphi: S^{2n+1}(r) \to S^{2n+2}(1)$ is given by

$$\varphi: S^{2n+1}(r) \ni x' = (x_1, x_2, \cdots, x_{2n+2}) \mapsto (x', \sqrt{1-r^2}) \in S^{2n+2}(1).$$

This φ is a standard isometric with constant principal curvature $\lambda_1 = \cot[\cos^{-1} t]$, $(-1 < t < 1)$, with the multiplicity $m_1 = \dim M = 2n+1$.

Due to Theorem 6.2, it is *pseudo biharmonic* if and only if

$$(\lambda_1)^2 \times 2n = 2n \qquad \Leftrightarrow \qquad \lambda_1 = \cot[\cos^{-1} t] = \pm 1.$$

$$\Leftrightarrow \quad t = \cos\left(\pm\frac{\pi}{4}\right) = \frac{1}{\sqrt{2}}. \tag{8.1}$$

This is just the example which is biharmonic but not minimal given by C. Oniciuc ([**121**]). Note that $\varphi: S^{2n+1}(r) \to S^{2n+2}(1)$ is *pseudo harmonic* if and only if

$$\mathrm{Trace}(B_\varphi|_{H(M) \times H(M)}) = 0 \quad \Leftrightarrow \quad \lambda_1 = 0$$

$$\Leftrightarrow \quad t = \cos\left(\frac{\pi}{2}\right) = 1. \tag{8.2}$$

This $t = 1$ gives a great hypersphere which is also minimal.

Example 8.2.   Let the Hopf fibration $\pi: S^{2n+3}(1) \to \mathbb{P}^{n+1}(4)$, and, let $\widehat{M} := S^1(\cos u) \times S^{2n+1}(\sin u) \subset S^{2n+3}(1)$ $\left(0 < u < \frac{\pi}{2}\right)$. Then, we have $\varphi: M^{2n+1} = \pi(\widehat{M}) \subset \mathbb{P}^{n+1}(4)$ which is a homogeneous real hypersurface of $\mathbb{P}^{n+1}(4)$ of type A_1 in the table of R. Takagi ([**129**]) whose principal curvatures and their multiplicities are given as follows ([**129**]):

$$\begin{cases} \lambda_1 = \cot u, & \text{multiplicity } m_1 = 2n, \\ \lambda_2 = 2\cot(2u), & \text{multiplicity } m_2 = 1. \end{cases} \tag{8.3}$$

Since $2\cot(2u) = \cot u - \tan u$, the mean curvature H and $\|B_\varphi\|^2$ are given by

$$H = \frac{1}{2n+1}\{(2n+1)\cot u - \tan u\}, \tag{8.4}$$

$$\|B_\varphi\|^2 = m_1\lambda_1{}^2 + m_2\lambda_2{}^2 = \tan^2 u + (2n+1)\cot^2 u - 2. \tag{8.5}$$

R. Takagi showed ([**129**]) to this example, that $\varphi: M^{2n+1} \to \mathbb{P}^{n+1}(4)$ is the geodesic sphere S^{2n+1}, and $J(-\xi)$ is the mean curvature vector of the principal curvature λ_2 (cf. Remark 1.1 in [**129**], p. 48), where ξ is a unit normal vector field along φ.

In the case (1) of Theorem 7.1, i.e., $(M^{2n+1}, g_\theta) = (S^{2n+1}, g_\theta)$ is a strictly pseudoconvex CR manifold and $J \, d\varphi(T)$ is tangential, we have

$$0 = h(J \, d\varphi(T), \xi) = h(J^2 \, d\varphi(T), J \, \xi) = h(d\varphi(T), J(-\xi)),$$

and $h(d\varphi(T), d\varphi(H(M))) = 0$. Then, the principal curvature vector field $J(-\xi)$ with principal curvature $\lambda_2 = 2 \cot(2u)$ coincides with $d\varphi(X)$ for some $X \in H(M)$. Since

$$\|X\| = \|d\varphi(X)\| = \|J(-\xi)\| = \|\xi\| = 1,$$

we can choose an orthonormal basis $\{X_i\}_{i=1}^{2n}$ of $H(M)$ in such a way $X_1 = X$. Then, $\{d\varphi(T), d\varphi(X_2), \cdots, d\varphi(X_{2n})\}$ give principal curvature vector fields along φ with principal curvature $\lambda_1 = \cot u$. Then,

$$\tau_b(\varphi) = \sum_{i=1}^{2n} B_\varphi(X_i, X_i) = 2 \cot(2u) + (2n - 1) \cot u$$

$$= 2n \cot u - \tan u. \tag{8.6}$$

Therefore, φ is *pseudo harmonic* if and only if

$$\tau_b(\varphi) = 0 \qquad \Leftrightarrow \qquad \tan u = \sqrt{2n}. \tag{8.7}$$

By Theorem 7.1, (1), φ is pseudo biharmonic if and only if

$$\|B_\varphi|_{H(M) \times H(M)}\|^2 = \frac{c(2n + 3)}{4} = 2n + 3. \tag{8.8}$$

Since the left hand side of (8.8) coincides with

$$\|B_\varphi|_{H(M) \times H(M)}\|^2 = (2 \cot(2u))^2 + (2n - 1) \cot^2 u$$

$$= (\cot u - \tan u)^2 + (2n - 1) \cot^2 u$$

$$= 2n \cot^2 u - 2 + \tan^2 u, \tag{8.9}$$

we have that (8.8) holds if and only if

$$2n \cot^2 u + \tan^2 u = 2n + 5 \qquad \Leftrightarrow \qquad x^2 - (2n + 5)x + 2n = 0, \tag{8.10}$$

where $x = \tan^2 u$. Therefore, φ is *pseudo biharmonic* if and only if $\tan u$ is $\sqrt{\alpha}$ or $\sqrt{\beta}$, where α and β are positive roots of (8.10).

In the case (2) of Theorem 7.1, i.e., $(M^{2n+1}, g_\theta) = S^{2n+1}, g_\theta)$ is a strictly pseudoconvex CR manifold, and $J \, d\varphi(T)$ is normal, i.e., $J \, d\varphi(T) = h(J \, d\varphi(T) \, \xi) \, \xi$. Then, we have that

$$0 \neq d\varphi(T) = h(d\varphi(T), J(-\xi)) \, J(-\xi).$$

And $J(-\xi)$ is the principal curvature vector field along φ with the principal curvature λ_2, and $d\varphi(H(M))$ is the space spanned by the principal curvature vectors along φ with the principal curvature λ_1

since $h(d\varphi(H(M)), J(-\xi)) = 0$. Then the pseudo tension field $\tau_b(\varphi)$ is given by

$$\tau_b(\varphi) = \sum_{i=1}^{2n} B_\varphi(X_i, X_i) = (2n \cot u)\, \xi \neq 0, \qquad (8.11)$$

so that φ is *not pseudo harmonic*. Due to the case (2) of Theorem 7.1 that $J\, d\varphi(T)$ is normal, φ is pseudo biharmonic if and only if

$$\begin{aligned}
2n &= \|B_\varphi|_{H(M)\times H(M)}\|^2 \\
&= m_1 \lambda_1{}^2 \\
&= 2n \cot^2 u
\end{aligned} \qquad (8.12)$$

occurs. Thus, we obtain

$$\tan^2 u = 1. \qquad (8.13)$$

Therefore, if $\tan u = 1$ $(u = \frac{\pi}{4})$, then the corresponding isometric immersion $\varphi : (M^{2n+1}, g_\theta) \to \mathbb{P}^{n+1}(4)$ is *pseudo biharmonic*, but *not pseudo harmonic*.

REMARK 8.1. *Let us recall our previous work* ([**63**], [**64**]) *that* $\varphi : (M^{2n+1}, g_\theta) \to \mathbb{P}^{n+1}(4)$ is biharmonic *if and only if*

$$\|B_\varphi\|^2 = \tan^2 u + (2n+1)\cot^2 u - 2 = \frac{n+2}{2}\, 4$$

$$\Leftrightarrow\ x^2 - 2(n+3)x + 2n + 1 = 0, \qquad (x = \tan^2 u).$$
$$(8.14)$$

The equation (8.14) has two positive solutions α, β, and if we put $\tan u = \sqrt{\alpha}$ or $\sqrt{\beta}$ $(0 < u < \frac{\pi}{2})$, then $\varphi : (M^{2n+1}, g_\theta) \to \mathbb{P}^{n+1}(4)$ is biharmonic, *and vice versa. Since the mean curvature is given by (8.4), $\varphi : (M^{2n+1}, g_\theta) \to \mathbb{P}^{n+1}(4)$ is* harmonic *(i.e., minimal) if and only if* $\tan u = \sqrt{2n+1}$ $(0 < u < \frac{\pi}{2})$.

CHAPTER 16

Harmonic Maps and Biharmonic Maps on the Principal Bundles and Warped Products

1

ABSTRACT. In this chapter, we study harmonic maps and biharmonic maps on the principal G-bundle in Kobayashi and Nomizu [82] and also the warped product $P = M \times_f F$ for a $C^\infty(M)$ function f on M studied by Bishop and O'Neill [12], and Ejiri [45].

1. Introduction

Variational problems play central roles in geometry; Harmonic map is one of important variational problems which is a critical point of the energy functional $E(\varphi) = \frac{1}{2} \int_M |d\varphi|^2 \, v_g$ for smooth maps φ of (M, g) into (N, h). The Euler-Lagrange equations are given by the vanishing of the tension filed $\tau(\varphi)$. In 1983, J. Eells and L. Lemaire [40] extended the notion of harmonic map to biharmonic map, which are, by definition, critical points of the bienergy functional

$$E_2(\varphi) = \frac{1}{2} \int_M |\tau(\varphi)|^2 \, v_g. \qquad (1.1)$$

After G.Y. Jiang [74] studied the first and second variation formulas of E_2, extensive studies in this area have been done (for instance, see [16], [90], [102], [123], [131], [63], [64], [73], etc.). Notice that harmonic maps are always biharmonic by definition. B.Y. Chen raised ([21]) so called B.Y. Chen's conjecture and later, R. Caddeo, S. Montaldo, P. Piu and C. Oniciuc raised ([16]) the generalized B.Y. Chen's conjecture.

B.Y. Chen's conjecture:

Every biharmonic submanifold of the Euclidean space $\mathbb{R}^n$ must be harmonic (minimal).

---

[1]This chapter is due to a part of [159]: H. Urakawa, *Biharmonic maps on principal G-bundles over Riemannian manifolds of non-positive Ricci curvature*, accepted in Michigan Math. J., (2017), and also
[160]: H. Urakawa, *Harmonic maps and biharmonic maps on the principal bundles and warped products*, accepted in J. Korean Math. Soc., (2018).

The generalized B.Y. Chen's conjecture:

Every biharmonic submanifold of a Riemannian manifold of non-positive curvature must be harmonic (minimal).

For the generalized Chen's conjecture, Ou and Tang gave ([**123**], [**124**]) a counter example in a Riemannian manifold of negative curvature. For the Chen's conjecture, affirmative answers were known for the case of surfaces in the three dimensional Euclidean space ([**21**]), and the case of hypersurfaces of the four dimensional Euclidean space ([**57**], [**34**]). K. Akutagawa and S. Maeta gave ([**1**]) showed a supporting evidence to the Chen's conjecture: *Any complete regular biharmonic submanifold of the Euclidean space $\mathbb{R}^n$ is harmonic (minimal).* The affirmative answers to the generalized B.Y. Chen's conjecture were shown ([**108**], [**109**], [**111**]) under the L^2-condition and completeness of (M, g).

In this paper, we first treat with a principal G-bundle over a Riemannian manifold, and show the following two theorems:

Theorem 3.2 *Let $\pi : (P, g) \to (M, h)$ be a principal G-bundle over a Riemannian manifold (M, h) with non-positive Ricci curvature. Assume P is compact so that M is also compact. If the projection π is biharmonic, then it is harmonic.*

Theorem 4.1 *Let $\pi : (P, g) \to (M, h)$ be a principal G-bundle over a Riemannian manifold with non-positive Ricci curvature. Assume that (P, g) is a non-compact complete Riemannian manifold, and the projection π has both finite energy $E(\pi) < \infty$ and finite bienergy $E_2(\pi) < \infty$. If π is biharmonic, then it is harmonic.*

We give two comments on the above theorems: For the generalized B.Y. Chen's conjecture, non-positivity of the sectional curvature of the ambient space of biharmonic submanifolds is necessary. However, it should be emphasized that for the principal G-bundles, we need not the assumption of non-positivity of the sectional curvature. We only assume *non-positivity of the Ricci curvature* of the domain manifolds in the proofs of Theorems 3.2 and 4.1. Second, finiteness of the energy and bienergy is necessary in Theorem 4.1. Otherwise, one have the counter examples due to Loubeau and Ou (cf. Sect. Four, Examples 1, 2 [**91**])

Next, we consider the warped products. For two Riemannian manifolds (M, h), (F, k) and a C^∞ function f on M, $f \in C^\infty(M)$, the warping function on M, let us consider the warped product (P, g) where

$\pi : P = M \times F \ni (x, y) \mapsto x \in M$ and $g = \pi^* h + f^2 k$. Let us consider the following two problems:

Problem 1. When $\pi : (P, g) \to (M, h)$ is harmonic?

Problem 2. In the case $(M, h) = (\mathbb{R}, dt^2)$, a line, can one choose $f \in C^\infty(\mathbb{R})$ such that $\pi : (P, g) \to (M, h)$ is biharmonic but not harmonic?

In this paper, we answer these two problems as follows.

Theorem 5.2 *Let $\pi : (P, g) \to (M, h)$ be the warped product with a warping function $f \in C^\infty(M)$. Then, the tension field $\tau(\pi)$ is given by*

$$\tau(\pi) = \ell \, \frac{\operatorname{grad} f}{f} = \ell \, \frac{\nabla f}{f}, \qquad (1.2)$$

where $\ell = \dim F$. Therefore, π is harmonic if and only if f is constant.

Theorem 6.2 *For the warped product $\pi : (P, g) \to (M, h)$, the bitension field $\tau_2(\pi)$ is given by*

$$\tau_2(\pi) = \overline{\Delta}(\tau(\pi)) - \rho^h(\tau(\pi)) - \ell \, \overline{\nabla}_{\frac{\nabla f}{f}} \tau(\pi), \qquad (1.3)$$

where $\overline{\Delta}$ is the rough Laplacian and $\overline{\nabla}$ is the induced connection from the Levi-Civita connection ∇^h of (M, h). Therefore, π is biharmonic if and only if

$$\overline{\Delta}(\tau(\pi)) - \rho^h(\tau(\pi)) - \ell \, \overline{\nabla}_{\frac{\nabla f}{f}} \tau(\pi) = 0. \qquad (1.4)$$

Here, ρ^h is the Ricci transform $\rho^h(u) := \sum_{i=1}^{m} R^h(u, e_i') e_i'$, $u \in T_x M$ for an locally defined orthonormal field $\{e_i'\}_{i=1}^{m}$ on (M, h).

Theorems 7.1 (1) *In the case $(M, h) = (\mathbb{R}, dt^2)$, a line, the warped product $\pi : (P, g) \to (\mathbb{R}, dt^2)$ is biharmonic if and only if $f \in C^\infty(\mathbb{R})$ satisfies the following ordinary equation:*

$$f''' f^2 + (\ell - 3) f'' f' f + (-\ell + 2) f'^3 = 0. \qquad (1.5)$$

(2) *All the solutions f of (1.5) are given by*

$$f(t) = c \, \exp \left(\int_{t_0}^{t} a \, \tanh \left[\frac{\ell}{2} \, a \, r + b \right] \, dr \right), \qquad (1.6)$$

where a, b, $c > 0$ are arbitrary constants.

(3) *In the case $(M, h) = (\mathbb{R}, dt^2)$, a line, let $f(t)$ be C^∞ function defined by (1.6) with $a \neq 0$ and $c > 0$. Then, the warped product $\pi : (P, g) \to (M, h)$ is biharmonic but not harmonic.*

Acknowledgement. We would like to express our gratitude to the referee who pointed to improve Theorem 3.2 in this version.

2. Preliminaries

2.1. Harmonic maps and biharmonic maps.
We first prepare the materials for the first and second variational formulas for the bienergy functional and biharmonic maps. Let us recall the definition of a harmonic map $\varphi : (M, g) \to (N, h)$, of a compact Riemannian manifold (M, g) into another Riemannian manifold (N, h), which is an extremal of the *energy functional* defined by

$$E(\varphi) = \int_M e(\varphi)\, v_g,$$

where $e(\varphi) := \frac{1}{2}|d\varphi|^2$ is called the energy density of φ. That is, for any variation $\{\varphi_t\}$ of φ with $\varphi_0 = \varphi$,

$$\frac{d}{dt}\bigg|_{t=0} E(\varphi_t) = -\int_M h(\tau(\varphi), V)v_g = 0, \tag{2.1}$$

where $V \in \Gamma(\varphi^{-1}TN)$ is a variation vector field along φ which is given by $V(x) = \frac{d}{dt}\big|_{t=0}\varphi_t(x) \in T_{\varphi(x)}N$, $(x \in M)$, and the *tension field* is given by $\tau(\varphi) = \sum_{i=1}^m B(\varphi)(e_i, e_i) \in \Gamma(\varphi^{-1}TN)$, where $\{e_i\}_{i=1}^m$ is a locally defined orthonormal frame field on (M, g), and $B(\varphi)$ is the second fundamental form of φ defined by

$$\begin{aligned} B(\varphi)(X, Y) &= (\widetilde{\nabla}d\varphi)(X, Y) \\ &= (\widetilde{\nabla}_X d\varphi)(Y) \\ &= \overline{\nabla}_X(d\varphi(Y)) - d\varphi(\nabla_X Y), \end{aligned} \tag{2.2}$$

for all vector fields $X, Y \in \mathfrak{X}(M)$. Here, ∇, and ∇^h, are Levi-Civita connections on TM, TN of (M, g), (N, h), respectively, and $\overline{\nabla}$, and $\widetilde{\nabla}$ are the induced ones on $\varphi^{-1}TN$, and $T^*M \otimes \varphi^{-1}TN$, respectively. By (2.1), φ is *harmonic* if and only if $\tau(\varphi) = 0$.

The second variation formula is given as follows. Assume that φ is harmonic. Then,

$$\frac{d^2}{dt^2}\bigg|_{t=0} E(\varphi_t) = \int_M h(J(V), V)v_g, \tag{2.3}$$

where J is an elliptic differential operator, called the *Jacobi operator* acting on $\Gamma(\varphi^{-1}TN)$ given by

$$J(V) = \overline{\Delta}V - \mathcal{R}(V), \tag{2.4}$$

where $\overline{\Delta}V = \overline{\nabla}^*\overline{\nabla}V = -\sum_{i=1}^m \{\overline{\nabla}_{e_i}\overline{\nabla}_{e_i}V - \overline{\nabla}_{\nabla_{e_i}e_i}V\}$ is the *rough Laplacian* and $\mathcal{R}$ is a linear operator on $\Gamma(\varphi^{-1}TN)$ given by $\mathcal{R}(V) = \sum_{i=1}^m R^N(V, d\varphi(e_i))d\varphi(e_i)$, and R^N is the curvature tensor of (N,h) given by $R^h(U,V) = \nabla^h_U\nabla^h_V - \nabla^h_V\nabla^h_U - \nabla^h_{[U,V]}$ for $U, V \in \mathfrak{X}(N)$.

J. Eells and L. Lemaire [40] proposed polyharmonic (k-harmonic) maps and Jiang [74] studied the first and second variation formulas of biharmonic maps. Let us consider the *bienergy functional* defined by

$$E_2(\varphi) = \frac{1}{2}\int_M |\tau(\varphi)|^2 v_g, \tag{2.5}$$

where $|V|^2 = h(V,V)$, $V \in \Gamma(\varphi^{-1}TN)$.

The first variation formula of the bienergy functional is given by

$$\frac{d}{dt}\Big|_{t=0} E_2(\varphi_t) = -\int_M h(\tau_2(\varphi), V)v_g. \tag{2.6}$$

Here,

$$\tau_2(\varphi) := J(\tau(\varphi)) = \overline{\Delta}(\tau(\varphi)) - \mathcal{R}(\tau(\varphi)), \tag{2.7}$$

which is called the *bitension field* of φ, and J is given in (2.4).

A smooth map φ of (M,g) into (N,h) is said to be *biharmonic* if $\tau_2(\varphi) = 0$. By definition, every harmonic map is biharmonic. We say, for an immersion $\varphi : (M,g) \to (N,h)$ to be *proper biharmonic* if it is biharmonic but not harmonic (minimal).

2.2. The principal G-bundle. Recall several notions on principal G-bundles. A manifold $P = P(M,G)$ is a principal fiber bundle over M with a compact Lie group G, where $p = \dim P$, $m = \dim M$, and $k = \dim G$. By definition, a Lie group G acts on P by right hand side denoted by $(G,P) \ni (a,u) \mapsto u \cdot a \in P$, and, for each point $u \in P$, the tangent space T_uP admits a subspace $G_u := \{A^*{}_u |\, A \in \mathfrak{g}\}$, the vertical subspace at u, and each $A \in \mathfrak{g}$ defines the fundamental vector field $A^* \in \mathfrak{X}(P)$ by

$$A^*{}_u := \frac{d}{dt}\Big|_{t=0} u\exp(t\,A) \in T_uP.$$

A Riemannian metric g on P is called *adapted* if it is invariant under all the right action of G, i.e., $R_a{}^*g = g$ for all $a \in G$. An adapted Riemannian metric on P always exists because for every Riemannian metric g' on P, define a new metric g on P by

$$g_u(X_u, Y_u) = \int_G g'(R_{a*}X_u, R_{a*}Y_u)\,d\mu(a),$$

where $d\mu(a)$ is a bi-invariant Haar measure on G. Then, $R_a{}^*g = g$ for all $a \in G$. Each tangent space T_uP has the orthogonal direct decomposition of the tangent space $T_u P$,

$$\text{(a)} \qquad\qquad T_u P = G_u \oplus H_u,$$

where the subspace G_u of P_u satisfies

$$\text{(b)} \qquad\qquad G_u = \{A^*_u | \, A \in \mathfrak{g}\},$$

and the subspace H_u of P_u satisfies that

$$\text{(c)} \qquad\qquad H_{u \cdot a} = R_{a*} H_u, \qquad a \in G, \ u \in P,$$

where the subspace H_u of P_u is called *horizontal subspace* at $u \in P$ with respect to g.

In the following, we fix a locally defined orthonormal frame field $\{e_i\}_{i=1}^p$ corresponding (a), (b) in such a way that

$\{e_i\}_{i=1}^m$ is a locally defined orthonormal basis of the horizontal subspace H_u $(u \in P)$, and

$\{e_i = A^*_{m+i}\}_{i=1}^k$ is a locally defined orthonormal basis of the vertical subspace G_u $(u \in P)$ for an orthonormal basis $\{A_{m+i}\}_{i=1}^k$ of the Lie algebra $\mathfrak{g}$ of a Lie group G with respect to the $\mathrm{Ad}(G)$ invariant inner product $\langle \, \cdot \, , \, \cdot \, \rangle$.

For each decomposition (a), one can define a $\mathfrak{g}$-valued 1-form ω on P by

$$\omega(X_u) = A, \qquad X_u = X_u{}^{\mathrm{V}} + X_u{}^{\mathrm{H}},$$

where

$$X_u{}^{\mathrm{V}} \in G_u, \qquad X_u{}^{\mathrm{H}} \in H_u, \qquad X_u{}^{\mathrm{V}} = A_u{}^*$$

for $u \in P$ and a unique $A \in \mathfrak{g}$. This 1-form ω on P is called a *connection form* of P.

Then, there exist a unique Riemannian metric h on M and an $\mathrm{Ad}(G)$-invariant inner product $\langle \, \cdot \, , \, \cdot \, \rangle$ on $\mathfrak{g}$ such that

$$g(X_u, Y_u) = h(\pi_* X_u, \pi_* Y_u) + \langle \omega(X_u), \omega(Y_u) \rangle, \quad X_u, \ Y_u \in T_u P, \ u \in P,$$

namely,

$$g = \pi^* h + \langle \omega(\cdot), \omega(\cdot) \rangle.$$

We call this Riemannian metric g on P, an *adapted* Riemannian metric on P.

Then, let us recall the following definitions for our question:

Definition 2.1. (1) The projection $\pi : (P, g) \to (M, h)$ is to be *harmonic* if the tension field vanishes, $\tau(\pi) = 0$, and

(2) the projection $\pi : (P, g) \to (M, h)$ is to be *biharmonic* if, the bitension field vanishes, $\tau_2(\pi) = J(\tau(\pi)) = 0$.

Here, J is the Jacobi operator for the projection π given by

$$J(V) := \overline{\Delta} V - \mathcal{R}(V), \qquad V \in \Gamma(\pi^{-1}TM),$$

where

$$\overline{\Delta} V := -\sum_{i=1}^{p} \left\{ \overline{\nabla}_{e_i}(\overline{\nabla}_{e_i} V) - \overline{\nabla}_{\nabla_{e_i} e_i} V \right\}$$

$$= -\sum_{i=1}^{m} \left\{ \overline{\nabla}_{e_i}(\overline{\nabla}_{e_i} V) - \overline{\nabla}_{\nabla_{e_i} e_i} V \right\}$$

$$- \sum_{i=1}^{k} \left\{ \overline{\nabla}_{A_{m+i}^*}(\overline{\nabla}_{A_{m+i}^*} V) - \overline{\nabla}_{\nabla_{A_{m+i}^*} A_{m+i}^*} V \right\},$$

for $V \in \Gamma(\pi^{-1}TM)$, i.e., $V(x) \in T_{\pi(x)}M$ ($x \in P$). Here, $\{e_i\}_{i=1}^{p}$ is a local orthonormal frame field on (P, g) which is given by that: $\{e_i\}_{i=1}^{m}$ is an orthonormal horizontal field on the principal G-bundle $\pi : (P, g) \to (M, h)$ and $\{e_{m+i, u} = A_{m+i, u}^*\}_{i=1}^{k}$ ($u \in P$) is an orthonormal frame field on the vertical space $G_u = \{A_u^* | A \in \mathfrak{g}\}$ ($u \in P$) corresponding to an orthonormal basis $\{A_{m+i}\}_{i=1}^{k}$ of $(\mathfrak{g}, \langle\,,\,\rangle)$.

2.3. The warped products. On the product manifold $P = M \times F$ for two Riemannian manifolds (M, h) and (F, k), and a C^∞ function, $f \in C^\infty(M)$ on M, let us consider the Riemannian metric

$$g = \pi^* h + f^2 k, \tag{2.8}$$

where the projection $\pi : P = M \times F \ni (x, y) \mapsto x \in M$. The Riemannian submersion $\pi : (P, g) \to (M, h)$ is called the *warped product* of (M, h) and (F, k) with a *warping function* $f \in C^\infty(M)$. In this section, we prepare several notions in order to calculate the tension field and bitension field.

We first construct a locally defined orthonormal frame field $\{e_i\}_{i=1}^{m+\ell}$ on (P, g) where $m = \dim M$ and $\ell = \dim F$ as follows: For $i = 1, \ldots, m$,

$$e_{i(x,y)} := (e'_{i\,x}, 0_y) \in T_{(x,y)}P = T_x M \times T_y F,$$

and for $i = m+1, \ldots, p$,

$$e_{i(x,y)} := \frac{1}{f(x)} (0_x, e''_{i,y}) \in T_{(x,y)}P = T_x M \times T_y F.$$

where $p = m + \ell$.

Recall the O'Neill's formulas on the warped product (cf. [12], [45]). For a C^∞ vector field $X \in \mathfrak{X}(M)$ on M, $X^* \in \mathfrak{X}(P)$, the *horizontal lift* of X which satisfies for $z \in P$,

$$X^*{}_z \in \mathcal{H}_z, \quad \text{and} \quad \pi_*(X^*{}_z) = X_{\pi(z)}, \tag{2.9}$$

where recall the *vertical subspace* $\mathcal{V}_z$ and *horizontal subspace* $\mathcal{H}_z$ of the tangent space $T_z P$:

$$\mathcal{V}_z = \mathrm{Ker}(\pi_{*(x,y)}), \tag{2.10}$$

$$T_z P = \mathcal{V}_z \oplus \mathcal{H}_z, \quad g(\mathcal{V}_z, \mathcal{H}_z) = 0, \tag{2.11}$$

where $\pi_{*(x,y)} : T_{(x,y)} P \to T_x M$ is the differential of the projection $\pi : P \to M$ at $(x,y) \in P$.

Let $q : P = M \times F \ni (x,y) \mapsto y \in F$ be the projection of P onto F. For a vector field V on F, there exists a unique vector field $\widetilde{V}$ on P satisfying that $\widetilde{V} \in \mathcal{V}$ and $q_*(\widetilde{V}) = V$. We identify $V \in \mathfrak{X}(F)$ with $\widetilde{V} \in \mathcal{V}$ denoting by the same letter V in the following.

LEMMA 2.1. *Let X, $Y \in \mathfrak{X}(M)$ be vector fields on M, and V, $W \in \mathfrak{X}(F)$, vector fields on F, and ∇^g, ∇^h, ∇^k, the Levi-Civita connections of (P,g), (M,h), and (F,k), respectively. Then,*

(1) $\mathrm{grad}(f \circ \pi) = \mathrm{grad} f$.

(2) $\pi_*(\nabla^g{}_{X^*} Y^*) = \nabla^h{}_X Y$,

where X^* and Y^* are the horizontal lifts of X and Y, respectively.

(3) $\nabla^g{}_{X^*} V = \nabla^g{}_V X^* = \dfrac{Xf}{f} V.$

(4) $\mathcal{H}(\nabla^g{}_V W) = -f\, k(V, W)\, G = -\dfrac{1}{f}\, g(V, W)\, G,$

where G is the gradient of f and $f \circ \pi$.

(5) $\mathcal{V}(\nabla^g{}_V W) = \nabla^k{}_V W,$

where $\mathcal{H}A$, and $\mathcal{V}A$ are the horizontal part, and the vertical part of A, respectively.

LEMMA 2.2. *(O'Neill's formulas)*

(1) $g_{(x,y)}(X^*{}_{(x,y)}, Y^*{}_{(x,y)}) = h_x(X_x, Y_x), \quad x \in M,$

(2) $\pi_*([X^*, Y^*]) = [X, Y],$

(3) $\pi_*(\nabla^g{}_{X^*} Y^*) = \nabla^h{}_X Y.$

LEMMA 2.3. *For a vector field $X \in \mathfrak{X}(M)$ whose $h(X, X)$ is constant, $\nabla^g{}_{X^*} X^*$ is the horizontal lift of $\nabla^h{}_X X$.*

Proof. By (3) of Lemma 2.2, we only have to see $\nabla^g{}_{X^*}X^*$ is a horizontal vector field. Due to Lemma 2.3 (1), for every vertical vector field $X \in \mathfrak{X}(M)$, we have

$$
\begin{aligned}
2g(\nabla^g{}_{X^*}X^*, V) &= X^*(g(X^*, V)) + X^*(g(V, X^*)) - V(g(X^*, X^*)) \\
&\quad + g(V, [X^*, X^*]) + g(X^*, [V, X^*]) - g(X^*, [X^*, V]) \\
&= 2g(X^*, [X^*, V]) \\
&= 0.
\end{aligned}
\tag{2.12}
$$

Here, the last equality of (2.12) follows as:

$$
\begin{aligned}
[X^*, V] &= \nabla^g{}_{X^*}V - \nabla^g{}_V X^* \\
&= \frac{Xf}{f}V - \frac{Xf}{f}V \\
&= 0.
\end{aligned}
\tag{2.13}
$$

by using Lemma 2.1 (3). □

Then, we can choose a locally defined orthonormal vector field

$$
\{e_1, \dots, e_m, e_{m+1}, \dots, e_{m+\ell}\}
$$

on (P, g) in such a way that $\{e_1, \dots, e_m\}$ are orthonormal vector fields which are horizontal lifts of the orthonormal vector fields $e'_1, \dots, e'_m$ on (M, h) and $e_{m+1} = \frac{1}{f}e''_{m+1}, \dots, e_{m+\ell} = \frac{1}{f}e''_{m+\ell}$. Then, by Lemma 2.3, $\nabla^g{}_{e_i}e_i$, $i = 1, \dots, m$, are the horizontal lifts of $\nabla^h{}_{e'_i}e'_i$.

For $i = m+1, \dots, m+\ell$, we have the following decomposition:

$$
\nabla^g{}_{e_i}e_i = \frac{1}{f^2}\left\{-(e''_i f)\, e_i + \nabla^k{}_{e''_i}e''_i - f\,\nabla(f \circ \pi)\right\}.
\tag{2.14}
$$

We first note that $\nabla(f \circ \pi)$ is a horizontal vector field on P. Because,

$$
g(\nabla(f \circ \pi), V) = Vf = 0
$$

for every $V \in \mathfrak{X}(F)$. And the first two terms of (2.14) are vertical since $\nabla^k{}_{e''_i}e''_i$, $i = m+1, \dots, m+\ell$, are vertical.

To prove (2.14), for $i = m+1, \dots, m+\ell$, we have

$$
\begin{aligned}
\nabla^g{}_{e_i}e_i &= \nabla^g{}_{\frac{1}{f}e''_i}\frac{1}{f}e''_i \\
&= \frac{1}{f}\left\{e''_i\left(\frac{1}{f}\right)e''_i + \frac{1}{f}\nabla^g{}_{e''_i}e''_i\right\} \\
&= \frac{1}{f^2}\left\{-\frac{e''_i f}{f}e''_i + \nabla^g{}_{e''_i}e''_i\right\}.
\end{aligned}
\tag{2.15}
$$

We decompose $\nabla^g_{e''_i} e''_i$ into the vertical and horizontal components:

$$\nabla^g_{e''_i} e''_i = \mathcal{V}\left(\nabla^g_{e''_i} e''_i\right) + \mathcal{H}\left(\nabla^g_{e''_i} e''_i\right). \tag{2.16}$$

Here, by Lemma 2.1 (5), we have

$$\mathcal{V}\left(\nabla^g_{e''_i} e''_i\right) = \nabla^k_{e''_i} e''_i. \tag{2.17}$$

By Lemma 2.1 (4) and $k(e''_i, e''_j) = \delta_{ij}$, we have

$$\begin{aligned} \mathcal{H}\left(\nabla^g_{e''_i} e''_i\right) &= -f\, k(e''_i, e''_i)\, G \\ &= -f\, G \\ &= -f\, \nabla f \\ &= -f\, \nabla (f \circ \pi) \end{aligned} \tag{2.18}$$

by Lemma 2.1 (1). We obtain (2.14).

3. Proof of Theorem 3.2

If the principal G-bundle $\pi : (P, g) \to (M, h)$ is harmonic, then it is clearly biharmonic. Our main interest is to ask the reverse holds under what conditions:

Problem 3.1. *If the projection π of a principal G-bundle $\pi : (P, g) \to (M, h)$ is biharmonic, is π harmonic or not.*

In this paper, we show that this problem is affirmative when the Ricci curvature of the base manifold (M, h) is negative definite. Indeed, we show that

Theorem 3.2. *Let $\pi : (P, g) \to (M, h)$ be a principal G-bundle over a Riemannian manifold (M, h) with non-positive Ricci curvature. Assume P is compact so that M is also compact. If the projection π is biharmonic, then it is harmonic.*

In this section, we give a proof of Theorem 3.2 in case of a compact Riemannian manifold (M, h) and the Ricci tensor of (M, h) is negative definite. We will give the proof of Theorem 4.1 in case of a non-compact complete Riemannian manifold (M, h) in the Section Four.

Let us first consider a principal G-bundle $\pi : (P, g) \to (M, h)$ whose the total space P is compact. Assume that the projection $\pi : (P, g) \to (M, h)$ is biharmonic, which is by definition, $J(\tau(\pi)) \equiv 0$, where $\tau(\pi)$ is the tension field of π which is defined by

$$\tau(\pi) := \sum_{i=1}^{p} \{\nabla^h_{e_i} \pi_* e_i - \pi_*(\nabla_{e_i} e_i)\}, \tag{3.1}$$

the Jacobi operator J is defined by

$$JV := \overline{\Delta}V - \mathcal{R}(V) \qquad (V \in \Gamma(\pi^{-1}TM)), \tag{3.2}$$

$\overline{\Delta}$ is the rough Laplacian defined by

$$\overline{\Delta}V := -\sum_{i=1}^{p} \{\overline{\nabla}_{e_i}(\overline{\nabla}_{e_i}V) - \overline{\nabla}_{\nabla_{e_i}e_i}V\}, \tag{3.3}$$

and

$$\mathcal{R}(V) := R^h(V, \pi_*e_i)\pi_*e_i, \tag{3.4}$$

where $\{e_i\}_{i=1}^{p}$ is a locally defined orthonormal frame field on (P, g).

The tangent space P_u ($u \in P$) is canonically decomposed into the orthogonal direct sum of the vertical subspace $G_u = \{A_u{}^* | A \in \mathfrak{g}\}$ and the horizontal subspace H_u: $P_u = G_u \oplus H_u$. Then, we have

$$\begin{aligned}
\tau_2(\pi) &= \overline{\Delta}\tau(\pi) - \sum_{i=1}^{p} R^h(\tau(\pi), \pi_*e_i)\pi_*e_i \\
&= \overline{\Delta}\tau(\pi) - \sum_{i=1}^{m} R^h(\tau(\pi), \pi_*e_i)\pi_*e_i \\
&\qquad - \sum_{i=1}^{k} R^h(\tau(\pi), \pi_*A_{m+i}^*)\pi_*A_{m+i}^* \\
&= \overline{\Delta}\tau(\pi) - \sum_{i=1}^{m} R^h(\tau(\pi), \pi_*e_i)\pi_*e_i,
\end{aligned}$$

where $p = \dim P$, $m = \dim M$, $k = \dim G$, respectively. Then, we obtain

$$\begin{aligned}
0 &= \int_M \langle J(\tau(\pi)), \tau(\pi) \rangle \, v_g \\
&= \int_M \langle \overline{\nabla}^* \overline{\nabla} \tau(\pi), \tau(\pi) \rangle \, v_g - \int_M \sum_{i=1}^{m} \langle R^h(\tau(\pi), \pi_*e_i)\pi_*e_i, \tau(\pi) \rangle \, v_g \\
&= \int_M \langle \overline{\nabla} \tau(\pi), \overline{\nabla} \tau(\pi) \rangle \, v_g - \int_M \sum_{i=1}^{m} \langle R^h(\tau(\pi), \pi_*e_i)\pi_*e_i, \tau(\pi) \rangle \, v_g.
\end{aligned}$$

Therefore, we obtain

$$\begin{aligned}
\int_M \langle \overline{\nabla} \tau(\pi), \overline{\nabla} \tau(\pi) \rangle \, v_g &= \int_M \sum_{i=1}^{m} \langle R^h(\tau(\pi), \pi_*e_i)\pi_*e_i, \tau(\pi) \rangle \, v_g \\
&= \int_M \sum_{i=1}^{m} \langle R^h(\tau(\pi), e_i')e_i', \tau(\pi) \rangle \, v_g \\
&= \int_M \mathrm{Ric}^h(\tau(\pi)) \, v_g, \tag{3.5}
\end{aligned}$$

where $\{e_i'\}_{i=1}^m$ is a locally defined orthonormal frame field on (M,h) satisfying $\pi_* e_i = e_i'$, and $\mathrm{Ric}(X)$ is the Ricci curvature of (M,h) along $X \in T_x M$. The left hand side of (3.5) is non-negative, and then, the both hand sides of (3.5) must vanish if the Ricci curvature of (M,h) is non-positive. Therefore, we obtain

$$\begin{cases} \overline{\nabla}_X \tau(\pi) = 0 \quad (\forall X \in \mathfrak{X}(P)), \text{ i.e., } \tau(\pi) \text{ is parallel, and} \\ \mathrm{Ric}^h(\tau(\pi)) = 0. \end{cases} \tag{3.6}$$

Let us define a 1-form $\alpha \in A^1(P)$ on P by $\alpha(X) = \langle d\pi(X), \tau(\pi)\rangle$, $X \in \mathfrak{X}(P)$. Then, we have

$$-\delta\alpha = \sum_{i=1}^p (\nabla_{e_i}\alpha)(e_i) = \langle \tau(\pi), \tau(\pi)\rangle + \langle d\pi, \overline{\nabla}\tau(\pi)\rangle. \tag{3.7}$$

Integrate the above (3.7) over P since P is compact without boundary. By (3.6), $\overline{\nabla}_X \tau(\pi) = 0$, $X \in \mathfrak{X}(P)$, we have

$$0 = -\int_P \delta\alpha\, v_g = \int_P \langle \tau(\pi), \tau(\pi)\rangle\, v_g \tag{3.8}$$

which implies that $\tau(\pi) = 0$, i.e., $\pi : (P,g) \to (M,h)$ is harmonic. $\quad\square$

4. Proof of Theorem 4.1

In this section, we will show

Theorem 4.1. *Let $\pi : (P,g) \to (M,h)$ be a principal G-bundle over a Riemannian manifold with non-positive Ricci curvature. Assume that (P,g) is a non-compact complete Riemannian manifold, and the projection π has both finite energy $E(\pi) < \infty$ and finite bienergy $E_2(\pi) < \infty$. If π is biharmonic, then it is harmonic.*

Here, we first recall the following examples:

Example 1 (cf. [91], p. 62) The inversion in the unit sphere $\phi : \mathbb{R}^n \backslash \{o\} \ni \mathbf{x} \mapsto \frac{\mathbf{x}}{|\mathbf{x}|^2} \in \mathbb{R}^n$ is a biharmonic morphism if $n = 4$. It is not harmonic since $\tau(\phi) = -\frac{4\mathbf{x}}{|\mathbf{x}|^4}$.

Here, a C^∞ map $\phi : (M,g) \to (N,h)$ is called to be a *biharmonic morphism* if, for every biharmonic function $f : U \subset N \to \mathbb{R}$ with $\phi^{-1}(U) \neq \emptyset$, the composition $f \circ \phi : \phi^{-1}(U) \subset M \to \mathbb{R}$ is biharmonic.

Example 2 (cf. [91], p. 70) Let (M^2, h) be a Riemannian surface, and let $\beta : M^2 \times \mathbb{R} \to \mathbb{R}^*$ and $\lambda : \mathbb{R} \to \mathbb{R}^*$ be two positive C^∞ functions. Consider the projection $\pi : (M^2 \times \mathbb{R}^*, g = \lambda^{-2} h + \beta^2 dt^2) \ni (p, t) \mapsto p \in (M^2, h)$. Here, we take $\beta = c_2 \, e^{\int f(x) \, dx}$, $f(x) = \frac{-c_1 (1 + e^{c_1 x})}{1 - e^{c_1 x}}$ with $c_1, c_2 \in \mathbb{R}^*$, and $(M^2, h) = (\mathbb{R}^2, dx^2 + dy^2)$. Then,

$$\pi : (\mathbb{R}^2 \times \mathbb{R}^*, dx^2 + dy^2 + \beta^2(x) \, dt^2) \ni (x, y, t) \mapsto (x, y) \in (\mathbb{R}^2, dx^2 + dy^2)$$

gives a family of *proper biharmonic* (i.e., biharmonic but not harmonic) Riemannian submersions.

For a non-compact and complete Riemannian manifold (N, h) with non-positive Ricci curvature, we will give a proof of Theorem 4.1.

(*The first step*) We first take a cut off function η on (P, g) for a fixed point $p_0 \in P$ as follows:

$$\begin{cases} 0 \leq \eta \leq 1 & \text{(on } P\text{)}, \\ \eta = 1 & \text{(on } B_r(p_0)\text{)}, \\ \eta = 0 & \text{(outside } B_{2r}(p_0)\text{)}, \\ |\nabla \eta| \leq \dfrac{2}{r} & \text{(on } P\text{)}, \end{cases} \tag{4.1}$$

where $B_r(p_0)$ is the ball in (P, g) of radius r around p_0.

Now assume that the projection $\pi : (P, g) \to (N, h)$ is biharmonic. Namely, we have, by definition,

$$0 = J_2(\pi) = J_\pi(\tau(\pi))$$

$$= \overline{\Delta} \, \tau(\pi) - \sum_{i=1}^p R^h(\tau(\pi), \pi_* e_i) \pi_* e_i, \tag{4.2}$$

where $\{e_i\}_{i=1}^p$ is a local orthonormal frame field on (P, g) and $\overline{\Delta}$ is the rough Laplacian which is defined by

$$\overline{\Delta} V := \overline{\nabla}^* \overline{\nabla} V = -\sum_{i=1}^p \left\{ \overline{\nabla}_{e_i} (\overline{\nabla}_{e_i} V) - \overline{\nabla}_{\nabla_{e_i} e_i} V \right\}, \tag{4.3}$$

for $V \in \Gamma(\pi^{-1} TM)$.

(*The second step*) By (4.2), we have

$$\int_P \langle \overline{\nabla}^* \overline{\nabla}\, \tau(\pi), \eta^2\, \tau(\pi) \rangle\, v_g = \int_P \eta^2 \left\langle \sum_{i=1}^p R^h(\tau(\pi), \pi_* e_i)\pi_* e_i, \tau(\pi) \right\rangle v_g$$

$$= \int_P \eta^2 \sum_{i=1}^p \left\langle R^h(\tau(\pi), \pi_* e_i)\pi_* e_i, \tau(\pi) \right\rangle v_g$$

$$= \int_P \eta^2 \sum_{i=1}^m \left\langle R^h(\tau(\pi), e'_i)e'_i, \tau(\pi) \right\rangle v_g$$

$$= \int_P \eta^2\, \mathrm{Ric}^h(\tau(\pi))\, v_g, \qquad (4.4)$$

where $\{e'_i\}_{i=1}^m$ is a local orthonormal frame field on (M, h), and $\mathrm{Ric}^h(u)$ $u \in T_y M$, $(y \in M)$ is the Ricci curvature of (M, h) which is non-positive by our assumption.

(*The third step*) Therefore, we obtain

$$0 \geq \int_P \langle \overline{\nabla}^* \overline{\nabla}\, \tau(\pi), \eta^2\, \tau(\pi) \rangle\, v_g$$

$$= \int_P \langle \overline{\nabla}\, \tau(\pi), \overline{\nabla}(\eta^2\, \tau(\pi)) \rangle\, v_g$$

$$= \int_P \sum_{i=1}^p \langle \overline{\nabla}_{e_i}\, \tau(\pi), \overline{\nabla}_{e_i}(\eta^2\, \tau(\pi)) \rangle\, v_g$$

$$= \int_P \sum_{i=1}^p \left\{ \eta^2 \langle \overline{\nabla}_{e_i}\tau(\pi), \overline{\nabla}_{e_i}\tau(\pi) \rangle + e_i(\eta^2) \langle \overline{\nabla}_{e_i}\tau(\pi), \tau(\pi) \rangle \right\} v_g$$

$$= \int_P \eta^2 \sum_{i=1}^p |\overline{\nabla}_{e_i}\tau(\pi)|^2\, v_g$$

$$+ 2 \int_P \sum_{i=1}^p \langle \eta\, \overline{\nabla}_{e_i}\tau(\pi), e_i(\eta)\, \tau(\pi) \rangle\, v_g. \qquad (4.5)$$

Therefore, we obtain by (4.5),

(*The fourth step*) Then, we have

$$\int_P \eta^2 \sum_{i=1}^p \left|\overline{\nabla}_{e_i}\tau(\pi)\right|^2\, v_g \leq -2 \int_P \sum_{i=1}^p \langle \eta\, \overline{\nabla}_{e_i}\tau(\pi), e_i(\eta)\, \tau(\pi) \rangle\, v_g$$

$$= -2 \int_P \sum_{i=1}^p \langle V_i, W_i \rangle\, v_g, \qquad (4.6)$$

where $V_i := \eta\, \overline{\nabla}_{e_i}\tau(\pi)$, and $W_i := e_i(\eta)\, \tau(\pi)$ $(i = 1, \ldots, p)$. Then, the right hand side of (4.6) is estimated by the Cauchy-Schwarz inequality,

$$\pm 2 \langle V_i, W_i \rangle \leq \epsilon |V_i|^2 + \frac{1}{\epsilon} |W_i|^2 \qquad (4.7)$$

since

$$0 \le |\sqrt{\epsilon}\, V_i \pm \frac{1}{\sqrt{\epsilon}} W_i|^2 = \epsilon |V_i|^2 \pm 2 \langle V_i, W_i \rangle + \frac{1}{\epsilon} |W_i|^2,$$

so that

$$\mp 2 \langle V_i, W_i \rangle \le \epsilon |V_i|^2 + \frac{1}{\epsilon} |W_i|^2.$$

Therefore, the right hand side of (4.6) is estimated as follows:

$$\text{RHS of (4.6)} := - \int_P \sum_{i=1}^p \langle V_i, W_i \rangle\, v_g$$

$$\le \epsilon \int_P \sum_{i=1}^p |V_i|^2\, v_g + \frac{1}{\epsilon} \int_P \sum_{i=1}^p |W_i|^2\, v_g. \qquad (4.8)$$

(*The fifth step*) By putting $\epsilon = \frac{1}{2}$, we have

$$\int_P \eta^2 \sum_{i=1}^p |\overline{\nabla}_{e_i} \tau(\pi)|^2\, v_g \le \frac{1}{2} \int_P \sum_{i=1}^p \eta^2\, |\overline{\nabla}_{e_i} \tau(\pi)|^2\, v_g$$

$$+ 2 \int_P \sum_{i=1}^p e_i(\eta)^2\, |\tau(\pi)|^2\, v_g. \qquad (4.9)$$

Therefore, we obtain

$$\frac{1}{2} \int_P \eta^2 \sum_{i=1}^p |\overline{\nabla}_{e_i} \tau(\pi)|^2\, v_g \le 2 \int_P |\nabla \eta|^2\, |\tau(\pi)|^2\, v_g. \qquad (4.10)$$

Substituting (4.1) into (4.10), we obtain

$$\int_P \eta^2 \sum_{i=1}^p |\overline{\nabla}_{e_i} \tau(\pi)|^2\, v_g \le 4 \int_P |\nabla \eta|^2\, |\tau(\pi)|^2\, v_g$$

$$\le \frac{16}{r^2} \int_P |\tau(\pi)|^2\, v_g. \qquad (4.11)$$

(The sixth step) Tending $r \to \infty$ by the completeness of (P, g) and $E_2(\pi) = \frac{1}{2} \int_P |\tau(\pi)|^2\, v_g < \infty$, we obtain that

$$\int_P \sum_{i=1}^p |\overline{\nabla}_{e_i} \tau(\pi)|^2\, v_g = 0, \qquad (4.12)$$

which implies that

$$\overline{\nabla}_X \tau(\pi) = 0 \qquad (\forall\, X \in \mathfrak{X}(P)). \qquad (4.13)$$

(*The seventh step*) Therefore, we obtain

$$|\tau(\pi)| \quad \text{is constant, say } c \qquad (4.14)$$

because

$$X |\tau(\pi)|^2 = 2 \langle \overline{\nabla}_X \tau(\pi), \tau(\pi) \rangle = 0 \qquad (\forall X \in \mathfrak{X}(M))$$

by (4.13).

(*The eighth step*) In the case that $\mathrm{Vol}(P, g) = \infty$ and $E_2(\pi) < \infty$, c must be zero. Because, if $c \neq 0$,

$$E_2(\pi) = \frac{1}{2} \int_P |\tau(\pi)|^2 \, v_g = \frac{c}{2} \, \mathrm{Vol}(P, g) = \infty$$

which is a contradiction.

Thus, if $\mathrm{Vol}(P, g) = \infty$, then $c = 0$, i.e., $\pi : (P, g) \to (M, h)$ is harmonic.

(*The ninth step*) In the case $E(\pi) < \infty$ and $E_2(\pi) < \infty$, let us define a 1-form $\alpha \in A^1(P)$ on P by

$$\alpha(X) := \langle d\pi(X), \tau(\pi) \rangle, \qquad (X \in \mathfrak{X}(P)). \tag{4.15}$$

Then, we obtain

$$\int_P |\alpha| \, v_g = \int_P \left(\sum_{i=1}^p |\alpha(e_i)|^2 \right)^{1/2} \leq \int_P |d\pi| \, |\tau(\pi)| \, v_g$$

$$\leq \left(\int_P |d\pi|^2 \, v_g \right)^{1/2} \left(\int_P |\tau(\pi)|^2 \, v_g \right)^{1/2}$$

$$= 2 \sqrt{E(\pi) \, E_2(\pi)} < \infty. \tag{4.16}$$

For the function $\delta\alpha := - \sum_{i=1}^p (\nabla_{e_i} \alpha)(e_i) \in C^\infty(P)$, we have

$$-\delta\alpha = \sum_{i=1}^p (\nabla_{e_i} \alpha)(e_i) = \sum_{i=1}^p \left\{ e_i(\alpha(e_i)) - \alpha(\nabla_{e_i} e_i) \right\}$$

$$= \sum_{i=1}^p \left\{ e_i \langle d\pi(e_i), \tau(\pi) \rangle - \langle d\pi(\nabla_{e_i} e_i), \tau(\pi) \rangle \right\}$$

$$= \sum_{i=1}^p \left\{ \left\langle \overline{\nabla}_{e_i} d\pi(e_i), \tau(\pi) \right\rangle + \left\langle d\pi(e_i), \overline{\nabla}_{e_i} \tau(\pi) \right\rangle - \left\langle d\pi(\nabla_{e_i} e_i, \tau(\pi) \right\rangle \right\}$$

$$= \left\langle \sum_{i=1}^p \left\{ \overline{\nabla}_{e_i} d\pi(e_i) - d\pi(\nabla_{e_i} e_i) \right\}, \tau(\pi) \right\rangle + \sum_{i=1}^p \left\langle d\pi(e_i), \overline{\nabla}_{e_i} \tau(\pi) \right\rangle$$

$$= \langle \tau(\pi), \tau(\pi) \rangle + \langle d\pi, \overline{\nabla}\tau(\pi) \rangle$$

$$= |\tau(\pi)|^2 \tag{4.17}$$

since $\overline{\nabla}\tau(\pi) = 0$. By (4.17), we obtain

$$\int_P |\delta\alpha| \, v_g = \int_P |\tau(\pi)|^2 \, v_g = 2 \, E_2(\pi) < \infty. \tag{4.18}$$

By (4.16), (4.18) and the completeness of (P, g), we can apply Gaffney's theorem which implies that

$$0 = \int_P (-\delta\,\alpha)\, v_g = \int_P |\tau(\pi)|^2\, v_g. \tag{4.19}$$

Thus, we obtain

$$\tau(\pi) = 0, \tag{4.20}$$

that is, $\pi : (P, g) \to (M, h)$ is harmonic. We obtain Theorem 4.1. $\square$

5. The tension fields of the warped products

In this section, we calculate the tension field $\tau(\pi)$. Let us recall the definition of the tension field:

DEFINITION 5.1.

$$\tau(\pi) = \sum_{i=1}^{m+\ell} \left\{ \bar{\nabla}_{e_i} \pi_* e_i - \pi_* \left(\nabla^g_{e_i} e_i \right) \right\}$$

$$= \sum_{i=1}^{m+\ell} \left\{ \nabla^h_{\pi_* e_i} \pi_* e_i - \pi_* \left(\nabla^g_{e_i} e_i \right) \right\}. \tag{5.1}$$

Since $\nabla^g_{e_i} e_i$ are the horizontal lifts of $\nabla^h_{e'_i} e'_i$ for $i = 1, \ldots, m$, and (2.14), we have

$$\tau(\pi) = \sum_{i=1}^{m} \left\{ \nabla^h_{\pi_* e_i} \pi_* e_i - \pi_* \left(\nabla^g_{e_i} e_i \right) \right\} + \sum_{i=m+1}^{m+\ell} \left\{ \nabla^h_{\pi_* e_i} \pi_* e_i - \pi_* \left(\nabla^g_{e_i} e_i \right) \right\}$$

$$= \sum_{i=1}^{m} \left\{ \nabla^h_{e'_i} e'_i - \nabla^h_{e'_i} e'_i \right\} + \sum_{i=m+1}^{m+\ell} \left\{ 0 - \left(-\frac{1}{f} \nabla \left(f \circ \pi \right) \right) \right\}$$

$$= \frac{\ell}{f} \nabla \left(f \circ \pi \right). \tag{5.2}$$

Indeed, we obtain the second equality of (5.2) as follows: The first sum vanishes since $\pi_* e_i = e'_i$ and $\pi_* \nabla^g_{e_i} e_i = \nabla^h_{e'_i} e'_i$, $(i = 1, \ldots, m)$. The second sum coincides with $\frac{\ell}{f} \nabla \left(f \circ \pi \right)$ since $\pi_* e_i = 0$ and also $\pi_* \nabla^g_{e_i} e_i = -\frac{1}{f} \nabla \left(f \circ \pi \right)$ $(i = m+1, \ldots, m+\ell)$. Therefore, we obtain

THEOREM 5.1. *Let* $\pi : (P, g) \to (M, h)$ *be the warped product. Then, we have*

$$\tau(\pi) = \frac{\ell}{f} \nabla \left(f \circ \pi \right). \tag{5.3}$$

Then, π is harmonic if and only if f is constant.

6. The bitension fields of the warped products

Let us recall the definition of the bitension field for a C^∞ mapping $\varphi : (P, g) \to (M, h)$ which is given by

$$\tau_2(\varphi) := \overline{\Delta}\,\tau(\varphi) - \mathcal{R}^h(\tau(\varphi)). \tag{6.1}$$

Here, recall, for $V \in \Gamma(\varphi^{-1}TM)$,

$$\overline{\Delta}V := -\sum_{i=1}^{p} \left\{ \overline{\nabla}_{e_i}(\overline{\nabla}_{e_i}V) - \overline{\nabla}_{\nabla^g_{e_i}e_i}V \right\}, \tag{6.2}$$

$$\mathcal{R}^hV := \sum_{i=1}^{p} R^h(V, \varphi_*e_i)\varphi_*e_i, \tag{6.3}$$

where $\{e_i\}_{i=1}^{p}$ is a locally defined orthonormal frame field on (P, g), $p = \dim P$, $\overline{\nabla}$ is the induced connection on the induced bundle $\varphi^{-1}TM$, and the curvature tensor of (N, h) is given by $R^h(U, V)W := \nabla^h_U(\nabla^h_VW) - \nabla^h_V(\nabla^h_UW) - \nabla^h_{[U,V]}W$, for $U, V, W \in \mathfrak{X}(M)$.

DEFINITION 6.1. $\pi : (P, g) \to (M, h)$ *is biharmonic if* $\tau_2(\pi) = 0$.

Let us $\pi : (P, g) \to (M, h)$ be the warped product whose Riemannian metric g is given by (2.8). For $V = \tau(\pi)$, then,

$$\begin{aligned}
\mathcal{R}^hV &= \sum_{i=1}^{p} R^h(\tau(\pi), \pi_*e_i)\pi_*e_i \\
&= \sum_{i=1}^{m} R^h(\tau(\pi), e'_i)e'_i \\
&= \rho^h(\tau(\pi)),
\end{aligned} \tag{6.4}$$

where $m = \dim M$ and ρ^h is Ricci transform of (M, h) given by $\rho^h(u) := \sum_{i=1}^{m} R^h(u, e'_i)e'_i$, $u \in T_xM$, and $\{e'_i\}_{i=1}^{m}$ is a locally defined orthonormal field on (M, h).

In the following, we calculate the rough Laplacian $\overline{\Delta}$ for $V = \tau(\pi)$. (*The first step*) We calculate $\overline{\nabla}_{e_i}\tau(\pi)$ and $\overline{\nabla}_{e_i}(\overline{\nabla}_{e_i}\tau(\pi))$ as follows:

$$\overline{\nabla}_{e_i}\tau(\pi) = \nabla^h_{\pi_*e_i}\tau(\pi) = \begin{cases} \nabla^h_{e'_i}\tau(\pi) & (i = 1, \dots, m = \dim M), \\ 0 & (i = m+1, \dots, m+\ell), \end{cases} \tag{6.5}$$

where $p := \dim P = m + \ell$, $m = \dim M$, and $\ell = \dim F$. Furthermore,

$$\overline{\nabla}_{e_i}(\overline{\nabla}_{e_i}\tau(\pi)) = \begin{cases} \nabla^h_{e'_i}(\nabla^h_{e'_i}\tau(\pi)) & (i = 1, \dots, m), \\ 0 & (i = m+1, \dots, m+\ell = p). \end{cases} \tag{6.6}$$

(*The second step*) We calculate $\overline{\nabla}_{\nabla^g_{e_i} e_i} \tau(\pi)$ by the similar way as the first step:

For $i = 1, \ldots, m,$

$$\overline{\nabla}_{\nabla^g_{e_i} e_i} \tau(\pi) = \nabla^h_{\pi_*(\nabla^g_{e_i} e_i)} \tau(\pi) = \nabla^h_{\nabla^h_{e'_i} e'_i} \tau(\pi), \tag{6.7}$$

and for $i = m+1, \ldots, m+\ell$, by (5.1),

$$\overline{\nabla}_{\nabla^g_{e_i} e_i} \tau(\pi) = \nabla^h_{\pi_*(\nabla^g_{e_i} e_i)} \tau(\pi) = \nabla^h_{-\frac{1}{f} \nabla(f \circ \pi)} \tau(\pi). \tag{6.8}$$

(*The third step*) Therefore, we calculate (6.2) for $V = \tau(\pi)$ as follows.

$$\overline{\Delta}\tau(\pi) := -\sum_{i=1}^{p} \left\{ \overline{\nabla}_{e_i}(\overline{\nabla}_{e_i} \tau(\pi)) - \overline{\nabla}_{\nabla^g_{e_i} e_i} \tau(\pi) \right\}$$

$$= -\sum_{i=1}^{m} \left\{ \overline{\nabla}_{e_i}(\overline{\nabla}_{e_i} \tau(\pi)) - \overline{\nabla}_{\nabla^g_{e_i} e_i} \tau(\pi) \right\} - \sum_{i=m+1}^{m+\ell} \left\{ \overline{\nabla}_{e_i}(\overline{\nabla}_{e_i} \tau(\pi)) - \overline{\nabla}_{\nabla^g_{e_i} e_i} \tau(\pi) \right\}$$

$$= -\sum_{i=1}^{m} \left\{ \nabla^h_{e'_i}(\nabla^h_{e'_i} \tau(\pi)) - \nabla^h_{\nabla^h_{e'_i} e'_i} \tau(\pi) \right\} - \sum_{i=m+1}^{m+\ell} \left\{ 0 - \nabla^h_{-\frac{1}{f} \nabla(f \circ \pi)} \tau(\pi) \right\}$$

$$= -\sum_{i=1}^{m} \left\{ \nabla^h_{e'_i} \left(\nabla^h_{e'_i} \tau(\pi) \right) - \nabla^h_{\nabla^h_{e'_i} e'_i} \tau(\pi) \right\} - \ell \nabla^h_{\frac{1}{f} \nabla(f \circ \pi)} \tau(\pi). \tag{6.9}$$

(*The fourth step*) Therefore, by (6.1), (6.4) and (6.9), we obtain

$$\tau_2(\pi) = -\sum_{i=1}^{m} \left\{ \overline{\nabla}_{e'_i} \left(\overline{\nabla}_{e'_i} \tau(\pi) \right) - \overline{\nabla}_{\nabla^h_{e'_i} e'_i} \tau(\pi) \right\} - \ell \nabla^h_{\frac{1}{f} \nabla f} \tau(\pi) - \rho^h(\tau(\pi))$$

$$= \ell J_{\mathrm{id}} \left(\frac{1}{f} \nabla f \right) - \ell^2 \nabla^h_{\frac{1}{f} \nabla f} \left\{ \frac{\nabla f}{f} \right\} \tag{6.10}$$

$$= \ell J_{\mathrm{id}} \left(\frac{1}{f} \nabla f \right) + \frac{\ell^2}{f^3} h(\nabla f, \nabla f) \nabla f - \frac{\ell^2}{f^2} \nabla^h_{\nabla f} \nabla f. \tag{6.11}$$

Therefore, we can summarize the above by recalling the following definitions:

$$J_{\mathrm{id}} := \overline{\Delta}^h - \rho^h, \tag{6.12}$$

is the Jacobi operator of the identity of (M, h), id $: (M, h) \to (M, h)$ acting on the space $\mathfrak{X}(M)$ of C^∞ vector fields on M, and the operator $\overline{\Delta}^h$ is defined by

$$\overline{\Delta}^h(X) := -\sum_{i=1}^{m} \left(\nabla^h_{e'_i} \nabla^h_{e'_i} - \nabla^h_{\nabla^h_{e'_i} e'_i} \right) X \qquad (X \in \mathfrak{X}(M)), \tag{6.13}$$

and ρ^h is the Ricci operator of (M, h) given by

$$\rho^h(X) = R^h(X, e_i')e_i' \qquad (X \in \mathfrak{X}(M)). \qquad (6.14)$$

Therefore, due to (6.1) and (6.2), we have

THEOREM 6.1.   *For the warped product* $\pi : (P, g) \to (M, h)$, *the bitension field* $\tau_2(\pi)$ *is given by*

$$\tau_2(\pi) = \overline{\Delta}(\tau(\pi)) - \rho^h(\tau(\pi)) - \ell \overline{\nabla}_{\frac{\nabla f}{f}} \tau(\pi), \qquad (6.15)$$

where $\overline{\Delta}$ *is the rough Laplacian and* $\overline{\nabla}$ *is the induced connection from the Levi-Civita connection* ∇^h *of* (M, h). *Therefore, the warped product* $\pi : (P, g) \to (M, h)$ *is biharmonic, i.e.,* $\tau_2(\pi) = 0$, *if and only if the following hold:*

$$J_{\mathrm{id}}\left(\frac{\nabla f}{f}\right) = \ell \nabla^h_{\frac{1}{f}\nabla f}\left\{\frac{\nabla f}{f}\right\} = -\frac{\ell}{f^3} h(\nabla f, \nabla f) \nabla f + \frac{\ell}{f^2} \nabla^h_{\nabla f} \nabla f. \qquad (6.16)$$

COROLLARY 6.1.   *For a positive* C^∞ *function* f *on* M, *let* $\pi : (P, g) = (M \times_f F, g) \to (M, h)$ *be the warped product with* $g = \pi^* h + f^2 k$ *over a Riemannian manifold* (M, h) *whose Ricci curvature is non-positive. If* π *is biharmonic, then*

$$\int_M (\nabla f)\left(h(\frac{\nabla f}{f}, \frac{\nabla f}{f})\right) v_h = 2 \int_M h\left(\nabla^h_{\frac{\nabla f}{f}} \frac{\nabla f}{f}, \frac{\nabla f}{f}\right) v_h \geq 0. \qquad (6.17)$$

Proof of Corollary 6.3.   If $\pi : (P, g) \to (M, h)$ is biharmonic, by (6.15), it holds that

$$0 = \tau_2(\pi) = J_{\mathrm{id}}\left(\ell \frac{\nabla f}{f}\right) - \ell^2 \nabla^h_{\frac{\nabla f}{f}} \frac{\nabla f}{f}. \qquad (6.18)$$

which implies that

$$0 \leq \int_M h\left(J_{\mathrm{id}}\left(\frac{\nabla f}{f}\right), \frac{\nabla f}{f}\right) v_h = \ell \int_M h\left(\nabla^h_{\frac{\nabla f}{f}} \frac{\nabla f}{f}, \frac{\nabla f}{f}\right) v_h. \qquad (6.19)$$

Because all the eigenvalues of J_{id} are non-negative since $J_{\mathrm{id}} = \overline{\Delta}^h - \rho^h$ and the Ricci transform ρ^h are non-positive (cf. [4], [6, p.161]).   □

7. The solutions of the ordinary differential equation

Assume that $(M, h) = (\mathbb{R}, dt^2)$, a line, and $(P, g) = F \times_f \mathbb{R}$, the warped product of a Riemannian manifold (F, k) and the line $(\mathbb{R}, dt^2)$, that is,

$$g = \pi^*(dt^2) + f^2 k, \tag{7.1}$$

for a C^∞ function $f \in C^\infty(\mathbb{R})$.

In this case, it holds that

$$
\begin{cases}
J_{\mathrm{id}}(\pi) = J_{\mathrm{id}}\left(\ell\,\dfrac{\nabla f}{f}\right) = -\ell\left(\dfrac{f'}{f}\right)''\dfrac{\partial}{\partial t}, \\[2mm]
\ell^2\,\nabla^h_{\frac{\nabla f}{f}}\dfrac{\nabla f}{f} = \ell^2\,\dfrac{f'}{f}\,\nabla^h_{\frac{\partial}{\partial t}}\left(\dfrac{f'}{f}\dfrac{\partial}{\partial t}\right) = \ell^2\,\dfrac{f'}{f}\dfrac{\partial}{\partial t}\left(\dfrac{f'}{f}\right)\dfrac{\partial}{\partial t} \\[2mm]
\qquad = \ell^2\left(\dfrac{f'f''}{f^2} - \dfrac{f'^3}{f^3}\right)\dfrac{\partial}{\partial t}.
\end{cases} \tag{7.2}
$$

Therefore, $\pi : (F \times_f \mathbb{R}, g) \to (\mathbb{R}, dt^2)$ is biharmonic, i.e.,

$$\tau_2(\pi) = J_{\mathrm{id}}\left(\ell\,\dfrac{\nabla f}{f}\right) - \ell^2\,\nabla^h_{\frac{\nabla f}{f}}\dfrac{\nabla f}{f} = 0 \tag{7.3}$$

if and only if

$$
\begin{aligned}
0 &= -\ell\left(\dfrac{f'}{f}\right)'' - \ell^2\left(\dfrac{f'f''}{f^2} - \dfrac{f'^3}{f^3}\right) \\[2mm]
&= -\ell\left(\dfrac{f''f - f'^2}{f^2}\right)' - \ell^2\left(\dfrac{f'f''}{f^2} - \dfrac{f'^3}{f^3}\right) \\[2mm]
&= -\ell\,\dfrac{f'''f^2 - 3f''f'f + 2f'^3}{f^3} - \ell^2\left(\dfrac{f'f''}{f^2} - \dfrac{f'^3}{f^3}\right) \\[2mm]
&= -\dfrac{\ell}{f^3}\left\{f'''f^2 + (\ell - 3)f''f'f + (-\ell + 2)f'^3\right\}
\end{aligned} \tag{7.4}
$$

if and only if

$$f'''f^2 + (\ell - 3)f''f'f + (-\ell + 2)f'^3 = 0. \tag{7.5}$$

Therefore, we have

THEOREM 7.1. *Let (F, k) be a Riemannian manifold.* (1) *the warped product $\pi : (F \times_f \mathbb{R}, g) \to (\mathbb{R}, dt^2)$ is biharmonic if and only if* (7.5) *holds.*

(2) *All the positive C^∞ solution f of* (7.5) *on $\mathbb{R}$ are given by*

$$f(t) = c\,\exp\left(\int_{t_0}^t a\,\tanh\left[a\,\dfrac{\ell}{2}r + b\right]\,dr\right), \tag{7.6}$$

where $a \neq 0$, b, $c > 0$ are arbitrary constants.

(3) In the case $(M, h) = (\mathbb{R}, dt^2)$, let $f(t)$ be a C^∞ function defined by (7.6) with $a \neq 0$, b any real number and $c > 0$. Then, the warped product $\pi : (\mathbb{R} \times_f F, g) \to (\mathbb{R}, dt^2)$ with the Riemannian metric

$$g = \pi^* dt^2 + f^2 k \tag{7.7}$$

is biharmonic but not harmonic.

In order to solve (7.5), we put $u := (\log f)' = \frac{f'}{f}$. Then (7.5) turns into the ordinary differential equation on u:

$$u'' + \frac{\ell}{2} (u^2)' = 0. \tag{7.8}$$

A general solution u of (5.6) is given by

$$u(t) = a \tanh \left[a \frac{\ell}{2} t + b \right], \tag{7.9}$$

where a and b are arbitrary constants. Thus, every positive solution $f(t)$ is given by

$$f(t) = c \exp \left(\int_{t_0}^{t} a \tanh \left[a \frac{\ell}{2} r + b \right] dr \right), \tag{7.10}$$

where a, b, $c > 0$ are arbitrary constants.

Therefore, we obtain Theorem 7.1 together with Theorem 6.2. $\quad \square$

Bibliography

[1] K. Akutagawa and S. Maeta, *Properly immersed biharmonic submanifolds in the Euclidean spaces*, Geometriae Dedicata, **164** (2013), 351–355.

[2] J. A. Alvarez López, *The basic component of the mean curvature of Riemannian mfoliations*, Ann. Global Anal. Geom., **10** (1992), 179–194.

[3] A. Aribi, *Le spectre du sous-laplacien sur les variétés CR strictement pseudoconvexes*, Univ. Tours, France, Thèse, 2012.

[4] A. Balmus, S. Montaldo and C. Oniciuc, *Classification results for biharmonic submanifolds in spheres*, Israel J. Math., **168** (2008), 201–220.

[5] A. Balmus, S. Montaldo and C. Oniciuc, *Biharmonic hypersurfaces in 4-dimensional space forms*, Math. Nachr., **283** (2010), 1696–1705.

[6] A. Balmus, S. Montaldo and C. Oniciuc, *Biharmonic PNMC submanifolds in spheres*, to appear in Ark. Mat.

[7] P. Baird and J. Eells, *A conservation law for harmonic maps*, Lecture Notes in Math., Springer, **894** (1981), 1–25.

[8] P. Baird, A. Fardoun and S. Ouakkas, *Liouville-type theorems for biharmonic maps between Riemannian manifolds*, Adv. Calc. Var., **3** (2010), 49–68.

[9] P. Baird and D. Kamissoko, *On constructing biharmonic maps and metrics*, Ann. Global Anal. Geom. **23** (2003), 65–75.

[10] P. Baird and J. Wood, *Harmonic Morphisms Between Riemannian Manifolds*, Oxford Science Publication, 2003, Oxford.

[11] E. Barletta, S. Dragomir and H. Urakawa, *Pseudoharmonic maps from a nondegenerate CR manifold into a Riemannian manifold*, Indiana Univ. Math. J., no. 2 **50** (2001), 719–746.

[12] R.L. Bishop and B.O'Neill, *Manifolds of negative curvature*, Trans. Amer. Math. Soc., **145** (1969), 1–49.

[13] C. Boyer and K. Galicki, *Sasakian Geometry*, Oxford Sci. Publ., 2008.

[14] F.E. Burstall and M.A. Guest, *Harmonic two-spheres in compact symmetric spaces, revisited*, Math. Ann., **309** (1997), 541–572.

[15] F.E. Burstall and F. Pedit, *Harmonic maps via Adler-Kostant-Symes theory*, In: *Harmonic Maps and Integrable Systems*, eds. by A. P. Fordy and J. C. Wood, Aspect of Mathematics, Vol. E 23, Vieweg (1993), 221–272.

[16] R. Caddeo, S. Montaldo, P. Piu, *On biharmonic maps*, Contemp. Math., **288** (2001), 286–290.

[17] R. Caddeo, S. Montaldo and C. Oniciuc, *Biharmonic submanifolds of* $\mathbb{S}^3$, Intern. J. Math., **12** (2001), 867–876.

[18] I. Castro and F. Urbano, *Twistor holomorphic Lagrangian surfaces in complex projective and hyperbolic planes*, Ann. Global. Anal. Geom. **13** (1995), 59-67.

[19] I. Castro, H.Z. Li and F. Urbano, *Hamiltonian-minimal Lagrangian submanifolds in complex space forms*, Pacific J. Math., **227** (2006), 43–63.

[20] S-Y. A. Chang, L. Wang and P.C. Yang, *A regularity theory of biharmonic maps*, Commun. Pure Appl. Math., **52** (1999), 1113–1137.

[21] B.-Y. Chen, *Some open problems and conjectures on submanifolds of finite type*, Soochow J. Math., **17** (1991), 169–188.

[22] B. -Y. Chen, *Total mean curvature and submanifolds of finite type*, Series in Pure Mathematics, **1**. World Scientific Publishing Co., Singapore, (1984).

[23] B.Y. Chen, *A report on submanifolds of finite type*, Soochow J. Math., **22** (1996), 117–337.

[24] B. -Y. Chen, *Complex extensors and Lagrangian submanifolds in complex Euclidean spaces*, Tohoku. Math. J. **49** (1997), 277-297.

[25] B. -Y. Chen, *Interaction of Legendre curves and Lagrangian submanifolds*, Israel J. Math. **99** (1997), 69-108.

[26] B.-Y. Chen, *Representation of flat Lagrangian H-umbilical submanifolds in complex Euclidean spaces*, Tohoku. Math J. **51** (1999), 13-21.

[27] B. Y. Chen, *Surfaces with parallel normalized mean curvature vector*, Monatsh. Math. **90** (1980), 185-194.

[28] B.-Y. Chen and S. Ishikawa, *Biharmonic pseudo-Riemannian submanifolds in pseudo-Euclidean spaces*, Kyushu J. Math. **52** (1998), no. 1-3, 101-108.

[29] S.S. Chern and J.K. Moser, *Real hypersurfaces in complex manifolds*, Acta Math., **133** (1974), 48–69.

[30] B.-Y. Chen and K. Ogiue, *On totally real submanifolds*, Trans. Amer. Math. Soc. **193** (1974), 257-266.

[31] Y.-J. Chiang and R.A. Wolak, *Transversally biharmonic maps between foliated Riemannian manifolds*, Intern. J. Math., **19** (2008), 981–996.

[32] Y. L. Xin and Q. Chen, *Gauss maps of hypersurfaces in the unit sphere* In: Conference of Differential Geometry in Shanghai (1983).

[33] Y-J. Dai, M. Shoji and H. Urakawa, *Harmonic maps into Lie groups and homogeneous spaces*, Differ. Geom. Appl., **7** (1997), 143–160.

[34] F. Defever, *Hypersurfaces in $\mathbb{E}^4$ with harmonic mean curvature vetor*, Math. Nachr., **196** (1998), 61–69.

[35] I. Dimitric, *Submanifolds of E^n with harmonic mean curvature vector*, Bull. Inst. Math. Acad. Sinica. **20** (1992), 53-65.

[36] J. Dorfmeister, F. Pedit and H. Wu, *Weierstrass type representation of harmonic maps into symmetric spaces*, Commun. Anal. Geom., **6** (1998), 633–668.

[37] S. Dragomir and S. Montaldo, *Subelliptic biharmonic maps*, J. Geom. Anal. **24** (2014), 223–245.

[38] S. Dragomir and G. Tomassini, *Differential Geometry and Analysis on CR Manifolds* , Progress in Math. 246, Birkhouser, 2006.

[39] J. Eells, M.J. Ferreira, *On representing homotopy classes by harmonic maps*, Bull. London Math. Soc., **23** (1991), 160–162.

[40] J. Eells, L. Lemaire, *Selected topics in harmonic maps*, CBMS, **50**, Amer. Math. Soc, 1983.

[41] J. Eells, L. Lemaire, *A report on harmonic maps*, Bull. London Math. Soc., **10** (1978), 1–68.

[42] J. Eells, L. Lemaire, *Another Report on Harmonic Maps*, Bull. London Math. Soc., **20** (1988), 385–524.

[43] J. Eells, J.C. Polking, *Removable singularities of harmonic maps*, Indiana U. Math. J., **33** (1984), 243–255.

[44] J. Eells and J.H. Sampson, *Harmonic mappings of Riemannian manifolds*, Amer. J. Math., **86** (1964), 109–160.

[45] N. Ejiri, *A construction of non-flat, compact irreducible Riemannian manifolds which are isospectral but not isometric*, Math. Z., **168** (1979), 207–212.

[46] Encyclopedia of Mathematics, ed. by Japan Math. Soc., the fourth edition, Iwanami Shoten, Tokyo, 2007.

[47] L.C. Evans, *Partial regularity for stationary harmonic maps into the sphere*, Arch. Rational Mech. Anal., **116** (1991), 1-1–113.

[48] M.J. Ferreira, B.A. Simões and J.C. Wood, *All harmonic 2-spheres in the unitary group, completely explicitly*, Math. Z., **266** (2010), 953–978.

[49] D. Fetcu, E. Loubeau, S. Montaldo and C. Oniciuc, *Biharmonic submanifolds of $\mathbb{C}P^n$*, Math. Z. **266** (2010), no. 3, 505-531.

[50] D. Fetcu and C. Oniciuc, *Biharmonic integral $\mathcal{C}$-parallel submanifolds in 7-dimensional Sasakian space forms*, Tohoku Math. J., **64** (2012), 195–222.

[51] D.S. Freed, K. Uhlenbeck, *Instantons and Four-Manifolds*, Springer, 1991.

[52] M.F. Gaffney, *A special Stokes' theorem for complete Riemannian manifold*, Ann. Math., **60** (1954), 140–145.

[53] O. Goertsches and G. Thorbergsson, *On the Geometry of the orbits of Hermann action*, Geom. Dedicata, **129** (2007), 101–118.

[54] S. Gudmundsson, *The Bibliography of Harmonic Morphisms*, http://matematik.lu.se/ matematiklu/personal/sigma/harmonic/bibliography.html

[55] M.A. Guest and Y. Ohnita, *Group actions and deformations for harmonic maps*, J. Math. Soc. Japan, **45** (1993), 671–704.

[56] P. Hartman, *Ordinary Differential Equations*, John Wiley & Sons, Inc., New York-London-Sydney, 1964.

[57] T. Hasanis and T. Vlachos *Hypersurfaces in $\mathbb{E}^4$ with harmonic mean curvaturer vector field*, Math. Nachr., **172** (1995), 145–169.

[58] F. Helein *Régularié des applications faiblement harmoniques entre une surface et une variét'e riemanienne*, C. R. Acad. Sci. Paris Sér I Math, **312** (1991), 591–596.

[59] S. Helgason, *Differential Geometry, Lie Group, and Symmetric Spaces*, Academic Press, 1978.

[60] S. Hiepko, *Eine innere Kennzeichnung der verzerrten Produkte*, Math. Ann. **241** (1979), 209-215.

[61] S. Hildebrant, H. Kaul, K.O. Widman, *An existence theorem for harmonic mappings of Riemannian manifolds*, Acta Math., **138** (1977), 1–15.

[62] D. Hirohashi, H. Tasaki, H.J. Song and R. Takagi, *Minimal orbits of the isotropy groups of symmetric spaces of compact type*, Differential Geom. Appl. **13** (2000), no. 2, 167–177.

[63] T. Ichiyama, J. Inoguchi, H. Urakawa, *Biharmonic maps and bi-Yang-Mills fields*, Note di Mat., **28**, (2009), 233–275.

[64] T. Ichiyama, J. Inoguchi, H. Urakawa, *Classifications and isolation phenomena of biharmonic maps and bi-Yang-Mills fields*, Note di Mat., **30**, (2010), 15–48.

[65] J. Inoguchi *Submanifolds with harmonic mean curvature in contact 3-manifolds.* Colloq. Math., **100**(2004), 163–179.

[66] J. Inoguchi and T. Sasahara, *Biharmonic hypersurfaces in Riemannian symmetric spaces I*, Hiroshima Math. J. **46** (2016), no. 1, 97–121.

[67] O. Ikawa, *The geometry of symmetric triad and orbit spaces of Hermann actions.*, J. Math. Soc. Japan **63** (2011), 79–136.

[68] O. Ikawa, *A note on symmetric triad and Hermann actions*, Proceedings of the workshop on differential geometry and submanifolds and its related topics, Saga, August 4–6, (2012), 220–229.

[69] O. Ikawa, *σ-actions and symmetric triads*, to appear in Tôhoku Math. J.

[70] O. Ikawa, T. Sakai and H. Tasaki, *Orbits of Hermann actions*, Osaka J. Math., **38** (2001), 923–930.

[71] O. Ikawa, T. Sakai and H. Tasaki, *Weakly reflective submanifolds and austere submanifolds*, J. Math. Soc. Japan, **61** (2009), 437–481.

[72] H. Iriyeh, *Hamiltonian minimal Lagrangian cones in $\mathbb{C}^m$*, Tokyo J. Math., **28** (2005), 91–107.

[73] S. Ishihara and S. Ishikawa, *Notes on relatively harmonic immersions*, Hokkaido Math. J., **4** (1975), 234–246.

[74] G.Y. Jiang, *2-harmonic maps and their first and second variational formula*, Chinese Ann. Math., **7A** (1986), 388–402; Note di Matematica, **28** (2009), 209–232, translated into English by H. Urakawa.

[75] J. Jost, *Two-Dimensional Geometric Variational Problems*, John Wiley & Sons Ltd., 1991.

[76] S. D. Jung, *The first eigenvalue of the transversal Dirac operator*, J. Geom. Phys., **39** (2001), 253–264.

[77] S. D. Jung *Eigenvalue estimates for the basic Dirac operator on a Riemannian foliation admitting a basic harmonic 1-form*, J. Geom. Phys., **57** (2007), 1239–1246.

[78] S. D. Jung, B.H. Kim and J.S. Pak, *Lower bounds for the eigenvalues of the basic Dirac operator on a Riemannian foliation*, J. Geom. Phys., **51**, (2004), 166–182.

[79] M.J. Jung and S. D. Jung, *On transversally harmonic maps of foliated Riemannian manifolds*, J. Korean Math. Soc., **49** (2012), 977–991.

[80] S. D. Jung, *Variation formulas for transversally harmonic and bi-harmonic maps*, J. Geometry Phys., **70** (2013), 9–20.

[81] T. Kajigaya, *Second variation formula and the stability of Legendrian minimal submanifolds in Sasakian manifolds*, Tohoku Math. J., **65** (2013), 523–543.

[82] S. Kobayashi and K. Nomizu, *Foundation of Differential Geometry, Vol. I, II*, Interscience Publ. 1963, 1969.

[83] S. Kobayashi, *Transformation Groups in Differential Geometry*, Springer, 1972.

[84] N. Koiso and H. Urakawa, *Biharmonic submanifolds in a Riemannian manifold*, arXiv: 1408.5494v1, 2014, accepted in Osaka J. Math.

[85] A. Kollross, *A classification of hyperpolar and cohomogeneity one actions*, Trans. Amer. Math. Soc. **354** (2002), no. 2, 571–612.

[86] J.J. Konderak and R. Wolak, *Transversally harmonic maps between manifolds with Riemannian foliations*, Quart. J. Math., **54** (3) (2003), 335–354.

[87] J.J. Konderak and R. Wolak, *Some remarks on transversally harmonic maps*, Glasgow Math. J., **50** (1) (2003), 1–16.

[88] Dominic S. P. Leung, *The reflection principle for minimal submanifolds of Riemannian symmetric spaces*, J. Differential Geom., **8** (1973), 153–160.

[89] E. Loubeau, C. Oniciuc, *The index of biharmonic maps in spheres*, Compositio Math., **141** (2005), 729–745.

[90] E. Loubeau and C. Oniciuc, *On the biharmonic and harmonic indices of the Hopf map*, Trans. Amer. Math. Soc., **359** (2007), 5239–5256.

[91] E. Loubeau and Y-L. Ou, *Biharmonic maps and morphisms from conformal mappings*, Tohoku Math. J., **62** (2010), 55–73.

[92] Y. Luo, *Weakly convex biharmonic hypersurfaces in nonpositive curvature space forms are minimal*, Results in Math. **65** (2014), 49–56.

[93] Y. Luo, *On biharmonic submanifolds in non-positively curved manifolds*, J. Geom. Phys. **88** (2015), 76–87.

[94] Y. Luo, *Remarks on the nonexistence of biharmonic maps*, Arch. Math. (Basel), **107** (2016), no 2, 191–200.

[95] S. Maeta and U. Urakawa, *Biharmonic Lagrangian submanifolds in Kähler manifolds*, Glasgow Math. J. , **55** (2013), 465–480.

[96] S. Maeta, N. Nakauchi and H. Urakawa, *Triharmonic isometric immersions into a manifold of non-positively constant curvature*, Monatsh. Math. **177** (2015), 551–567.

[97] M. Markellos and H. Urakawa, *The bienergy of unit vector fields*, Ann. Global Anal. Geom. **46** (2014), 431–457.

[98] M. Markellos and H. Urakawa, *The biharmonicity of sections of the tangent bundle*, Monatsh. Math. **178** (2015), 389–404.

[99] M. Markellos and H. Urakawa, *Biharmonic vector fields on pseudo-Riemannian manifolds*, a preprint, 2018.

[100] F. Matsuura and H. Urakawa, *On exponential Yang-Mills connections*, J. Geom. Phys. **17** (1995), 73–89.

[101] Sh. Mizohata, *Theory of Partial Differential Equations*, Iwanami, Tokyo, 1965.

[102] S. Montaldo, C. Oniciuc, *A short survey on biharmonic maps between Riemannian manifolds*, Rev. Un. Mat. Argentina **47** (2006), 1–22.

[103] M. Hidano-Mukai and Y. Ohnita, *Geometry of the moduli spaces of harmonic maps into Lie groups via gauge theory over Riemann surfaces*, International J. Math., **12** (2001), 339–371.

[104] M. Hidano-Mukai and Y. Ohnita, *Gauge-theoretic approach to harmonic maps and subspaces in moduli spaces*, In: Integrable Systems, Geometry and Topology, AMS/IP Study Advances Math., **36**, Amer. Math. Soc., Providence, (2006), 191–234.

[105] M. Mukai and Y. Ohnita, *Gauge-theoretic equations for harmonic maps into symmetric spaces*, In: The third Pacific Rim Geometry Conf., Monogr. Geometry Topology, **25**, Int. Press, Cambrdge, (1998), 195–209.

[106] Y. Nagatomo, *Harmonic maps into Grassmannians and a generalization of do Carmo-Wallach theorem*, Proc. the 16th OCU Intern. Academic Symp. 2008, OCAMI Studies, **3** (2008), 41–52.

[107] H. Naito and H. Urakawa, *Conformal change of Riemannian metrics and biharmonic maps*, Indiana Univ. Math. J., **63** (2014), 1631–1657.

[108] N. Nakauchi and H. Urakawa, *Biharmonic hypersurfaces in a Riemannian manifold with non-positive Ricci curvature*, Ann. Global Anal. Geom., **40** (2011), 125–131.

[109] N. Nakauchi and H. Urakawa, *Biharmonic submanifolds in a Riemannian manifold with non-positive curvature*, Results in Math.,**63** (2013), 467–474.

[110] N. Nakauchi and H. Urakawa, *Bubbling phenomena of biharmonic maps*, J. Geom. Phys. **98** (2015), 355–375.

[111] N. Nakauchi, H. Urakawa and S. Gudmundsson, *Biharmonic maps into a Riemannian manifold of non-positive curvature*, Geom. Dedicata, **169** (2014), 263–272.

[112] B. O'Neill, *The fundamental equation of a submersion*, Michigan Math. J., **13** (4) (1966), 459–469.

[113] S. Nishikawa and Ph. Tondeur and L. Vanhecke, *Spectral geometry for Riemannian foliations*, Ann. Global Anal. Geom., **10** (1992), 291–304.

[114] Y. Ohnita, *Group actions and deformations for harmonic maps into symmetric spaces*, Kodai Math. J., **17** (1994), 463–475.

[115] S. Ohno, *A sufficient condition for orbits of Hermann action to be weakly reflective*, to appear in Tokyo Journal Mathematics.

[116] S. Ohno, T. Sakai and H. Urakawa, *Biharmoic homogeneous hypersurfaces in compact symmetric spaces*, Differential Geom. Appl. **43** (2015), 155–179.

[117] S. Ohno, T. Sakai and H. Urakawa, *Rigidity of transversally biharmonic maps between foliated Riemannian manifolds*, accepted in Hokkaido Math. J., 2017.

[118] S. Ohno, T. Sakai and H. Urakawa, *Harmonic maps and bi-harmonic maps on CR-manifolds and foliated Riemannian manifolds*, J. Appl. Math. Physics, **4** (2016), 2272–2289.

[119] S. Ohno, T. Sakai and H. Urakawa, *Biharmonic homogeneous submanifolds in compact symmetric spaces and compact Lie groups*, accepted in Hiroshima Math. J., in 2018, January.

[120] C. Oniciuc, *On the second variation formula for biharmonic maps to a sphere*, Publ. Math. Debrecen., **67** (2005), 285–303.

[121] C. Oniciuc, *Biharmonic maps between Riemannian manifolds*, Ann. Stiint Univ. Al. I. Cuza Iasi, Mat. (N.S.), **68** No. 2, (2002), 237–248.

[122] Ye-Lin Ou, *Biharmonic hypersurfaces in Riemannian manifolds* , Pacific J. Math., **248** (2010), 217–237.

[123] Ye-Lin Ou and Liang Tang, *On the generalized Chen's conjecture on biharmonic submanifolds*, arXiv: 1006.1838.

[124] Ye-Lin Ou and Liang Tang, *On the generalized Chen's conjecture on biharmonic submanifolds* Michigan Math. J., **61** (2012), 531–542.

[125] S. Ouakkas, *Biharmonic maps, conformal deformations and the Hopf maps*, Differential Geom. Appl. **26** (2008), 495–502.

[126] J. H. Park, *The Laplace-Beltrami operator and Riemannian submersion with minimal and not totally geodesic fibers*, Bull. Korean Math. Soc., **27** (1990), 39–47.

[127] E. Park and K. Richardson, *The basic Laplacian of a Riemannian foliation*, Amer. J. Math., **118**, (1996), 1249–1275.

[128] H. Reckziegel, *Horizontal lights of isometric immersions into the budge space of a pseudo-Riemannian submersion*, Global differential geometry and global analysis (1984), 264-279, Lecture Notes in Math, **1156** (1985).

[129] R. Takagi, *Real hypersurfaces in a complex projective space with constant principal curvatures*, J. Math. Soc. Japan, **27** (1975), 43–53.

[130] J. Sack, K. Uhlenbeck, *The existence of minimal immersions of 2-spheres*, Ann. Math., **113** (1981), 1–24.

[131] T. Sasahara, *Legendre surfaces in Sasakian space forms whose mean curvature vectors are eigenvectors*, Publ. Math. Debrecen, **67** (2005), 285–303.

[132] T. Sasahara, *Biharmonic Lagrangian surfaces of constant mean curvature in complex space forms*, Glasg. Math. J. **49** (2007), 497–507.

[133] T. Sasahara, *Stability of biharmonic Legendrian submanifolds in Sasakian space forms*, Canad. Math. Bull. **51** (2008), 448–459.

[134] T. Sasahara, *A class of biminimal Legendrian submanifolds in Sasaki space forms*, a preprint, 2013, to appear in Math. Nach.

[135] T. Sasahara, *A classification result for biminimal Lagrangian surfaces in complex space forms*, J. Geom. Phys. **60** (2010), 884–895.

[136] T. Sasahara, *Biminimal Lagrangian H-umbilical submanifolds in complex space forms*, Geom. Dedicata **160**, (2012), 185–193.

[137] R.M. Schoen, *Analytic aspects of the harmonic map problem*, MSRI Publ., Springer, **2** (1984), 321–358.

[138] R.M. Schoen, K. Uhlenbeck, *A regurarity theory for harmonic maps*, J. Differ. Geom., **17** (1982), 307–335.

[139] R. Schoen and S.T. Yau, *Harmonic maps and the topology of stable hypersurfaces and manifolds with non-negative Ricci curvature*, Comment. Math. Helv. **51** (1976), 333–341.

[140] R.T. Smith, *The second variation for harmonic mappings*, Proc. Amer. Math. Soc., **47** (1975), 229–236.

[141] M. Struwe, *Variational Methods, Applications to Nonlinear Partial Differential Equations and Hamiltonian Systems*, Springer, 1990.

[142] M. Struwe, *Partial regularity for biharmonic maps, revisited*, Calculus Var., **33** (2008), 249–262.

[143] T. Takahashi, *Minimal immersions of Riemannian manifoplds*, J. Math. Soc. Japan, **18** (1966), 380–385.

[144] K. Tsukada, *Eigenvalues of the Laplacian of warped product*, Tokyo J. Math., **3** (1980), 131–136.

[145] Ph. Tondeur, *Foliations on Riemannian Manifolds*, Springer-Verlag, New York, 1988.

[146] Ph. Tondeur, *Geometry of Foliations*, Birkhäuser, Basel, 1997.

[147] K. Uhlenbeck, *Harmonic maps into Lie groups (classical solutions of the chiral model)*, J. Differ. Geom., **30** (1989), 1–50.

[148] H. Urakawa, *Spectral Geometry of the Laplacian–Spectral Analysis and Differential Geometry of the Laplacian*, World Scientific Publishing Co. Pte. Ltd. 2017.

[149] H. Urakawa, *Biharmonic maps into compact Lie groups and symmetric spaces*, "Alexandru Myller" Mathematical Seminar, 246–263, AIP Conf. Proc. **1329**, Amer. Snst. Phys., Melville, NY, 2011.

[150] H. Urakawa, *The geometry of biharmonic maps*, Harmonic maps and differential geometry, 159–175, Contemp. Math., **542**, Amer. Math. Soc., Providence, RI, 2011.

[151] H. Urakawa, *Geometry of harmonic maps and biharmonic maps*, Lecture Notes Seminario Interdiscipl. di Mat., **11**, 41–83.

[152] H. Urakawa, *Biharmonic maps into compact Lie groups and integrable systems*, Hokkaido Math. J., **43** (2014), 73–103.

[153] H. Urakawa, *Biharmonic maps into symmetric spaces and integrable systems*, Hokkaido Math. J., **43** (2014), 105–136.

[154] H. Urakawa, *Geometry of biharmonic maps: L^2-rigidity, biharmonic Lagrangian submanifolds of Kähler manifolds, and conformal change of metrics*, Prospects of differential geometry and its related topics, 1–14, World Sci. Publ., Hackensack, NJ, 2014.

[155] H. Urakawa, *Sasaki manifolds, Kähler cone manifolds and biharmonic submanifolds*, Illinois J. Math., **58** (2014), 521–535.

[156] H. Urakawa, *Harmonic maps and biharmonic maps*, Symmetry, **7** (2015), 651–674.

[157] H. Urakawa, *Geometry of Sasaki manifolds, Kähler cone manifolds and biharmonic submanifolds*, Topology and its Appl., **196, B** (2015), 1023–1032.

[158] H. Urakawa, *CR rigidity of pseudo harmonic maps and pseudo biharmonic maps*, Hokkaido Math. J., **46** (2017), 141–187.

[159] H. Urakawa, *Biharmonic maps on principal G-bundles over Riemannian manifolds of non-positive Ricci curvature*, accepted in Michigan Math. J., 2017.

[160] H. Urakawa, *Harmonic maps and biharmonic maps on the principal bundles and warped products*, accepted in J. Korean Math. Soc., 2018.

[161] Z-P Wang and Y-L Ou, *Biharmonic Riemannian submersions from 3-manifolds*, Math. Z., **269** (2011), 917–925.

[162] J.C. Wood, *On the explicit construction and parametrization of all harmonic maps from the two-sphere to a complex Grassmanian*, In: Harmonic Mappings, Twistors and σ-models, Adv. Ser. Math. Phys., **4**, World Sci. Publ., Singapore, (1988), 246–260.

[163] J. C. Wood, *Harmonic maps into symmetric spaces and integrable systems*, In: *Harmonic Maps and Integrable Systems*, eds. by A. P. Fordy and J. C. Wood, Aspect of Mathematics, Vol. E 23, Vieweg (1993), 29–55.

[164] J.C. Wood, Completely explicit formulae for harmonic 2-spheres in the unitary group and related spaces, In: Riemann Surfaces, Harmonic Maps and Visualization, OCAMI Studies, **3**, Osaka Municipal Univ. Press, Osaka, (2010), 53-65.

[165] Y. L. Xin and Q. Chen, *Gauss maps of hypersurfaces in the unit sphere*, In: Conference of Differential Geometry in Shanghai (1983).

[166] S.T. Yau, *Some function-theoretic properties of complete Riemannian manifold and their applications to geometry*, Indiana Univ. Math. J., **25** (1976), 659–670.

[167] S. Yorozu and T. Tanemura, *Green's theorem on a foliated Riemannian manifold and its applications*, Acta Math. Hungarica, **56** (1990), 239–245.

Printed in the United States
By Bookmasters